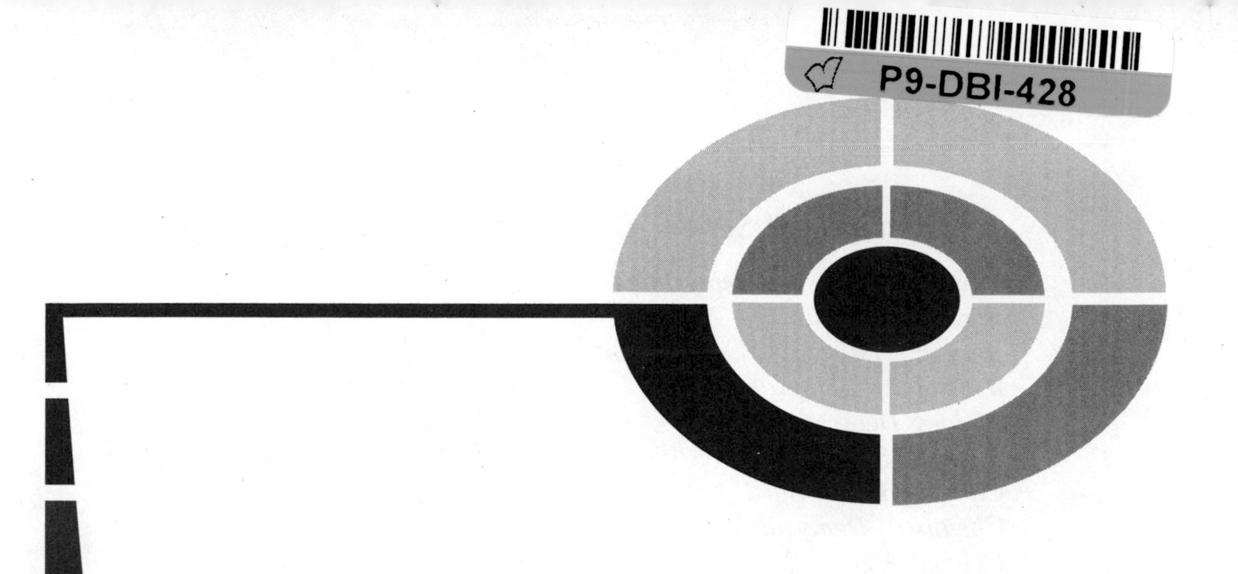

Digital Electronics Demystified

Demystified Series

Advanced Statistics Demystified
Algebra Demystified
Anatomy Demystified
Astronomy Demystified
Biology Demystified
Business Statistics Demystified
Calculus Demystified
Chemistry Demystified
College Algebra Demystified
Differential Equations Demystified
Earth Science Demystified
Electronics Demystified
Everyday Math Demystified
Geometry Demystified
Math Word Problems Demystified
Physics Demystified
Physiology Demystified
Pre-Algebra Demystified
Pre-Calculus Demystified
Project Management Demystified
Robotics Demystified
Statistics Demystified
Trigonometry Demystified

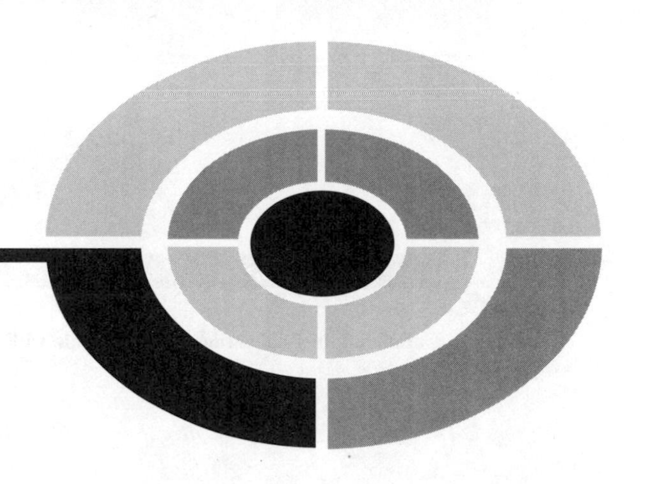

Digital Electronics Demystified

MYKE PREDKO

McGRAW-HILL

New York Chicago San Francisco Lisbon London
Madrid Mexico City Milan New Delhi San Juan
Seoul Singapore Sydney Toronto

The McGraw-Hill Companies

Cataloging-in-Publication Data is on file with the Library of Congress

2 3 4 5 6 7 8 9 0 DOC/DOC 0 1 0 9 8 7 6 5

ISBN 0-07-144141-7

The sponsoring editor for this book was Judy Bass and the production supervisor was Pamela A. Pelton. It was set in Times Roman by Keyword Publishing Services Ltd. The art director for the cover was Margaret Webster-Shapiro; the cover designer was Handel Low.

Printed and bound by RR Donnelley.

This book is printed on recycled, acid-free paper containing a minimum of 50% recycled, de-inked fiber.

CONTENTS

PREFACE

Philosophy is sometimes described as the study of what people take for granted. It examines the reasons why people make assumptions about things in their lives by understanding the relationships between the basic "truths" that are used to come up with these assumptions. This analysis takes a very precise logical path that is scientific in nature. For example, the following statement can be broken down into a set of simple truths and the relationships between them plotted out and understood to allow philosophers to carry on the natural thought process (such as what is a body that has three "extensions" with a "thinking substance").

> Thus, extension in length, breadth and depth, constitutes the nature of corporeal substance; and thought constitutes the nature of thinking substance. For all else that may be attributed to body presupposes extension, and is but a mode of this extended thing; as everything that we find in mind is but so many diverse forms of thinking.
>
> Descartes

Surprisingly enough, the rules that were developed for understanding philosophic statements like the one above were applied in the 1930s and 1940s to help define how electrical circuits could be designed that would be used in the first electronic computers. One of the elements of the success of this effort was to reduce the electronic logic "truths" into two simple electrical states.

These two electrical states are often represented as two numbers that can be manipulated using "binary arithmetic." Binary arithmetic was formally described by the English mathematician George Boole in the middle of the 19th century and is often referred to as "Boolean arithmetic" or "Boolean algebra" as a way to perform mathematical operations on numbers that only have two values ("0" or "1"). These two values are manipulated within electronic computers and other devices built from "digital electronics."

Over the past 60-plus years, digital logic circuits, processing binary signals have been miniaturized, sped up and integrated together to create the fantastic electronic gadgets that we take for granted. Despite their complexity, they operate using the basic rules and circuits that are explained in this book. After working through this book, not only will you understand how these products are designed but you will also have some experience in designing and working through the problems of implementing them on your own.

This book was written for people that would like to learn about digital electronics without taking a formal course. After working through this book, along with a reasonably good understanding of the subject as well as some of the background material needed to create electronic circuits, it can also serve as a supplemental text in a classroom, tutored or home-schooling environment. The book should also be useful for career changers who need to refresh their knowledge in electronics and would like to better understand what are the different facets of current digital electronic products.

This introductory work contains an abundance of practice quiz, test and exam questions. They are all multiple-choice and are similar to the sorts of questions used in standardized texts. There is a short quiz at the end of every chapter. The quizzes are "open-book." You may (and should) refer to the chapter texts when taking them. When you think you're ready, take the quiz, write down the answers and then give your list of answers to a friend. Have the friend tell you the score, but not which questions you got wrong. The answers are listed in the back of the book. Stick with a chapter until you get most of the answers correct.

This book is divided into two parts. At the end of each part is a multiple-choice test. Take these tests when you've completed with the respective sections and have taken all the chapter quizzes. The section tests are "closed-book", but the questions are not as difficult as those in the quizzes. A satisfactory score is three-quarters of the answers correct. Again, answers are in the back of the book.

There is a final exam at the end of this course. It contains questions drawn uniformly from all the chapters in the book. Take it when you have finished both sections, both section tests and all of the chapter quizzes. A satisfactory score is at least 75% correct answers.

With the section tests and the final exam, as with the quizzes, have a friend tell you your score without letting you know which questions you missed. That way, you will not subconsciously memorize the answers. You can check to see where your knowledge is strong and where it is not.

I recommend that you complete one chapter a week. An hour or two daily ought to be enough time for this. As part of this work, you should notice that I have given a number of suggestions on how you could implement the described circuits to see exactly how they work. When you've worked through this material, you can use this book as a permanent reference.

Now, work hard, but be sure to have fun and look to see where you can use the information provided here to help you to understand how the complex electronic devices of modern society are implemented using digital logic devices that are just capable of following simple rules of logic.

myke

ACKNOWLEDGMENT

I would like to thank my wife, Patience, for her love and support and willingness to become the first person to have worked through the material in this book. Without her support, suggestions, love, and willingness to understand what "fanout" means, this book and its material would never have been possible.

PART ONE

Introduction to Digital Electronics

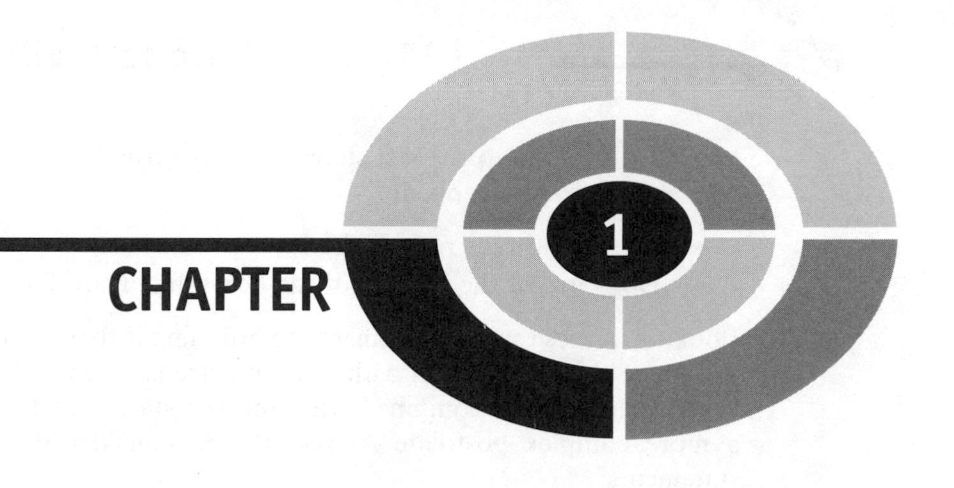

The Underpinnings of Digital Electronics

If you were asked to define what a bit is, chances are you would probably do a pretty good job, saying something like:

A bit is something that can only have two values: on or off.

Instead of "on or off", you might have used terms for two values like "one or zero", "high or low voltage", "up or down", "empty or full" or (if you fancy yourself as being erudite) "dominant or recessive". All of these terms are correct and imply that the two values are at opposite extremes and are easily differentiated.

When you think of "bits", you are probably thinking of something in a wire or an electronic device contained within a computer, but when the concept of binary (two state) logic was first invented, the values that were applied were tests to see if a statement was "true" or "false". Examples of true and false statements are:

- The sun always rises in the East. (true)
- Dogs live at the bottom of the ocean like fish. (false)

Looking at these simple statements determining if they are true or false seems to reduce the information within to an extreme degree. The truthfulness of a statement can be combined with another statement to help determine if a more complex postulate is true. If you consider the following "true" statements:

- A dog has fur over its body.
- A dog has four legs.
- Animals have four legs and fur.
- Humans have two legs.
- A snake has scales on its body.
- A reptile's body has scales or smooth skin.

and combining them together, you can make some surprisingly complex "assertions" from these data using three basic operations. These three basic operations consist of "AND" which is true if all the statements combined together by the AND are true, "OR" which is true if any of the combined statements are true and "NOT" which is true if a single statement is false. To differentiate these three operations from their prose synonyms, I will capitalize them (as well as other basic logic operations) throughout the book. These operations are often called "logic operations" because they were first used to understand the logic of complex philosophical statements.

From the seven statements above and using these three basic operations, you can make the following true assertions:

- Humans are not dogs.
- A dog is an animal.
- A snake is a reptile.

The first statement is true because we know that a human has two legs (violating a condition that is required for the definition of a dog to be true). This is an example of the "negation" or "NOT" operation; the assertion is true if the single input statement is false:

The room is dark because the lights are not on.

The NOT function is often called an "Inverter" because it changes the value of the input from high to low and vice versa.

The second assertion, "A dog is an animal", is true because both of the two statements concerning animals are true when applied to dogs (which have four legs and fur). This is an example of the "AND" operation; the assertion is true if and only if the input statements are true. The AND operation has two or more input statements. In trying to differentiate bicycles and motorcycles from cars, you might make the assertion which uses the AND operation:

A car has four wheels and a motor.

The last assertion, "A snake is a reptile", is true because one of the two statements giving the necessary characteristics for a reptile is true. This is an example of an "inclusive or" (usually referred to as just "OR") operation; the assertion is true if any of the input statements are true. Like the "and" operation, OR can have two or more input statements. If you're a parent, you will be familiar with the assertion:

During the course of a day, a baby eats, sleeps, cries or poops.

I use this example to illustrate an important point about the "OR" operation that is often lost when it is used in colloquial speech: if more than one input statement is true, the entire assertion is still true. As incredible as it sounds to someone who has not had children yet, a baby is very capable of performing all four actions listed above simultaneously (and seemingly constantly).

I'm making this point because when we speak, we usually use the "exclusive or" instead of "inclusive or" to indicate that only one of two actions can be true. An example statement in which an "exclusive or" is used in everyday speech could be:

Tom is at a restaurant or the movies.

This is an example of "exclusive OR" because Tom can only be at one of the two places at any given time. I will discuss the "exclusive or" operation in more detail later in this chapter, but for now try to remember that an assertion using the "OR" operation will be true if one or more of the input statements are true.

So far I have been working with "bits" of "binary" information contained in "statements" and "assertions". You are probably wondering why a term like "bit electronics" or "binary electronics" is used instead of "digital electronics". "Digital" comes from the Latin word for "fingers" and indicates that there are many discrete signals that are at one of two values. Naming the circuitry "bit electronics" or "binary electronics" would imply that it can only work with one piece of information; digital electronic circuits can process many bits of information simultaneously, either as separate pieces of information or collections of large amounts of data.

In the first few pages of this book, I have introduced you to the concept of the "bit", the "digit", the "NOT", "AND" and "OR" operations along with the "exclusive OR". Different combinations of these concepts are the basis for the majority of the material presented through the remainder of this book and any course in digital electronics. I suggest that you read over this chapter and make sure you are completely comfortable with the terms and how they work before going on.

Boolean Arithmetic, Truth Tables and Gates

In the introduction to this chapter, I demonstrated the operation of the three operations "AND", "OR" and "NOT", which can be used to test input values (which are in the form of two state "bits") and produce assertions based on the state of the input bits. The verbose method I used could be used with digital electronics, but you will find that it is cumbersome and not intuitively obvious when you are working with electronic circuits. Fortunately, a number of different tools have been developed to simplify working with logic operations.

The first tool that simplifies how logic operations are expressed is known as "Boolean arithmetic" (or sometimes as "Boolean logic"), a branch of mathematics where a mathematical expression is used to express how bit inputs can be transformed into an output using the three operations presented in the introduction. Boolean arithmetic was first described by the English mathematician Charles Lutwidge Dodgson, whom you may be familiar with by his nom de plume Lewis Carroll, and expanded upon by George Boole, in the mid 19th century, as a way of understanding, proving or disproving complex philosophical statements. Boole demonstrated that a statement, involving bits of data and the AND, OR or NOT operations could be written in the form:

```
Result = Data1 operation Data2  {operation Data3 ...}
```

The braces ("{" and "}") are often used to indicate that what's inside them is optional and the three periods ("...") indicate that the previous text can be repeated. Using these conventions you can see that a Boolean arithmetic statement is not limited to just one operation with two input bits – they can actually be very lengthy and complex with many bit inputs and multiple operations.

To demonstrate how a Boolean arithmetic statement could be articulated, I can write the proof that a dog is an animal in the form:

`Result = (Does Dog have 4 Legs) AND (Does Dog have Fur)`

If both statements within the parentheses are true, then the "Result" will be true.

This method of writing out assertions and the logic behind them is quite a bit simpler and much easier to understand, but we can do better. Instead of writing out the true or false statement as a condition, it can be expressed in terms of a simple "variable" (like "X"). So, if we assign "A" as the result of testing if dogs have four legs and "B" as the result of testing if dogs have fur, we can write out the Boolean arithmetic equation above as:

`Result = A AND B`

To further simplify how a logic operation is written out, the basic characters "·", "+" and "!" can be used instead of AND, OR and NOT, respectively. AND behaves like a binary multiplication, so along with the "·" character, you may see an "x" or "*". The OR operation may be represented as "|". The ampersand ("&") for AND and "|" for OR are often used because they are the same symbols as are used in most computer programming languages. When I write out Boolean arithmetic equations throughout the book, I will be using the "·", "+" and "!" characters for the three basic logic operations instead of the full words.

An important advantage of converting a statement into a simple equation is that it more clearly shows how the logic operation works. If the variables "A" and "B" were just given the values of "true" or "false", the "Result" of the equation above could be written out in the form shown in Table 1-1. This is known as a "truth table" and it is a very effective way of expressing how a Boolean operator works. The truth table is not limited to just three inputs,

Table 1-1 "AND" operation truth table using Gray code inputs.

Input "A"	Input "B"	"AND" Output
False	False	False
False	True	False
True	True	True
True	False	False

and a function with more than one Boolean operator can be modeled in this way. Functions with more than one output can be expressed using the truth table, but I don't recommend doing this because relationships between inputs and outputs (which I will discuss in greater detail later in the book) can be obscured.

One other thing to notice about the truth table is that I have expressed the inputs as a "Gray code", rather than incrementing inputs. Gray codes are a technique for sequencing multiple bits in such a way that only one bit changes from one state to the next. Incrementing inputs behave as if the inputs were bits of a larger binary number and the value of this number is increased by one when moving from one state to the next. The truth table above, for the "AND" gate could be written out using incrementing inputs as Table 1-2.

In many cases, truth tables are introduced with incrementing inputs, but I would like to discourage this. Incrementing inputs can obscure relationships between inputs that become obvious when you use Gray codes. This advantage will become more obvious as you work through more complex logic operations and are looking for ways to simplify the expression.

The OR operation's truth table is given in Table 1-3, while the NOT operation's truth table is shown in Table 1-4.

The OR operation would be written in Boolean arithmetic, using the "+" character to represent the OR operation as:

$$\texttt{Output} = \texttt{A} + \texttt{B}$$

and the NOT operation (using the "!" character) is written out in Boolean arithmetic as:

$$\texttt{Output} = !\texttt{A}$$

Table 1-2 "AND" operation truth table using incrementing inputs.

Input "A"	Input "B"	"AND" Output
False	False	False
False	True	False
True	False	False
True	True	True

Table 1-3 "OR" operation truth table using Gray code inputs.

Input "A"	Input "B"	"OR" Output
False	False	False
False	True	True
True	True	True
True	False	True

Table 1-4 "NOT" operation truth table using Gray code inputs.

Input	"NOT" Output
False	True
True	False

Sometimes, when a signal is NOTted, its symbol is given either a minus sign ("–") or an underscore ("_") as its first character to indicate that it has been inverted by a NOT operation.

The final way of expressing the three basic logic operations is graphically with the inputs flowing through lines into a symbol representing each operation and the output flowing out of the line. Figures 1-1 through 1-3 show the graphical representations of the AND, OR and NOT gates, respectively.

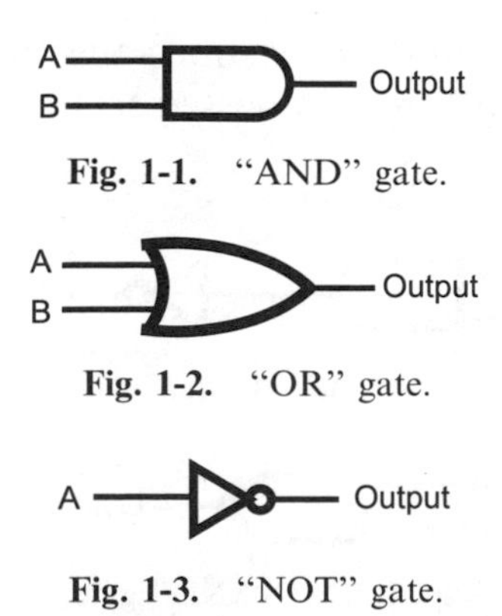

Fig. 1-1. "AND" gate.

Fig. 1-2. "OR" gate.

Fig. 1-3. "NOT" gate.

The graphical representation of the logic operations is a very effective way of describing and documenting complex functions and is the most popular way of representing logic operations in digital electronic circuits. When graphics are used to represent the logic operations, they are most often referred to as "gates", because the TRUE is held back until its requirements are met, at which point it is allowed out by opening the gate. "Gate" is the term I will use most often when describing Boolean arithmetic operations in this book.

If you were to replace the lines leading to each gate with a wire and the symbol with an electronic circuit, you can transfer a Boolean arithmetic design to a digital electronic circuit directly.

The Six Elementary Logic Operations

When you look at a catalog of digital electronics chips, you are going to discover that they are built from ANDs, ORs and NOTs as well as three other elementary gates. Two of these gates are critically important to understand because they are actually the basis of digital logic while the third is required for adding numbers together.

TTL logic is based on the "NAND" gate which can be considered a "NOTted AND" – the output of an AND gate is passed through a NOT gate as shown in Fig. 1-4. Instead of drawing the NAND gate as an AND gate and NOT gate connected together as in Fig. 1-4, they are contracted into the one symbol shown in Fig. 1-5. It's truth table is in Table 1-5.

When writing out the NAND function in Boolean arithmetic, it is normally in the form:

$$\texttt{Output} = !(\texttt{A} \cdot \texttt{B})$$

which is a literal translation of the operation – the inputs are ANDed together and the result is NOTted before it is passed to the Output.

Fig. 1-4. "NAND" gate made from "AND" and "OR" gates.

Fig. 1-5. "NAND" gate.

Table 1-5 "NAND" operation truth table.

Input "A"	Input "B"	"NAND" Output
False	False	True
False	True	True
True	True	False
True	False	True

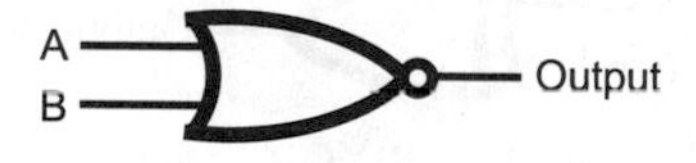

Fig. 1-6 "NOR" gate.

You will see the small circuit on various parts in different electronic devices, both on inputs and outputs. The small circle on the NAND gate is the conventional shorthand symbol to indicate that the input or output of a gate is NOTted.

In case you didn't note the point above, the NAND gate is the basis for TTL logic, as I will explain later in the book. Being very comfortable with NAND gates is very important to being able to design and use TTL electronics. This is a point that I find is not stressed enough in most electronics courses and by having a strong knowledge of how NAND gates work as well as how they are implemented you will better understand what is happening within your digital electronics circuits.

If you are going to be working with CMOS logic, in the same way you should be comfortable with the NAND gate for TTL, you should be familiar with the "NOR" gate (Fig. 1-6). The NOR gate can be considered to be a contraction of the OR and NOT gates (as evidenced by the circle on the output of the OR gate) and operates in the opposite manner as the OR gate, as shown in Table 1-6. When using NOR operations in Boolean arithmetic, a similar format to the NAND gate is used:

```
Output = ! (A + B)
```

The last elementary logic gate that you will have to work with is the "Exclusive OR" (Fig. 1-7) with Table 1-7 being its truth table. The

Table 1-6 "NOR" operation truth table.

Input "A"	Input "B"	"NAND" Output
False	False	True
False	True	False
True	True	False
True	False	False

Fig. 1-7. "XOR" gate.

Table 1-7 "Exclusive OR" operation truth table.

Input "A"	Input "B"	"Exclusive OR" Output
False	False	False
False	True	True
True	True	False
True	False	True

Exclusive OR (also referred to as "Ex-OR" or "XOR") only returns a True output if only one of its inputs is true. If both inputs are the same, then the Exclusive OR outputs False. The Boolean arithmetic symbol for Exclusive OR is often a caret ("^") as is used in computer programming languages or a circle with an "x" character in it ⊗. Writing a Boolean statement with the Exclusive OR would be in the format:

```
Output = A ^ B
```

Table 1-8 summarizes the six elementary gates along with their Boolean arithmetic symbols and formats, graphical symbols and truth tables.

Table 1-8 Summary of the six elementary logic operations.

<table>
<tr><th>Gate</th><th>Boolean arithmetic symbols</th><th>Boolean arithmetic equation</th><th>Graphic symbol</th><th>Truth table</th></tr>
<tr><td>AND</td><td>·, &, *, x</td><td>Out = A · B</td><td>A B Output</td><td>A B | Out
-- -- + --
F F | F
F T | F
T T | T
T F | F</td></tr>
<tr><td>OR</td><td>+, |</td><td>Out = A + B</td><td>A B Output</td><td>A B | Out
-- -- + --
F F | F
F T | T
T T | T
T F | T</td></tr>
<tr><td>NOT</td><td>!, _, -</td><td>Out = ! A</td><td>A Output</td><td>A | Out
-- + --
F | T
T | F</td></tr>
<tr><td>NAND</td><td>! ·</td><td>Out = ! (A · B)</td><td>A B Output</td><td>A B | Out
-- -- + --
F F | T
F T | T
T T | F
T F | T</td></tr>
<tr><td>NOR</td><td>! +</td><td>Out = ! (A + B)</td><td>A B Output</td><td>A B | Out
-- -- + --
F F | T
F T | F
T T | F
T F | F</td></tr>
<tr><td>Exclusive OR</td><td>^, ⊙</td><td>Out = A ^ B</td><td>A B Output</td><td>A B | Out
-- -- + --
F F | F
F T | T
T T | F
T F | T</td></tr>
</table>

Combinatorial Logic Circuits: Combining Logic Gates

As I hinted at in the previous section, multiple gate functions can be combined to form more complex or different Boolean logic functions. Wiring together multiple gates are used to build a complex logic function that only outputs a specific value when a specific combination of True and False inputs are passed to it is known as "combinatorial logic". The output of a combinatorial logic circuit is dependent on its input; if the input changes then the output will change as well.

When I wrote the preceding paragraph, I originally noted that combinatorial logic circuits produce a "True" output for a given set of inputs. This is incorrect, as there will be some cases where you will require a False output in your application. I made the definition a bit more ambiguous so that you do not feel like the output has to be a single, specific value when the input consists of the required inputs. It is also important to note that in a combinatorial logic circuit, data flows in one direction and outputs in logic gates cannot be used as inputs to gates which output back to themselves. These two points may seem subtle now, but they are actually critically important to the definition of combinatorial logic circuits and using them in applications.

An example of a combinatorial circuit is shown in Fig. 1-8. In this circuit, I have combined three AND gates, a NOR gate, a NOT gate and an XOR gate to produce the following logic function:

$$\texttt{Output} = ((\texttt{A} \cdot \texttt{B}) \cdot !(!(\texttt{A} + \texttt{C}))) \ \hat{}\ (!(\texttt{A} + \texttt{C}) \cdot \texttt{B})$$

This combinatorial circuit follows the convention that inputs to a gate (or a chip or other electronic component) are passed into the left and outputs

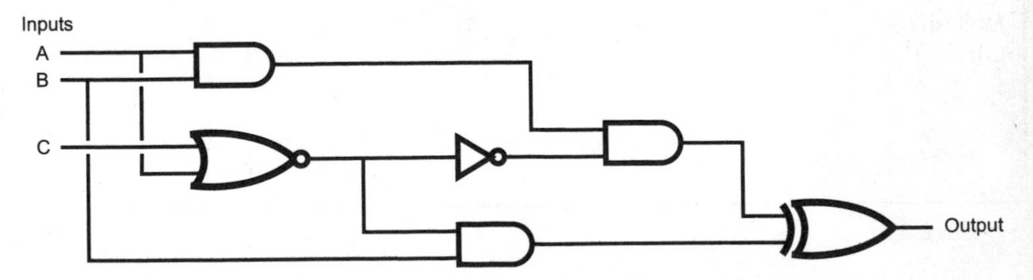

Fig. 1-8 Combinatorial circuit built from multiple logic gates.

exit from the right. This will help you "read" the circuit from left to right, something that should be familiar to you.

While seeing a series of logic gates, like the one in Fig. 1-8, seems to be overwhelming, you already have the tools to be able to work through it and understand how it works. In the previous section, I noted that gates could be connected by passing the output of one into an input of another; a combinatorial circuit (like Fig. 1-8) is simply an extension of this concept and, being an extension, you can use the same tools you used to understand single gates to understand the multiple gate operation.

I should point out that the two broken lines on the left side of Fig. 1-8 (leading down from "A" and "B") indicate that these lines are not connected to the lines that they intersect with. You will find that it can be very difficult to design logic circuits without connected and separate lines from becoming confused. In Fig. 1-9, I have shown a couple of the conventional ways of drawing intersecting lines, depending on whether or not they connect or bypass each other. Probably the most intuitively obvious way of drawing connecting and bypassing lines is to use the dot and arc, respectively. I tend not to because they add extra time to the logic (and circuit) diagram drawing process. As you see more circuit diagrams, you will see the different conventions used and you should be comfortable with recognizing what each means.

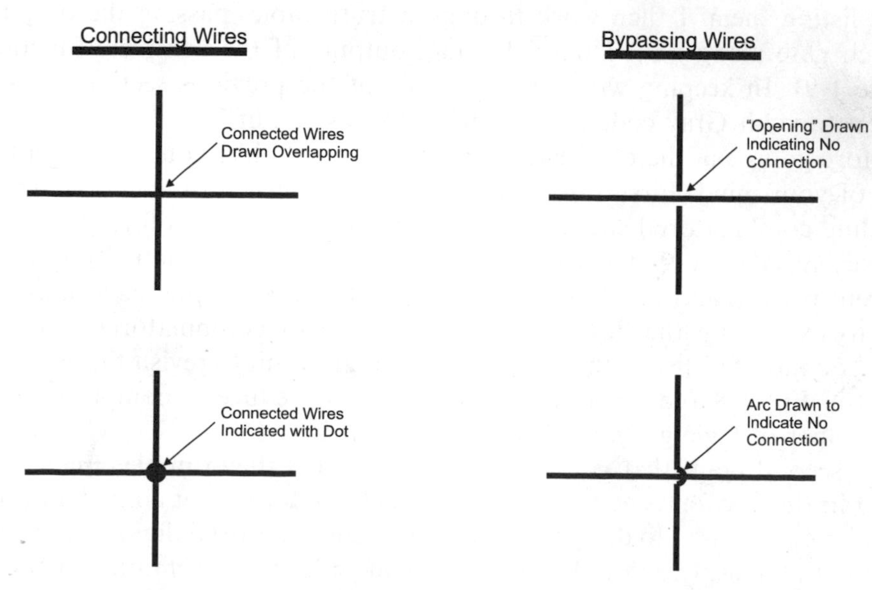

Fig. 1-9. Different representations for wires that connect or bypass.

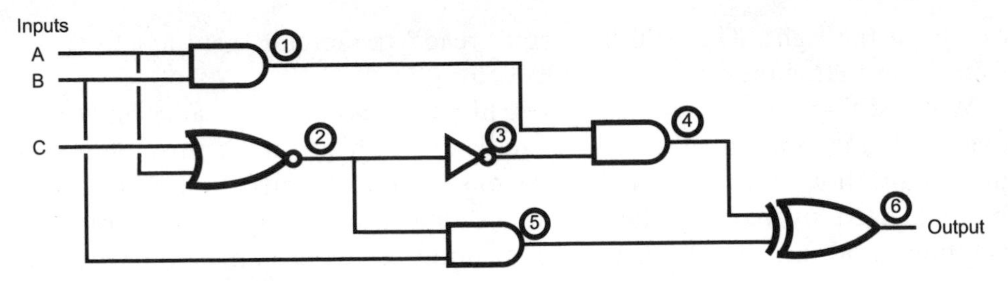

Fig. 1-10. Combinatorial circuit with logic gate outputs marked.

When I am faced with a complex combinatorial circuit, the first thing I do is to mark the outputs of each of the gates (Fig. 1-10) and then list them according to their immediate inputs:

```
Output 1 = A · B
Output 2 = !(A + C)
Output 3 = !2
Output 4 = 1 · 3
Output 5 = B · 2
Output 6 = 4 ^ 5
```

After listing them, I then work through a truth table, passing the outputs of each gate along until I have the final outputs of the complete function (Table 1-9). In keeping with my comments of the previous section, I have used a three bit Gray code for the inputs to this circuit.

Before going on, there are two points that I would like you to keep in the back of your mind. First, this is actually quite an efficient methodology for decoding combinatorial circuits that you are given the design for. Designing a logic gate circuit that responds in a specific manner is actually quite a different process and I will be devoting the rest of this chapter as well as the next to explaining the design and optimization of combinatorial circuits. Once you have finished with Chapter 2, you might want to revisit the example circuit in Fig. 1-8 and see how effectively you can reduce its complexity and the number of logic gates needed to implement it.

The second point that you should be aware of is the example circuit that I used in this section is actually quite unwieldy and does not conform to the typical methods used to design most practical combinatorial digital electronic circuits. In the next section, I will present you with the conventional methods for specifying and documenting combinatorial circuits.

Table 1-9 Decoding the response of the combinatorial circuit in Fig. 1-8.

Inputs			1 = A · B	2 = !(A + C)	3 = !2	4 = 1 · 3	5 = B · 2	6 = 4 ^ 5
A	B	C						
F	F	F	False	True	False	False	False	False
F	F	T	False	False	True	False	False	False
F	T	T	False	False	True	False	False	False
F	T	F	False	True	False	False	True	True
T	T	F	True	False	True	True	False	True
T	T	T	True	False	True	True	False	True
T	F	T	False	False	True	False	False	False
T	F	F	False	False	True	False	False	False

Sum of Products and Product of Sums

Presenting combinatorial circuits as a collection of gates wired together almost randomly, like the circuit shown in Fig. 1-8, is sub-optimal from a variety of perspectives. The first is, the function provided by the combinatorial circuit is not obvious. Secondly, using a variety of different gates can make your parts planning process difficult, with only one gate out of many available in a chip being used. Lastly, the arrangement of gates will be difficult for automated tools to combine on a printed circuit board ("PCB") or within a logic chip. What is needed is a conventional way of drawing combinatorial logic circuits.

The most widely used format is known as "sum of products". Earlier in the chapter, I presented the concept that the AND operation was analogous to multiplication just as the OR operation is to addition. Using this background, you can assume that a "sum of products" circuit is built from AND and OR gates. Going further, you might also guess that the final output is the "OR" (because addition produces a "sum") with the gates that

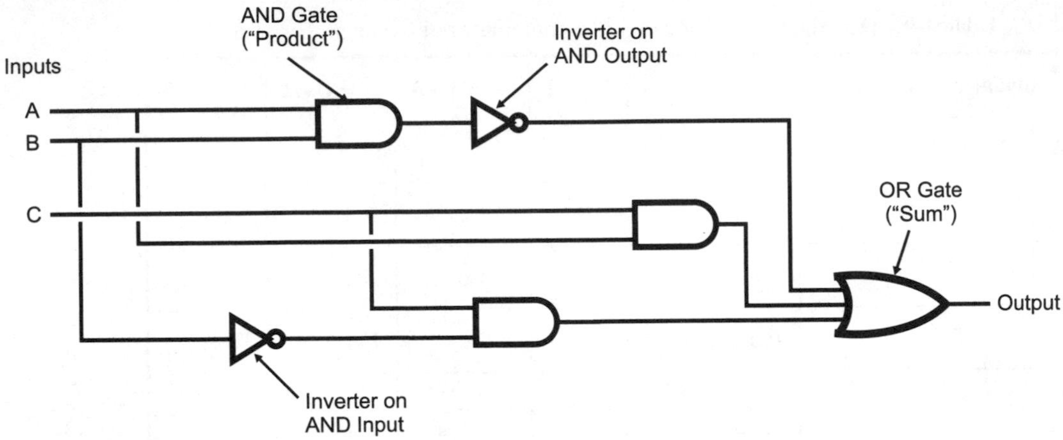

Inputs
A
B
C
AND Gate ("Product")
Inverter on AND Output
OR Gate ("Sum")
Output
Inverter on AND Input

Fig. 1-11. Example "sum of products" combinatorial logic circuit.

convert the inputs being "AND" gates (a "product" is the result of a multiplication operation). An example "sum of products" combinatorial logic circuit is shown in Fig. 1-11.

In this circuit, the inputs are ANDed together and the result is passed to an OR gate. In this circuit, the output will be "True" if any of the inputs to the OR gate (which are the outputs from the AND gates) are true. In some cases, to make sure that the inputs and outputs of the AND gates are in the correct state, they will be inverted using NOT gates, as I have shown in Fig. 1-11.

Figure 1-11 has one feature that I have not introduced to you yet and that is the three input OR gate on the right side of the diagram. So far, I have only discussed two input gates, but I should point out that three input gates can be built from multiple two input gates, as I show in Fig. 1-12, in which two, two input AND gates are combined to form a single three input AND gate. A three input OR gate could be built exactly the same way.

A three input NAND or NOR gate is a bit trickier, as Fig. 1-13 shows. For this case, the output of the NAND gate processing "A" and "B" must be inverted (which can be accomplished with a NAND gate and both inputs tied together as I show in Fig. 1-13) to make its output the same as an "AND". The NAND gate's function is to first AND its inputs together and then invert them before driving out the output signal. As I will explain in greater detail in the next chapter, an inverted output, when it is inverted, becomes a "positive" output and I use this rule to produce the three input NAND gate. A three input NOR gate would be built in exactly the same way as a three input NAND gate.

To Build a 3 Input AND:

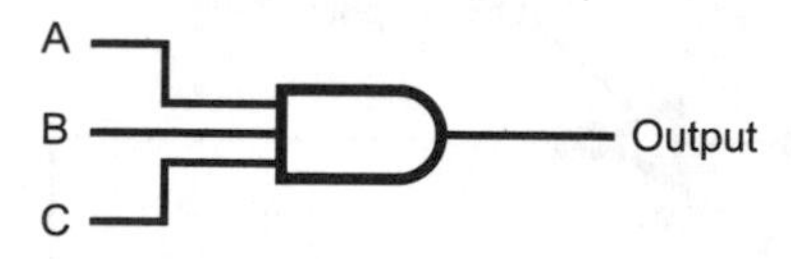

Use two, 2 Input ANDs:

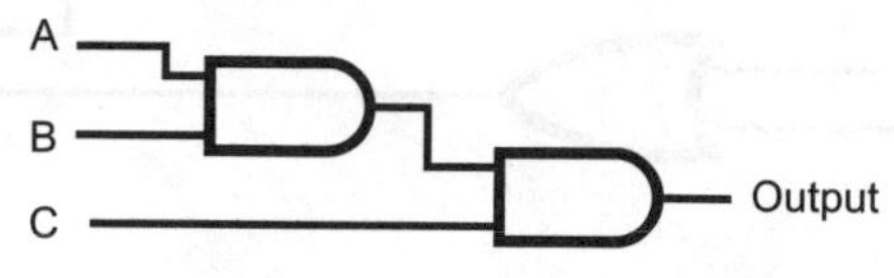

Fig. 1-12. 3 Input AND gate built from 2 input AND gates.

To Build a 3 Input NAND:

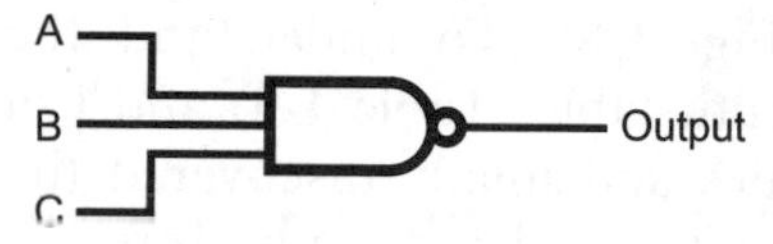

Use three, 2 Input NANDs:

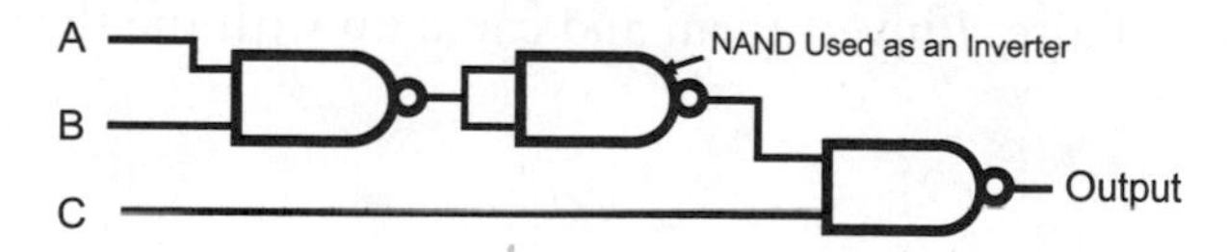

Fig. 1-13. 3 Input AND gate built from 2 input AND gates.

Along with having a "sum of products" combinatorial logic circuit that outputs a True when one of the intermediate AND gates outputs True, there is the complementary "product of sums" (Fig. 1-14), which outputs False when one of its intermediate OR gates outputs False.

While product of sums combinatorial circuits can produce the same functions as sum of product combinatorial circuits, you will not see as many product of sum combinatorial circuits in various designs because they rely on what I call "negative logic". Most people cannot easily visualize something happening because the inputs do not meet an expected case, which is exactly what happens in a product of sums combinatorial logic circuit.

To demonstrate how a sum of product combinatorial logic circuit is designed, consider the messy combinatorial logic circuit I presented

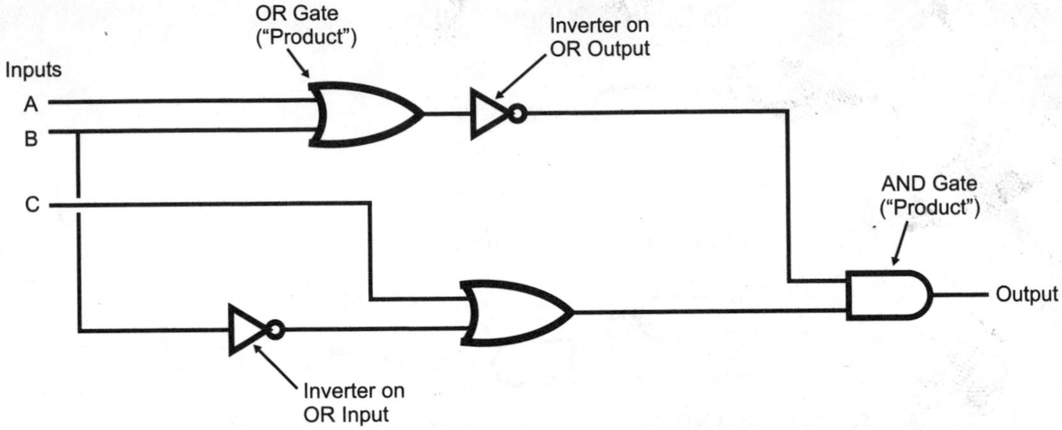

Fig. 1-14. Example "product of sums" combinatorial logic circuit.

in the previous section (see Fig. 1-8). To understand the operation of this circuit, I created a large truth table (Table 1-9) and listed the outputs of each of the intermediate gates and finally discovered that the function output True in three cases that can be directly translated into AND operations by assuming that in each case the output was true and the input conditions were all true. To make all three inputs True to the AND gates when the input is False, I invert them and came up with the three statements below:

A . B · !C

A · B · C

!A · B · !C

These three AND statements can be placed directly into a sum of products combinatorial circuit, as shown in Fig. 1-15.

Looking at Fig. 1-15, you'll probably notice that this circuit has the same total number of gates as the original circuit – and, assuming that each three input gate is made up of two, two input AND gates, it probably requires four *more* gates than the original circuit shown in Fig. 1-8. The only apparent advantage to the sum of product format for combinatorial logic circuit is that it is easier to follow through and see that the output is True for the three input conditions I listed above.

In the following chapters, I will introduce you to combinatorial logic circuit optimization as well as explain in more detail how digital electronic gates are actually built. It will probably be surprising to discover that the sum

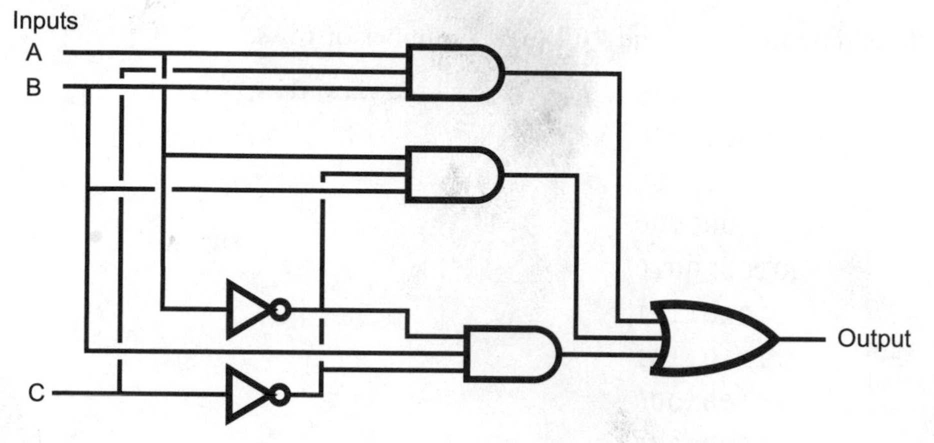

Fig. 1-15. Original combinatorial circuit built in "sum of products" format.

of product combinatorial logic circuit format leads to applications that are more efficient (in terms of total gate or transistor count along with speed and heat dissipation) than ones using less conventional design methodologies.

Waveform Diagrams

So far in this chapter, I have shown how logic functions can be presented as English text, mathematical equations (Boolean arithmetic), truth tables and graphical circuit diagrams. There are actually two more ways in which the logic data can be presented that you should be familiar with. The first method is not one that you will see a lot of except when debugging microprocessor instructions from a very low level, while the second is one that you will have to become very familiar with, especially when the digital electronic signals pass from the combinatorial logic shown here to more complex circuits that have the ability to "store" information.

The first method, the "state list" consists of a list of text columns for each state of the circuit. The state list is really a compressed form of the truth table and is best suited for displaying a fairly large amount of numerical data. Going back to the example circuit of Fig. 1-8, and Table 1-9, I could express the truth table as the series of columns below. Note that I have used the numeric values "1" for True and "0" for False because they are easier to differentiate than "T" and "F" over a number of rows.

```
ABC123456   < – Signals being displayed
000010000
001001000
011001000
010010010
110101101
111101101
101001000
100001000
```

As I said, not a lot of information is obvious from the state list. Some formatting could be done to make the inputs and outputs better differentiated, but for the most part, I don't recommend using state lists for most digital electronics applications. Where the state list is useful is in debugging state machine or microcontroller applications in which you have added hardware to the data, address and control busses to record how the device responds to specific inputs.

The state list is not ideal for this type of application, but it's better than nothing.

The other method, which is not only recommended as a circuit analysis and design tool but is also one you should be intimately familiar with is the "waveform diagram". Waveforms are electrical signals that have been plotted over time. The original waveform display tool was the oscilloscope; a drawing of a typical oscilloscope's display is shown in Fig. 1-16.

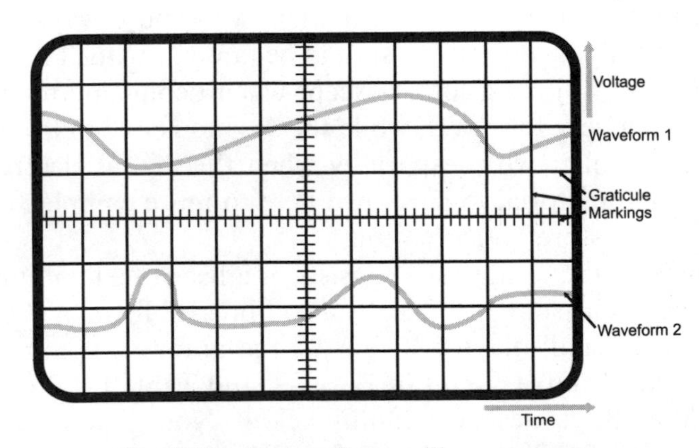

Fig. 1-16. Basics of an oscilloscope screen.

The features of the two "waveforms" displayed on the oscilloscope screen can be measured by placing them against the "graticule markings" on the display. These markings (usually just referred to as "graticules" and etched onto the glass screen of the oscilloscope) are indicators of a specific passage of time or change in voltage. Along with the "gross" graticules, most oscilloscopes have finer markings, to allow more accurate measurements by moving the waveforms over them.

Oscilloscopes are very useful tools for a variety of different applications, which contain varying voltage levels (which are known as "analog" voltage levels). They can be (and often are) used for digital logic applications but they are often not the best tool because digital waveforms only have two levels, when applied to electronics: digital signals are either a high voltage or a low voltage. The timing of the changes of these two voltage levels is more interesting to the designer.

So instead of thinking of digital waveforms in terms of voltage over time, in digital electronics, we prefer to think of them as states (High/Low, True/False, 1/0) over time and display them using a waveform diagram like the one shown in Fig. 1-17. When designing your digital electronics circuit, you will create a waveform diagram to help you understand how the logic states will be passed through the circuit; later, when you are debugging the circuit, you will be comparing what you actually see with this diagram to see if your assumptions about how the circuit would operate are correct. The different signals shown in Fig. 1-17 are samples of what you will see when you are designing your own application circuit.

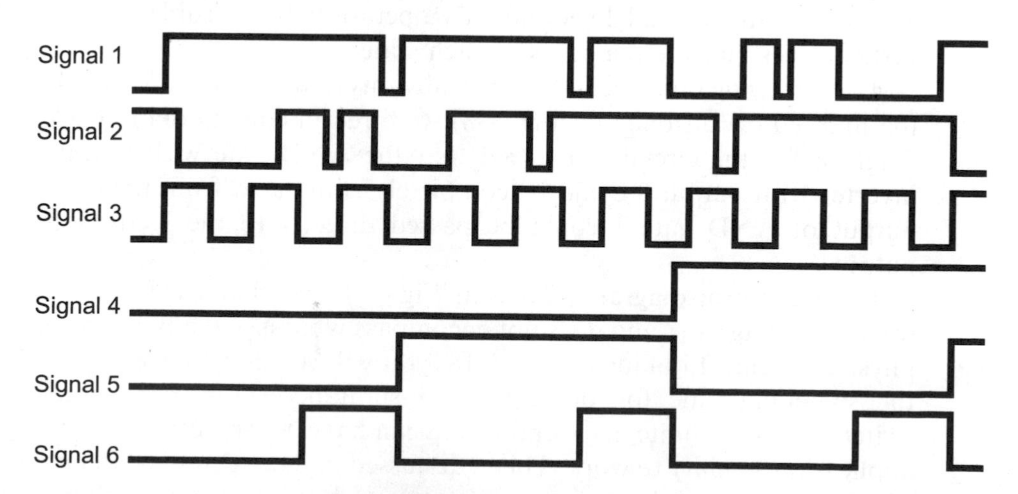

Fig. 1-17. Digital waveforms.

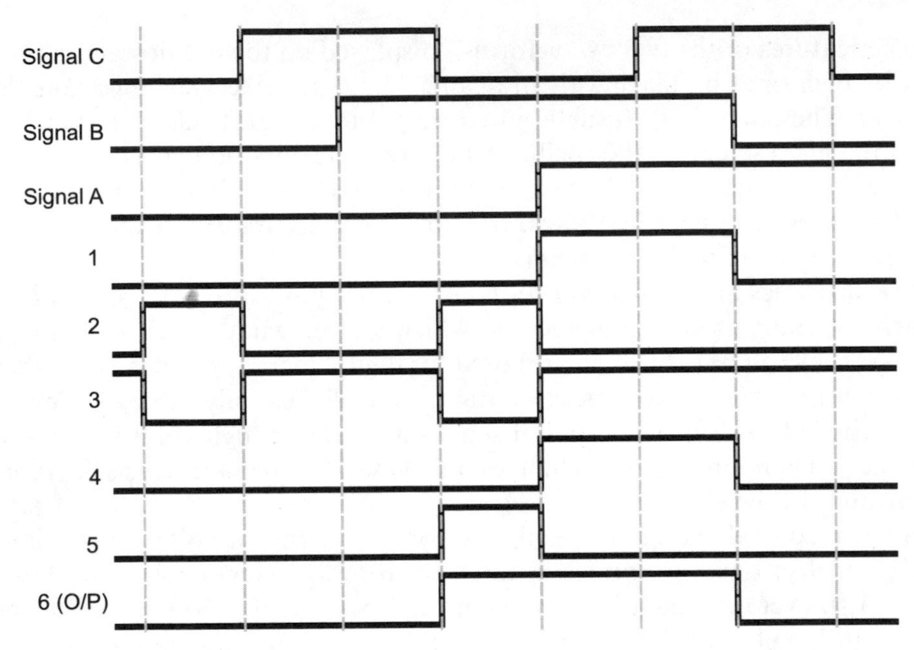

Fig. 1-18. Digital waveforms of Fig. 1-8 example combinatorial logic circuit.

The waveform diagram is the first tool that will help you optimize your circuit. Before writing up this section, I was planning on the diagrams I wanted to include with it and one was a waveform representation of the first example combinatorial logic circuit's operation from Table 1-9. The thin vertical lines indicate the edges of each state.

After drawing out Fig. 1-18, it was obvious that signals "1" and "4" (from the marked circuit diagram Fig. 1-8) were redundant. Looking back at the diagram for the circuit, I realized that the AND gate with output 4 and inverter with output 3 could be completely eliminated from the circuit – the output of AND gate 1 could be passed directly to the XOR gate (with output 6).

The waveform diagram shown in Fig. 1-18 is what I call an "idealized waveform diagram" and does not encompass what is actually happening in a physical circuit. Looking at Fig. 1-18, you will see that I have assumed that the switching time for the gates is instantaneous. In real components, switching time is finite, measurable and can have a significant impact to your application's ability to work. This is discussed in more detail in later chapters. Finally, this circuit does not allow for basic errors in understanding, such as what happens when multiple gate outputs are passed to a single gate input – your assumption of this type of case may not match what really happens in an actual circuit.

In this chapter, I have introduced you to the basic concepts of combinatorial logic circuits and the parts that make them up. In the next chapter, I will start working through some of the practical aspects of designing efficient digital electronic circuits.

Quiz

1. Which of the following statements is true?
   (a) Negative logic is the same as reverse psychology. You get somebody to do something by telling them to do what you don't want them to do
   (b) Using the logic definition, "A dog has four legs and fur", a cat could be accurately described as a dog
   (c) "High" and "Higher" are valid logic states
   (d) Assertions are the same as logic operations
2. Boolean arithmetic is a:
   (a) way to express logic statements in a traditional mathematical equation format
   (b) terrible fraud perpetrated by philosophers to disprove things they don't agree with
   (c) very difficult calculation used in astronomy
   (d) fast way to solve problems around the house
3. The truth table using "incrementing input" for the OR gate is correctly represented as:
   (a)

Input "A"	Input "B"	"OR" Output
False	True	True
True	False	True
False	False	False
True	True	True

(b)

Input "A"	Input "B"	"OR" Output
False	False	False
False	True	True
True	False	True
True	True	False

(c)

Input "A"	Input "B"	"OR" Output
False	False	False
False	True	True
True	False	True
True	True	True

(d)

Input "A"	Input "B"	"OR" Output
False	False	False
False	True	False
True	False	False
True	True	True

4. When writing a logic equation, which symbols are typically used to represent optional operations?
   (a) {and}
   (b) < and >
   (c) (and)
   (d) [and]

5. If the output of an Exclusive OR gate was passed to a NOT gate's input, the NOT gate output would be "True" if:
   (a) Input "A" was True and input "B" is False
   (b) There is only one input and the output would be True if the input was False
   (c) A dot was placed on the output of the Exclusive OR symbol
   (d) Both inputs were at the same state (either True or False)
6. Boolean arithmetic statements are similar to:
   (a) Verbal descriptions of what the logic is to do
   (b) HTML, the language used to program an internet web page
   (c) Simple mathematical equations
   (d) The holes punched into computer cards
7. When decoding a combinatorial logic circuit diagram, you
   (a) Write out the Boolean arithmetic equation for the function and list the output for each possible input
   (b) Start slamming your forehead on your desk
   (c) Give each gate's output a unique label and list their outputs for each changing circuit input as well as outputs for other gates in the circuit
   (d) Rearrange the gates in the diagram to help you understand what the function is doing
8. "Sum of product" combinatorial logic circuits are popular because:
   (a) They are the most efficient way of designing circuitry
   (b) Their operation can be quickly seen by looking at the circuit diagram
   (c) They dissipate less heat than other design methodologies
   (d) They are more robust and allow for more reliable product designs
9. When trying to debug a digital clock circuit, what tool is not recommended?
   (a) Truth tables
   (b) Boolean arithmetic
   (c) State lists
   (d) Graphical circuit diagrams

10. Waveform diagrams display:
    (a) Logic state changes over time
    (b) Switching times of digital electronic gates
    (c) Problems with line impedance
    (d) Voltage variances in a logic signal over time

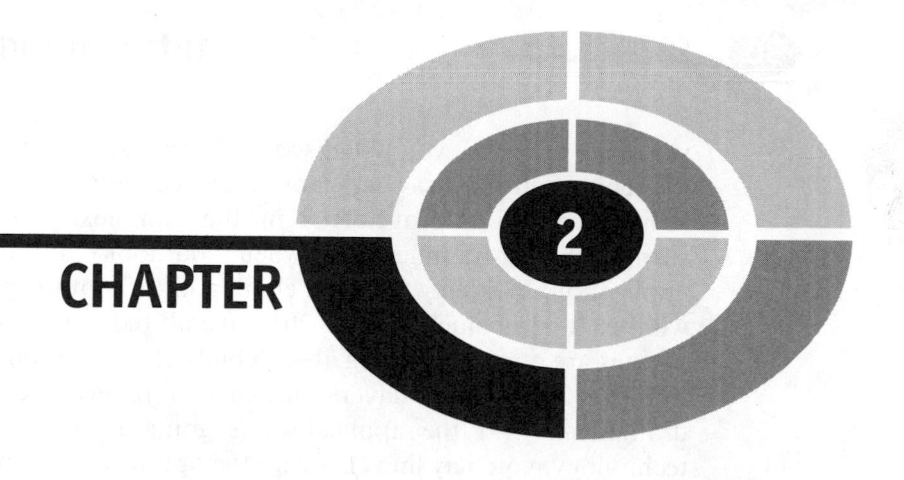

Effectively Optimizing Combinatorial Circuits

In the first chapter, I introduced you to the basic theory behind digital electronics: binary data is manipulated by six different simple operations. With this knowledge, you actually have enough information to be able to design very complex operations, taking a number of different bits as input. The problem with these circuits is that they will probably not be "optimized" in order to minimize the number of gates, the speed which the digital electronic circuit responds to the inputs and finally, whether or not the circuit is optimized for the technology that it will be implemented in.

These three parameters are the basic measurements used to determine whether or not a circuit is effectively optimized. The number of gates should be an obvious one and you should realize that the more gates, the higher the chip count and cost of implementing the circuit as well as the increased complexity in wiring it. Right now, connections between logic gates are just black lines on paper to you – but when you start trying to wire circuits that you have designed, you will discover first hand that simplifying the wiring of a circuit often reduces costs more than reducing the number of chips would indicate. Small improvements in the complexity of a circuit can have surprising cost ramifications when you look at the overall cost of the application. You may find that eliminating 1% of the gates in an application will result in as much as a 10–20% overall reductions in product cost. These savings are a result of being able to build the circuit on a smaller PCB or one which requires fewer layers (which can reduce the overall product cost dramatically). If the application is going to use a programmable logic technology, you may find that with the optimized circuit, lower cost chips can be substituted into the design. Fewer gates in an application also results in less power being dissipated by the circuit, requiring less cooling and a smaller power supply.

The speed that signals pass through gates is not infinite; standard TTL requires 8 billionths of a second (called a "nanosecond" and uses the abbreviation "ns") to pass a signal through a "NAND" gate. The term given to this time is known as the "gate delay". Halving the number of gates a signal has to pass through (which is halving the number of gate delays) will double the speed in which it can respond to a changing input. As you work with more complex circuits, you will discover that you will often have to optimize a circuit for speed or else you may have to use a faster (and generally more expensive) technology.

The last parameter, what I call "technology optimization", on the surface may seem more intangible than the other two parameters as well as have its measurements use the other two parameters, but when working with physical devices, it is the most important factor in properly optimizing your application. Before moving on and considering your circuit "done", you should look at how it will actually be implemented in the technology that you are using and look for optimizations that will reduce the actual number of gates and gate delays required by the application.

You can consider logic optimization to be a recursive operation, repeatedly optimizing all the different parameters and measurements. Once you have specified the required logic functions, you should look at how it will be implemented in the actual circuit. Once you have converted it to the actual circuit, you will then go back and look for opportunities for decreasing the number of gates, speeding up the time the signal passes through the gates and again look for technology optimizations. This will continue until you are satisfied with the final result.

To illustrate what I mean, in this chapter, I will look at a practical example, a simple home burglar alarm. In Fig. 2-1, I have drawn a very basic house, which has two windows, a door and power running to it. Sensors, on

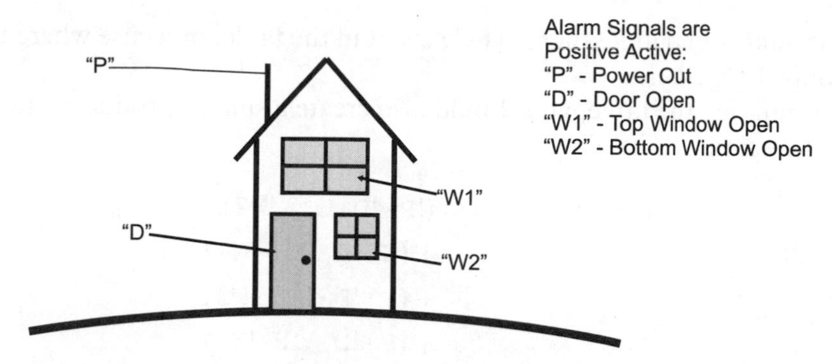

Fig. 2-1. Home alarm logic.

Table 2-1 Home alarm truth table.

P	D	W1	W2	Alarm Response
0	0	0	0	
0	0	0	1	
0	0	1	0	Sound Alarm
0	0	1	1	
0	1	0	0	
0	1	0	1	
0	1	1	0	Sound Alarm
0	1	1	1	
1	0	0	0	
1	0	0	1	Sound Alarm
1	0	1	0	Sound Alarm
1	0	1	1	Sound Alarm
1	1	0	0	
1	1	0	1	Sound Alarm
1	1	1	0	Sound Alarm
1	1	1	1	Sound Alarm

the windows, door and power are passed to an alarm system. When the alarm system was designed a table of the different possible combinations of inputs was generated (Table 2-1), with the combinations that would cause the alarm to sound indicated. As I have noted in Fig. 2-1, the alarm inputs are positive active, which means I can represent them as being active with a "1".

In this fictional house, I assumed that the upper window ("W1") should never be opened – if it were opened, then the alarm would sound. Along with this, I decided that if the power failed and either of the windows were opened, then the alarm failed; this would be the case where the power to the house was cut and somebody forced open the window. Table 2-1 shows the cases where the alarm should sound and you will notice that the cases where the alarm should sound are either a single event in the table, or a case where three are grouped together.

After building the table, you should also create a sum of products equation for the function:

```
Alarm State = (!P · !D · W1 · !W2)
            + (!P · D · W1 · !W2)
            + (P · !D · !W1 · W2)
            + (P · !D · W1 · !W2)
            + (P · !D · W1 · W2)
            + (P · D · !W1 · W2)
            + (P · D · W1 · !W2)
            + (P · D · W1 · W2)
```

You could also draw a logic diagram using the gate symbols that I introduced in the first chapter. I found that this diagram was very complex and very difficult to follow. If you were to try it yourself, you would discover that the logic diagram would consist of 12 NOTs, 24 two input ANDs (knowing that a single four input AND can be produced from three two input ANDs) and seven two input OR gates with the maximum gate delay being eleven (the number of basic TTL gates the signal has to pass through). At first take, this alarm function is quite complex.

Looking at Table 2-1 and the sum of products equation, you will be hard pressed to believe that this home alarm circuit can be significantly optimized, but in this chapter, I will show how these four alarm inputs and eight alarm events can be reduced to fit in the most basic TTL chip there is.

Truth Table Function Reduction

I like to tell new circuit designers to approach optimizing a logic circuit by first looking for opportunities in its truth table. This may not seem like a useful tool (especially in light of Table 2-1), but it can be as effective a tool as any of the others presented in this chapter. It can also be used as a useful verification tool for making sure that an optimized logic circuit will perform the desired function. The drawback to the truth table function reduction is that it tends to be the most demanding in terms of the amount of rote effort that you will have to put into it.

Table 2-2 Gray code home alarm truth table.

P	D	W1	W2	Alarm Response
0	0	0	0	
0	0	0	1	
0	0	1	1	
0	0	1	0	Sound Alarm
0	1	1	0	Sound Alarm
0	1	1	1	
0	1	0	1	
0	1	0	0	
1	1	0	0	
1	1	0	1	Sound Alarm
1	1	1	1	Sound Alarm
1	1	1	0	Sound Alarm
1	0	1	0	Sound Alarm
1	0	1	1	Sound Alarm
1	0	0	1	Sound Alarm
1	0	0	0	

In the introduction to this chapter, the initial truth table I came up with didn't seem very helpful. The reason for this is something that I will harp upon throughout this book – listing logic responses to binary input is not very effective, because of the large number of states that can change at any given time. If you look at Table 2-1, you will see that going from the state where $P=0$, $D=W1=W2=1$ to $P=1$, $D=W1=W2=0$ involves the changing of four bits. While this is a natural progression of binary numbers and probably an intuitive way of coming up with a number of different input states, it is not an effective way to look at how a logic circuit responds to varying inputs.

A much better method is to list the output responses in a truth table that is ordered using Gray codes, as I have shown in Table 2-2. Gray codes are a numbering system in which only one bit changes at a time: they are explained in detail along with how they are generated in Chapter 4. When you are listing data, regardless of the situation, you should *always* default to using Gray code inputs instead of incrementing binary inputs, as I have shown in Table 2-1.

Taking this advice, I recreated the home alarm system truth table using Gray codes in Table 2-2. When you look at Table 2-2, you should notice that the "discontinuities" of Table 2-1 have disappeared. The bit patterns which "Sound Alarm" group together quite nicely.

Looking at each value which "Sound Alarm", you'll notice that each pair has three bits in common. To illustrate this, in Table 2-3, I have circled the bit which is different between each of the four pairs. In each of these

Table 2-3 Uncommon bits in "Sound Alarm" pairs.

P	D	W1	W2	Alarm Response
0	0	0	0	
0	0	0	1	
0	0	1	1	
0	0	1	0	Sound Alarm
0	1	1	0	Sound Alarm
0	1	1	1	
0	1	0	1	
0	1	0	0	
1	1	0	0	
1	1	0	1	Sound Alarm
1	1	1	1	Sound Alarm
1	1	1	0	Sound Alarm
1	0	1	0	Sound Alarm
1	0	1	1	Sound Alarm
1	0	0	1	Sound Alarm
1	0	0	0	

Table 2-4 Home alarm truth table with don't care bits replaced with an "x".

P	D	W1	W2	Alarm Response
0	0	0	0	
0	0	0	1	
0	0	1	1	
0	x	1	0	Sound Alarm
0	1	1	1	
0	1	0	1	
0	1	0	0	
1	1	0	0	
1	1	x	1	Sound Alarm
1	x	1	0	Sound Alarm
1	0	x	1	Sound Alarm
1	0	0	0	

pairs, to sound the alarm we have very specific requirements for three bits, but the fourth bit can be in either state.

Another way of saying this is: for the alarm to sound, we don't care what the fourth bit is and it can be ignored when we are determining the sum of products equation for the logic function. To indicate the "don't care" bit, in Table 2-4, I have combined the bit pairs and changed the previously circled bits with an "x". This "x" indicates that the bit can be in either state for the output to be true. By replacing the two truth table entries with a single one with the don't care bit indicated by an "x" you should see that something magical is starting to happen.

The obvious observation is that the table is shorter, but you should notice that the number of events which "Sound Alarm" has been halved and they are less complex than the eight original events. The sum of products equation for the bits shown in Table 2-4 is:

$$\begin{aligned}\texttt{Alarm State} = & (!\texttt{P} \cdot \texttt{W1} \cdot !\texttt{W2}) \\ & + (\texttt{P} \cdot \texttt{D} \cdot \texttt{W2}) \\ & + (\texttt{P} \cdot \texttt{W1} \cdot !\texttt{W2}) \\ & + (\texttt{P} \cdot !\texttt{D} \cdot \texttt{W2})\end{aligned}$$

This sum of products expression will require four NOT gates, eight AND gates and three OR gates and the maximum gate delay will be nine. This has reduced the total gate count to less than 50% of the original total and this logic equation will operate somewhat faster than the original.

This is pretty good improvement in the logic circuit. You should be asking yourself if we can do better. To see if we can do better, I rearranged the data in Table 2-4 so that the "Sound Alarm" events with common don't care bits were put together and came up with Table 2-5.

When I put the "Sound Alarm" events that had the same don't care bits together, I noticed that in each of these cases, two of the remaining bits were in common and one bit changed in the two events (which I circled in Table 2.5).

In Table 2-5, you may have noticed that the single changing bit of the original Gray code input sequence has been lost; this is not a problem. The Gray code sequence has served its purpose – it has indicated the initial input patterns which are common with its neighbors. In complex truth tables, you may have to rearrange bit patterns multiple times to find different

Table 2-5 Optimized home alarm truth table rearranged with don't care event bits moved together.

P	D	W1	W2	Alarm Response
0	0	0	0	
0	0	0	1	
0	0	1	1	
0	x	1	0	Sound Alarm
1	x	1	0	Sound Alarm
0	1	1	1	
0	1	0	1	
0	1	0	0	
1	1	0	0	
1	1	x	1	Sound Alarm
1	0	x	1	Sound Alarm
1	0	0	0	

Table 2-6 Reoptimized home alarm truth table with "don't care" bits replaced with an "x".

P	D	W1	W2	Alarm Response
0	0	0	0	
0	0	0	1	
0	0	1	1	
x	x	1	0	Sound Alarm
0	1	1	1	
0	1	0	1	
0	1	0	0	
1	1	0	0	
1	x	x	1	Sound Alarm
1	0	0	0	

commonalities. When you do this, don't worry about "loosing data"; the important bit patterns are still saved in the active bit patterns.

Table 2-6 shows what happens when the second don't care bit is indicated. Since the two events which "Sound Alarm" do not have common don't care bits, we can't repeat this process any more. The two events from Table 2-6 can be written out as the sum of products:

$$\begin{aligned}\texttt{Alarm State} &= (\texttt{W1} \cdot \texttt{!W2}) \\ &\quad + (\texttt{P} \cdot \texttt{W2})\end{aligned}$$

This optimized "Alarm State" truth table has reduced our component count to one NOT gate, two AND gates and one OR gate and executes in five gate delays – quite an improvement from the original 43 gates and 11 gate delays!

Depending on how cynical you are, you might think that I "cooked up" this example to come up with such a dramatic improvement. Actually, the application shown here was my first attempt at coming up with a logic circuit to demonstrate how optimization operations of a logic circuit are performed; you will find similar improvements as this one when you start with a basic logic circuit and want to see how much you can reduce it.

Karnaugh Maps

Using truth tables is an effective but not very efficient method of optimizing digital logic circuits. A very clever French mathematician, Maurice Karnaugh (pronounced "carno") came up with a way to simplify the truth table optimization process by splitting the truth table inputs down the middle and arranging the two halves perpendicularly in order to display the relationships between bits more effectively. These modified truth tables are called "Karnaugh Maps" and are best suited for single bit output functions with three to six input bits.

My description of what a Karnaugh map is may sound cursory, but it is actually very accurate. A standard truth table can be considered to be a single dimensional presentation of a logic function and when it is properly manipulated, relationship between active outputs can be observed as I showed in the previous section. The problem with this method is that it is fairly labor intensive and will burn up a lot of paper. Karnaugh maps present the data in a two-dimensional "field" which allows for quick scanning of active output bits against their inputs, to find basic relationships between them.

An example of converting a three input logic function from a truth table to a Karnaugh map is shown in Fig. 2-2. The initial logic function would be:

$$\texttt{Output} = (!\texttt{A} \cdot !\texttt{B} \cdot \texttt{C}) + (!\texttt{A} \cdot \texttt{B} \cdot !\texttt{C}) + (\texttt{A} \cdot \texttt{B} \cdot !\texttt{C}) + (\texttt{A} \cdot \texttt{B} \cdot \texttt{C}) + (\texttt{A} \cdot !\texttt{B} \cdot \texttt{C})$$

To create the Karnaugh map, I created a two by four matrix, with the rows being given the two different values for "A" and the columns given the four different values for "B" and "C". Note that the columns are listed as a two bit Gray code – this is an important feature of the Karnaugh map and, as I have pointed out, an important tool to being able to optimize a function.

Once the two axes of the Karnaugh map are chosen, the outputs from the truth table are *carefully* transferred from the truth table to the Karnaugh map. When transferring the outputs, treat the Karnaugh map as a two-dimensional array, with the "X" dimension being the inputs which

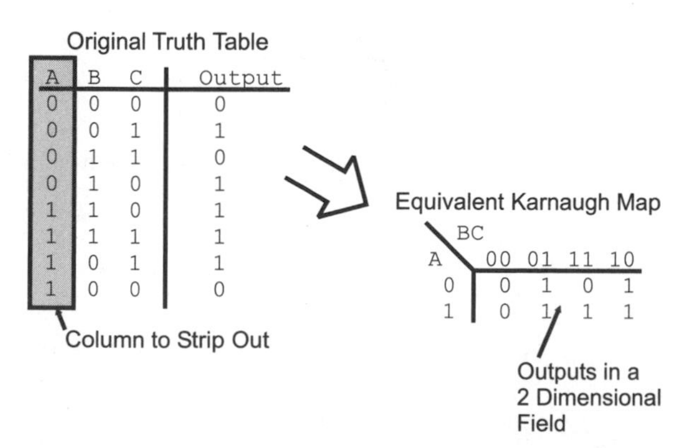

Fig. 2-2. Converting a truth table to a Karnaugh map.

weren't stripped out and the "Y" dimension being the inputs which were stripped out from the truth table. When you are first starting out, this will be an operation in which you will tend to make mistakes because it is unfamiliar to you. To make sure you understand the process, it is a good idea to go back and convert your Karnaugh map into a truth table and compare it to your original truth table.

When you have created the Karnaugh map for your function, it is a good idea to either photocopy it or write it out in pen before going on. I am suggesting this action because, just as you did with the truth table, you are going to circle outputs which have the same unchanging bit patterns. As you circle the outputs, chances are you are not going to see the most effective groups of bits to circle together, or you will find that you have made a mistake in circling the bits. A photocopy or list in ink will allow you to try again without having to redraw the Karnaugh map.

For the example shown in Fig. 2-2, the Karnaugh map has three circles put on it, as shown in Fig. 2-3. Each circle should result in combining two input sets together and making at least one bit into a "don't care".

Correctly circling bits can be difficult to understand, but there are a few rules that can be applied to it. First, each circle *must* be around a power of two number of bits – you cannot circle three bits (as shown in Fig. 2-4 for this example). Secondly, it is not a problem if circles overlap over specific bits. I should point out that there is the case for redundant circles (Fig. 2-5). If a circle is drawn and all the circled bits are enclosed in another circle, then the enclosed circle is redundant. Thirdly, remember that when you are circling bits that you want to circle a power of two number of bits, not just two. In Fig. 2-6, I have modified the three bit Karnaugh map with the outputs at A = 0 and B = C = 1 and A = 1 and B = C = 0 being a "1"

A \ BC	00	01	11	10
0	0	1	0	1
1	0	1	1	1

Fig. 2-3. Circling like bits in the example Karnaugh map.

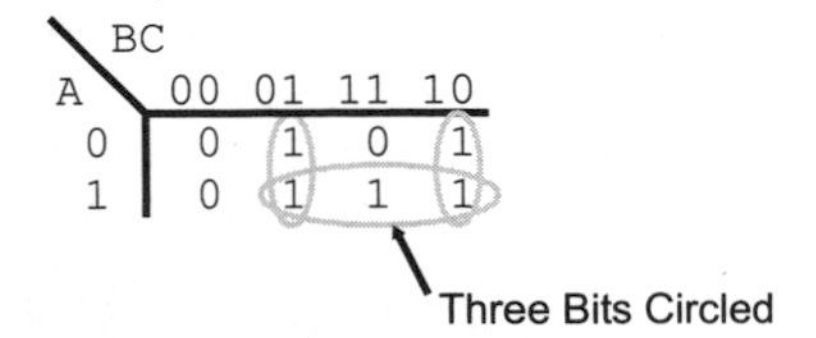

Fig. 2-4. Incorrectly circling an odd number of bits in the example Karnaugh map.

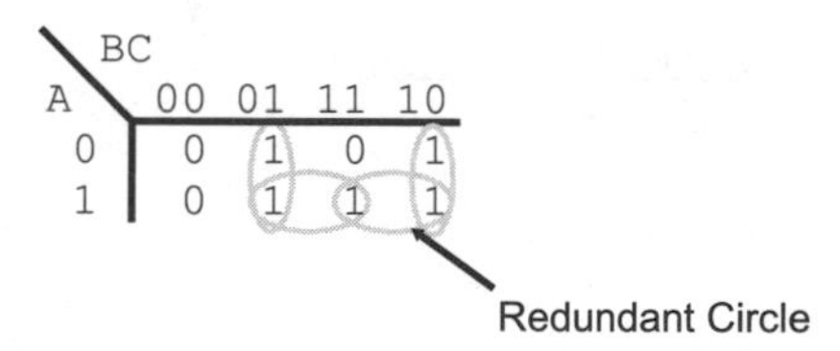

Fig. 2-5. Redundant circles on the example Karnaugh map.

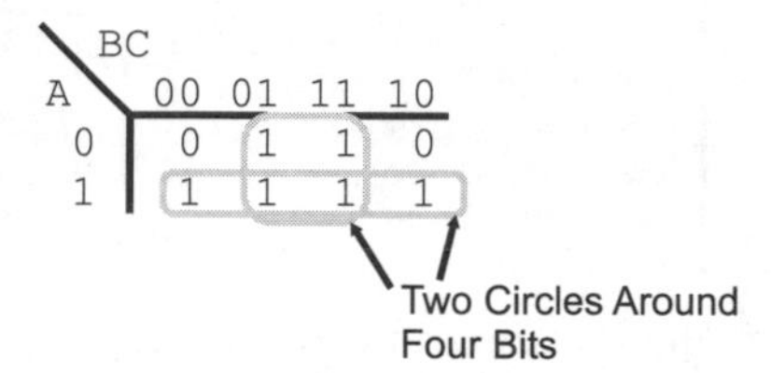

Fig. 2-6. Karnaugh map showing that more than two bits can be circled at the same time.

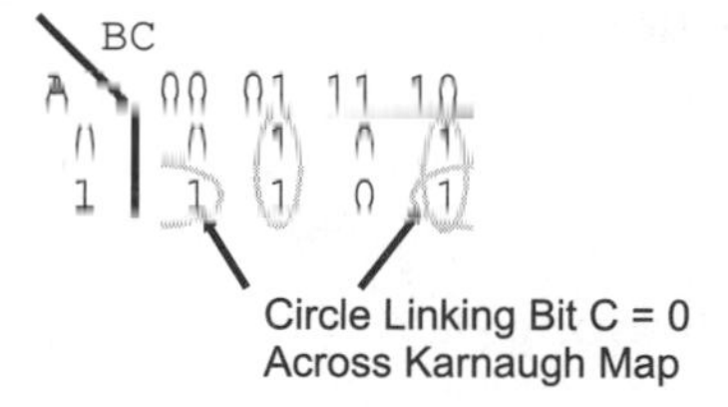

Fig. 2-7. Circle extending outside the apparent boundaries of the Karnaugh map.

and found that I could circle two groups of four bits. In each of these cases, I have made two bits "don't care".

Finally, saying that a Karnaugh map is like a two-dimensional array is inaccurate – it is actually a continuum unto itself, with the tops and sides being connected. When you draw out your Karnaugh map, you may find that the bits which can be circled (meaning ones with similar patterns) are on opposite ends of the Karnaugh map. This is not a problem as long as there are matching bits.

Once you have the outputs circled, you can now start writing out the optimized equation. As an exercise, you might want to look at the example Karnaugh maps in Figs. 2-3, 2-6 and 2-7. The output equations for these figures are:

$$\text{Output}_{2.03} = (!B \cdot C) + (B \cdot !C) + (A \cdot C)$$

$$\text{Output}_{2.06} = A \cdot C$$

$$\text{Output}_{2.07} = (!B \cdot C) + (B \cdot !C) + (A \cdot !C)$$

P	D	W1	W2	Alarm Response
0	0	0	0	
0	0	0	1	
0	0	1	1	
0	0	1	0	Sound Alarm
0	1	1	0	Sound Alarm
0	1	1	1	
0	1	0	1	
0	1	0	0	
1	1	0	0	
1	1	0	1	Sound Alarm
1	1	1	1	Sound Alarm
1	1	1	0	Sound Alarm
1	0	1	0	Sound Alarm
1	0	1	1	Sound Alarm
1	0	0	1	Sound Alarm
1	0	0	0	

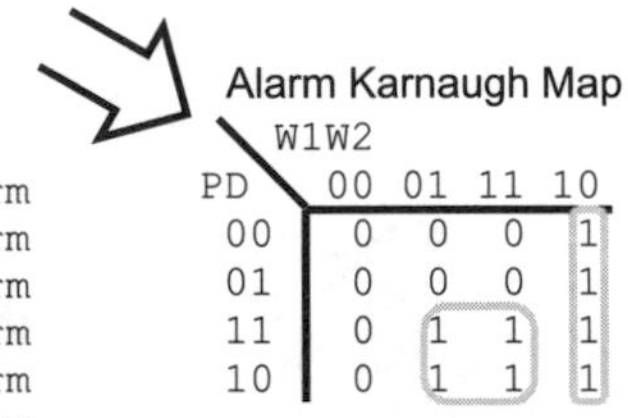

Alarm Karnaugh Map

PD \ W1W2	00	01	11	10
00	0	0	0	1
01	0	0	0	1
11	0	1	1	1
10	0	1	1	1

Fig. 2-8. Home alarm truth table to Karnaugh map.

In this chapter, I wanted to show how the different optimizing tools are used for the home alarm system presented in the chapter introduction. The alarm system's functions can be optimized using the Karnaugh map shown in Fig. 2-8. In Fig. 2-8, I have drawn the circles around the two groups of four active output bits which are in common and result in the logic equation

$$\texttt{Alarm Response} = (\texttt{P} \cdot \texttt{W2}) + (\texttt{W1} \cdot \texttt{!W2})$$

which is identical to the equation produced by the truth table reduction and a lot less work.

Before going on, I want to just say that once you are comfortable with Karnaugh maps, you will find them to be a fast and efficient method of optimizing simple logic functions. Becoming comfortable and being able to accurately convert the information from a truth table to a Karnaugh map will take some time, as will correctly circling active outputs to produce the optimized sum of products circuit. Once you have mastered this skill, you will find that you can go directly to the Karnaugh map from the requirements without the initial step of writing out the truth table.

Boolean Arithmetic Laws

One of the ways of optimizing circuits is look through their output equations and try to find relationships that you can take advantage of using the rules and laws in Table 2-7. These rules should be committed to memory as quickly as possible (or at least written down on a crib sheet) to help you with

Table 2-7 Boolean arithmetic laws and rules.

Rule/law	Boolean arithmetic example
AND identity function	$A \cdot 1 = A$
OR identity function	$A + 0 = A$
Output reset	$A \cdot 0 = 0$
Output set	$A + 1 = 1$
Identity law	$A = A$
AND complementary law	$A \cdot !A = 0$
OR complementary law	$A + !A = 1$
AND idempotent law	$A \cdot A = A$
OR idempotent law	$A + A = A$
AND commutative law	$A \cdot B = B \cdot A$
OR commutative law	$A + B = B + A$
AND associative law	$(A \cdot B) \cdot C = A \cdot (B \cdot C) = A \cdot B \cdot C$
OR associative law	$(A + B) + C = A + (B + C) = A + B + C$
AND distributive law	$A \cdot (B + C) = (A \cdot B) + (A \cdot C)$
OR distributive law	$A + (B \cdot C) = (A + B) \cdot (A + C)$
De Morgan's NOR theorem	$!(A + B) = !A \cdot !B$
De Morgan's NAND theorem	$!(A \cdot B) = !A + !B$

optimizing logic equations without the need of truth tables or Karnaugh maps. Many of these rules and laws will seem self-evident, but when you are working at optimizing a logic equation in an exam, it is amazing what you will forget or won't seem that obvious to you.

When I talk about using the laws and rules in Table 2-7 to simplify a logic equation, I normally use the term “reduce” instead of “optimize”. The reason for thinking of these operations as a reduction is due to how much the logic equation shrinks as you work through it, trying to find the most efficient sum of products expression.

The two identity functions are used to indicate the conditions where an input value can pass unchanged through an AND or OR gate. The output set, reset and complementary laws are used to output a specific state when a value is passing through an AND or OR gate. The idempotent laws can be summarized by saying that if an input passes through a non-inverting gate, its value is not changed.

The remaining laws – commutative, associative and distributive – and De Morgan’s theorems are not as trivial and are extremely powerful tools when you have a logic equation to optimize. The commutative laws state that the inputs to AND and OR gates can be reversed, which may seem obvious, but when you have a long logic equation that is written in an arbitrary format (not necessarily in sum of product format), you can get confused very easily as to what is happening. It’s useful to have a law like this in your back pocket to change the logic equation into something that you can more easily manipulate.

To demonstrate the operation of these laws, we can go back to some of the logic circuits described in the Karnaugh map examples of the previous section. Looking at Fig. 2-3, the initial sum of products logic equation would be:

$$\texttt{Output} = (!\texttt{A} \cdot !\texttt{B} \cdot \texttt{C}) + (!\texttt{A} \cdot \texttt{B} \cdot !\texttt{C}) + (\texttt{A} \cdot !\texttt{B} \cdot \texttt{C}) + (\texttt{A} \cdot \texttt{B} \cdot \texttt{C}) + (\texttt{A} \cdot \texttt{B} \cdot !\texttt{C})$$

Using the AND associative law, I can rewrite this equation with the A term separate from the B and C terms to see if there are any cases where the B and C terms are identical.

$$\texttt{Output} = !\texttt{A} \cdot (!\texttt{B} \cdot \texttt{C}) + !\texttt{A} \cdot (\texttt{B} \cdot !\texttt{C}) + \texttt{A} \cdot (!\texttt{B} \cdot \texttt{C}) + \texttt{A} \cdot (\texttt{B} \cdot \texttt{C}) + \texttt{A} \cdot (\texttt{B} \cdot !\texttt{C})$$

By doing this, I can see that the inside terms of the first and third products are identical. Along with this, I can see that the second and fifth products are also identical. Using the OR distributive law, I can combine the first and third terms like:

$$
\begin{aligned}
&!A \cdot (!B \cdot C) + A \cdot (!B \cdot C) \\
&\quad = (!A + A) \cdot (!B \cdot C)
\end{aligned}
$$

Using the OR complementary law, I know that A OR !A will always be true. This is actually a clear and graphic example of the "don't care" bit; regardless of the value of this bit, the output will be true so it can be ignored. The partial equation of the two terms reduces to:

$$
\begin{aligned}
&(!A \cdot !B \cdot C) + A \cdot (!B \cdot C) \\
&\quad = 1 \cdot (!B \cdot C)
\end{aligned}
$$

The 1 ANDed with !B AND C can be further reduced using the AND identity law (1 AND A equals A):

$$
\begin{aligned}
&!A \cdot (!B \cdot C) + A \cdot (!B \cdot C) \\
&\quad = (!B \cdot C)
\end{aligned}
$$

This can be repeated for the second and fifth terms:

$$
\begin{aligned}
&(!A \cdot B \cdot !C) + (A \cdot B \cdot !C) \\
&\quad = (B \cdot C)
\end{aligned}
$$

If you go back to the original logic equation, you will see that the fourth term $(A \cdot B \cdot C)$ has not been reduced by combining it with another term. It can actually be paired with the third term $(A \cdot !B \cdot C)$ by rearranging the two terms (using the AND commutative law) so that part of the terms operating on two bits are in common $(A \cdot C)$. Once this is done, the third and fourth terms can be reduced as:

$$
\begin{aligned}
&(A \cdot !B \cdot C) + (A \cdot B \cdot !C) \\
&\quad = (A \cdot !C)
\end{aligned}
$$

After doing this work, the optimized or reduced sum of product logic equation for this function is

$$
\text{Output} = (!B \cdot C) + (B \cdot !C) + (A \cdot C)
$$

which is identical to what was found using the Karnaugh map.

Looking at the reduced logic equation, you should have noticed that there are two terms that will output a "1" at the same time ($(!B \cdot C)$ and $(A \cdot C)$ with $A = 1$, $B = 0$ and $C = 1$). This is not a problem because the OR gate (even though the symbol that I use is a "+") will only output a 1, regardless of how many true inputs it has. This was mentioned when the

Karnaugh maps were presented, but I wanted to reinforce that the same issue is present when you are reducing logic equations.

Before moving on, let's go back to the home alarm logic equation and see if it can be reduced in the same way as the example above. Starting with the sum of products logic equation:

```
Alarm State = (!P · !D · W1 · !W2)
            + (!P · D · W1 · !W2)
            + (P · !D · !W1 · W2)
            + (P · !D · W1 · !W2)
            + (P · !D · W1 · W2)
            + (P · D · !W1 · W2)
            + (P · D · W1 · !W2)
            + (P · D · W1 · W2)
```

We can bring out the "P" values from the products and look for similarities in the remaining bracketed values and combine them using the associative, distributive, complementary and AND identity laws. I can see that the first and fourth, second and seventh can be combined, resulting in the logic equation:

```
Alarm State = (!D · W1 · !W2)
            + (D · W1 · !W2)
            + (P · !D · !W1 · W2)
            + (P · !D · W1 · W2)
            + (P · D · !W1 · W2)
            + (P · D · W1 · W2)
```

Bringing "W1" to the forefront allows the combination of the third and fourth and fifth and sixth terms of the logic equation above, resulting in the new equation:

```
Alarm State = (!D · W1 · !W2)
            + (D · W1 · !W2)
            + (P · !D · W2)
            + (P · D · W2)
```

We have eliminated half the terms and, of those remaining, they are 25% smaller. Looking at the new logic equation, we can see that by combining the first and second terms (making "D" a don't care bit in the process)

Table 2-8 Testing optimized home alarm logic equation.

P	D	W1	W2	Alarm	W1 ● !W2	P ● W2	OR
0	0	0	0	0	0	0	0
0	0	0	1	0	0	0	0
0	0	1	1	0	0	0	0
0	0	1	0	1	1	0	1
0	1	1	0	1	1	0	1
0	1	1	1	0	0	0	0
0	1	0	1	0	0	0	0
0	1	0	0	0	0	0	0
1	1	0	0	0	0	0	0
1	1	0	1	1	0	1	1
1	1	1	1	1	0	1	1
1	1	1	0	1	1	0	1
1	0	1	0	1	1	0	1
1	0	1	1	1	0	1	0
1	0	0	1	1	0	1	0
1	0	0	0	0	0	0	0

and combining the third and fourth terms ("D" again is the don't care bit) we end up with:

$$\text{Alarm State} = (\text{W1} \cdot !\text{W2}) + (\text{P} \cdot \text{W2})$$

which is, again, the logic equation found by optimizing the function using truth tables or Karnaugh maps.

Personally, I tend to optimize logic equations using the Boolean arithmetic laws and rules listed in Table 2-7. Once a reduced sum of products equation has been produced, I then go back and compare its outputs in a truth table with the required outputs. In doing this, I present the values for each product (AND) and the final sum (OR) in separate columns, as shown in Table 2-8.

Optimizing for Technology

If you review the laws in Table 2-7 and correlate them to the text in the previous section, you'll see that I missed the last two (De Morgan's theorem). These two laws are not typically used during basic logic equation reduction because they typically involve converting part of an equation into an NAND or NOR gate, which is important when finally implementing a logic function in actual electronics. Another important aspect of optimizing for technology is adding functions out of the leftover gates in your circuit; by looking at how differently a logic circuit could be implemented, you may be able to add functionality to your circuit, without adding any cost to it.

Table 2-9 XOR gate truth table.

A	B	A ^ B
0	0	0
0	1	1
1	1	0
1	0	1

So far in the book, I haven't discussed the "Exclusive OR" (XOR) gate in a lot of detail, but it is vital for implementing binary adders, as I will show you later in the book. In the first chapter, I presented the XOR gate with the truth table shown in Table 2-9.

You should probably be able to create the logic equation for the XOR table as:

$$\texttt{Output} = (!\texttt{A} \cdot \texttt{B}) + (\texttt{A} \cdot !\texttt{B})$$

which does not seem like a very likely candidate for optimization. Similarly, you probably would have a hard time believing that the following logic equation would perform the same function:

$$\texttt{Output} = !((\texttt{A} \cdot \texttt{B}) + !(\texttt{A} + \texttt{B}))$$

But, using De Morgan's theorem as well as the other rules and laws from Table 2-7, I can go through the manipulations shown in Table 2-10 to show that they are equal, as well as count out the gates required by intermediate steps to give you a list of different implementations of the XOR gate. Each intermediate step in Table 2-10 is an implementation of the XOR gate that you could implement using the number of gates listed to the right of the terms.

It's interesting to note that a total of five gates is required for each implementation – this is not something that you can count on when you are working at optimizing a circuit.

The basic gate used in TTL is the "NAND" gate: this means that the three basic gates (AND, OR and NOT) are built from multiples of it, as I've shown in Fig. 2-9. The basic gate for CMOS is the NOR gate, and Fig. 2-10 shows how the three basic gates are implemented for it. The three gate NAND and NOR equivalencies for the OR and AND gates, respectively, are perfect examples of De Morgan's theorem in operation. These implementations

Table 2-10 Different implementations of the XOR gate.

Terms	NOTs	ANDs	ORs	NANDs	NORs
$(!A \cdot B) + (A \cdot !B)$	2	2	1	0	0
$!(!(!A \cdot B) \cdot !(A \cdot !B))$	2	0	0	3	0
$!((A + !B) \cdot (!A + B))$	2	0	2	1	0
$!((A \cdot B) + (!A \cdot !B))$	2	2	0	0	1
$!((A \cdot B) + !(A + B))$	0	1	0	0	2

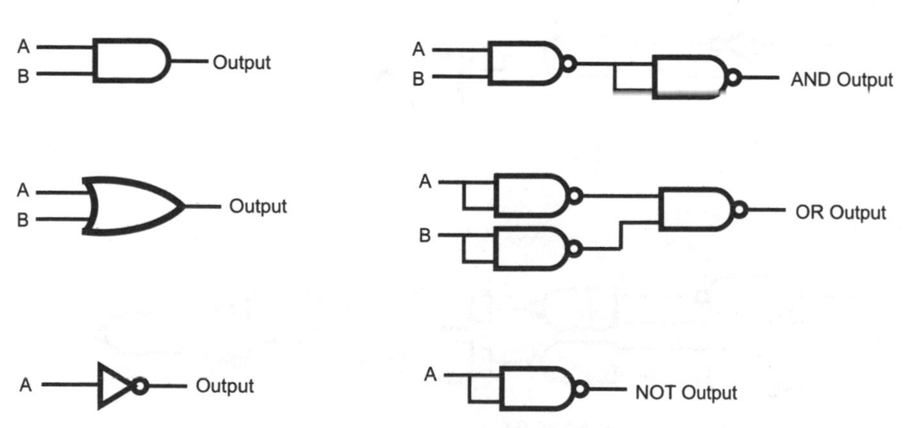

Fig. 2-9. Implementing the three basic gates using NAND gates.

can be checked against De Morgan's theorem and the rules and laws presented in Table 2-7.

By understanding how gates are implemented in chips, we can now look at how to optimize the gates to provide the fastest possible operation of the logic function. Using the example of the XOR gate, we can graphically show how the gate is implemented using ANDs, ORs and NOTs and how these gates are implemented as NAND gates in TTL chips (Fig. 2-11).

Looking at the bottom logic diagram of Fig. 2-11, you can see that there are two sets of NAND gates wired as inverters together. Going back to Table 2-7, we can see that a doubly inverted signal is the same signal, so we can eliminate these two sets of NOT gates, as shown in Fig. 2-12. The resulting XOR circuit will pass signals through three NAND gates, which

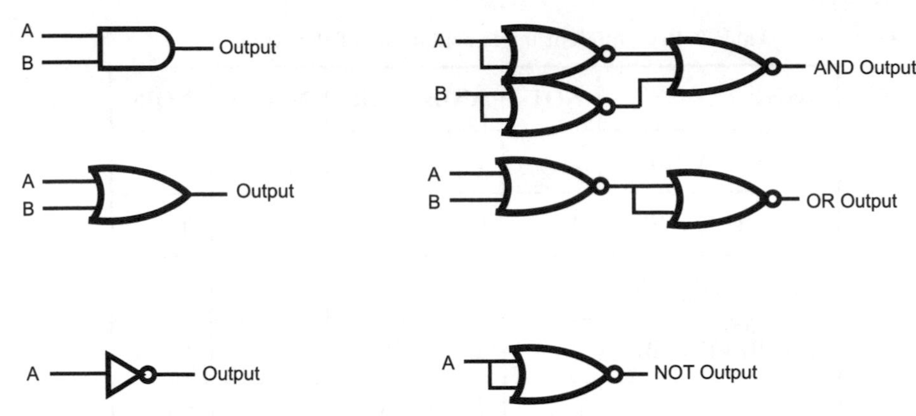

Fig. 2-10. Implementing the three basic gates using NOR gates.

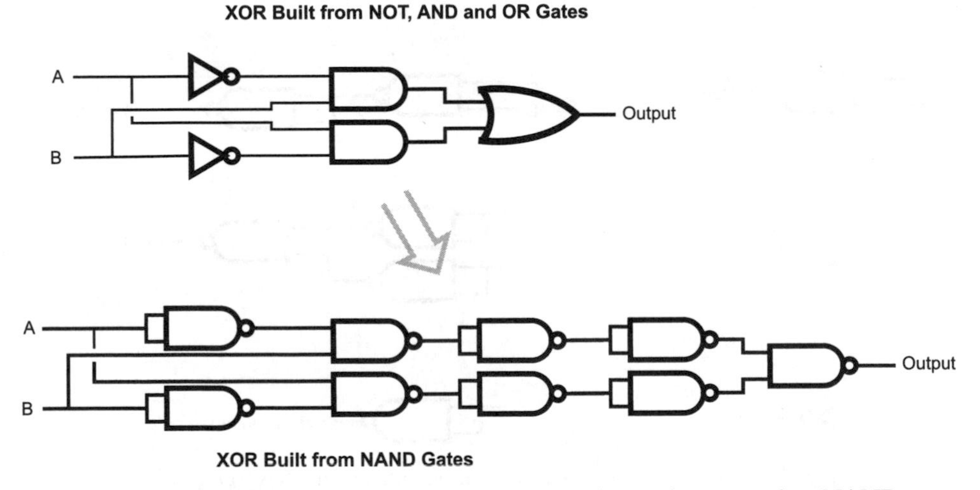

Fig. 2-11. XOR gate built from sum of products equation and converted to NAND gates.

counts as three "gate delays". This is an example of what I call "technology optimization": the logic circuit has been reduced to its bare minimum, taking advantage of the operation of the basic logic gates that make up the technology that it is implemented in.

Before moving on, I want to take one more look at the home alarm circuit that has been discussed throughout this chapter. I made a pretty bold statement at the start of the chapter, saying that it could be reduced to fit into the most basic TTL chip available – let's see how honest I was being.

The (repeatedly) optimized logic equation for the home alarm system was:

$$\texttt{Alarm State} = (\texttt{W1} \cdot \texttt{!W2}) + (\texttt{P} \cdot \texttt{W2})$$

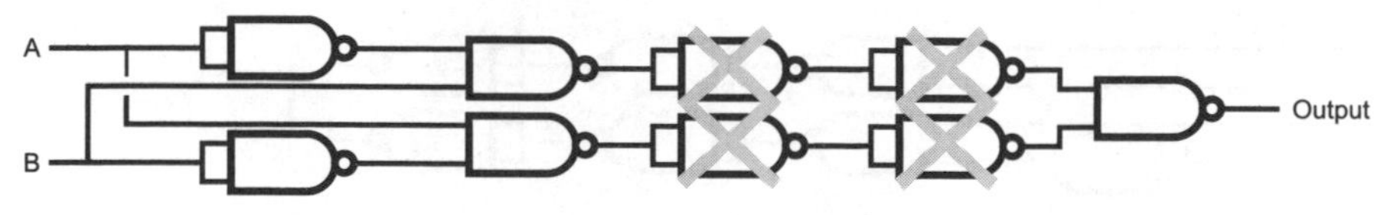

XOR Built from NAND Gates with Redundant Gates Marked and Removed Below

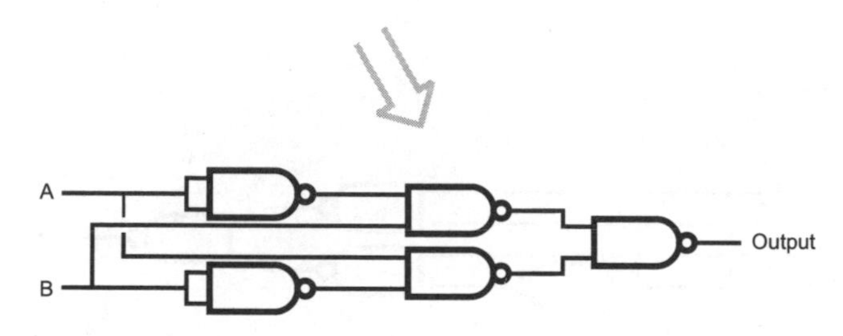

Fig. 2-12. Optimized XOR gate built from NAND gates.

Alarm Circuit Built from NOT, AND and OR Gates

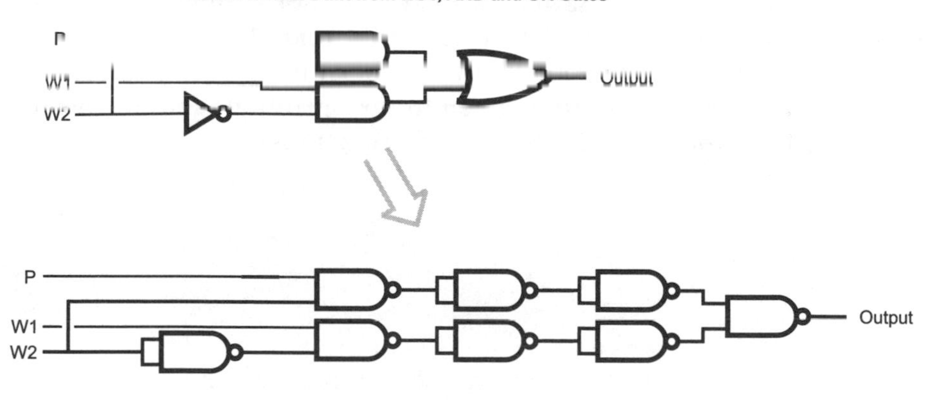

Alarm Circuit Directly Built from NAND Gates

Fig. 2-13. Home alarm logic circuit built using AND, OR and NOT gates and converted to NAND gates.

which could be first implemented in two AND, one OR and one NOT gate, as shown in Fig. 2-13 and converted to just NAND gates. You may have noted in Fig. 2-13 the remarkable similarity between the home alarm logic diagram and the XOR logic diagram – as I've shown in Fig. 2-14, the logic function reduces to just four NAND gates (one less than the XOR gate built out of NAND gates).

The final home alarm logic function requires four two input NAND gates – which is just what the 7400, the most basic TTL chip, provides. Every TTL chip, except for this one and a derivative revision, has more than four

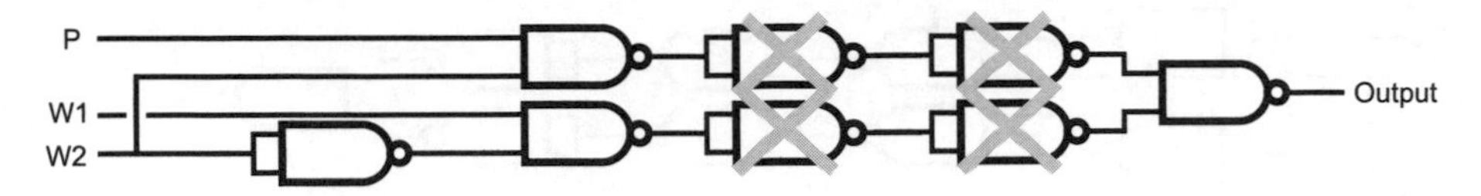

Home Alarm Logic Built from NAND Gates with Redundant Gates Marked and Removed Below

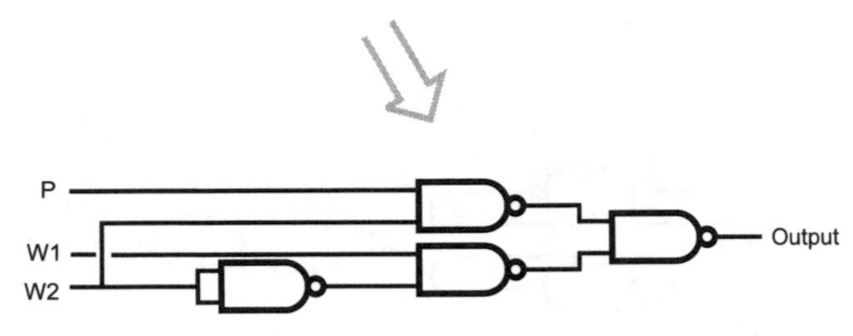

Fig. 2-14. Optimixed alarm circuit built from NAND gates.

gates built into them because they provide additional functions requiring multiple NAND gates. I was not exaggerating when I said that the home alarm logic function could be reduced to the most basic TTL chip available. In the next chapter, I will introduce you to the operation of TTL chips that provide the basis for digital electronic logic functions.

Quiz

1. The three parameters that are used to measure the optimization of a digital electronic circuit are:
   (a) Cost, speed and complexity
   (b) Gate delay, gate count and technology optimization
   (c) Gate count, number of gate delays a signal must pass through and technology optimization
   (d) Gate count, number of connections a signal must pass through and technology optimization
2. If TTL logic has a gate delay of 8 ns and the signal passing through an XOR gate built from NAND gates has to go through 9 gates and the shortest path is five gate delays, the time required for a signal to pass through the gates is:
   (a) 40 ns
   (b) 8 ns

(c) indeterminate
(d) 24 ns

3. When writing out a truth table, the inputs should be listed:
   (a) Using a "Gray code"
   (b) Using a "binary progression"
   (c) In alphabetical order
   (d) In order of importance

4. The "don't care" bit in a truth table is:
   (a) Indicated by a "dc" and replaces the common bits in two true sets of inputs
   (b) Indicated by an "x" and replaces the common bits in two true sets of inputs
   (c) Indicated by a "dc" and replaces the uncommon bits in two true sets of inputs
   (d) Indicated by an "x" and replaces the uncommon bits in two true sets of inputs

5. When optimizing a logic function you can expect:
   (a) That the number of chips that are required is reduced from the initial design
   (b) That the optimized function runs faster than the initial design
   (c) Cheaper chips can be used than in the initial design
   (d) Answers (a) through (c) are all possible and it might not be able to optimize the circuit from the initial sum of products equation

6. Karnaugh maps are:
   (a) Tools designed to help you find your way around a digital electronic circuit
   (b) A tool that will help you optimize a logic function
   (c) The most efficient method of optimizing logic fuctions
   (d) Hard to understand but must be used in every logic function design

7. The sum of products logic equation

   $$\texttt{Output} = (\texttt{A} \cdot \texttt{!B} \cdot \texttt{C}) + (\texttt{!A} \cdot \texttt{!B} \cdot \texttt{C})$$

   can be reduced to:
   (a) A · C
   (b) !A · !B
   (c) C · !B
   (d) C

8. Which of the following pairs of Boolean arithmetic laws cannot be used together?
   (a) Identity and De Morgan's theorem
   (b) Associative and idempotent
   (c) Complementary and commutative
   (d) All the laws and rules can be used together

9. The NAND equivalent to an AND gate is:
   (a) Built from two NAND gates and requires two gate delays for a signal to pass through
   (b) Built from three NAND gates and requires two gate delays for a signal to pass through
   (c) Built from three NAND gates and requires three gate delays for a signal to pass through
   (d) Built from one NAND gate as well as a NOT gate and requires two gate delays for a signal to pass through

10. Technology optimization is defined as:
    (a) Designing the circuit which uses the fewest number of chips and signals pass through it as fast as possible
    (b) Implementing logic functions to take advantage of the base logic of the logic technology used as well as using any leftover gates
    (c) Finding the most efficient digital electronic technology to use for the application
    (d) Designing circuitry that dissipates the least amount of heat to perform a desired function

CHAPTER 3

Creating Digital Electronic Circuits

In the previous chapters, I introduced you to the basic Boolean arithmetic theory behind decoding and design combinatorial circuits; binary data is manipulated by simple operations to produce a desired output. Before going on and showing you how these basic operations are extended to create complicated functions and products, I want to take a step back and look at basic electrical theory and semiconductor operation and how they are applied to digital electronics. While digital electronics work with "ones and zeros", it is still built from the basic electronic devices that are outlined in the beginning of this chapter. It is impossible to work successfully with digital electronics without understanding basic electrical theory and how simple electronic devices work.

For many people, this chapter will be a review, but I still urge you to read through this chapter and answer the quiz at the end of it. While you may be familiar with electrical rules and device operation, you may not be so comfortable understanding how they are used to create digital electronics.

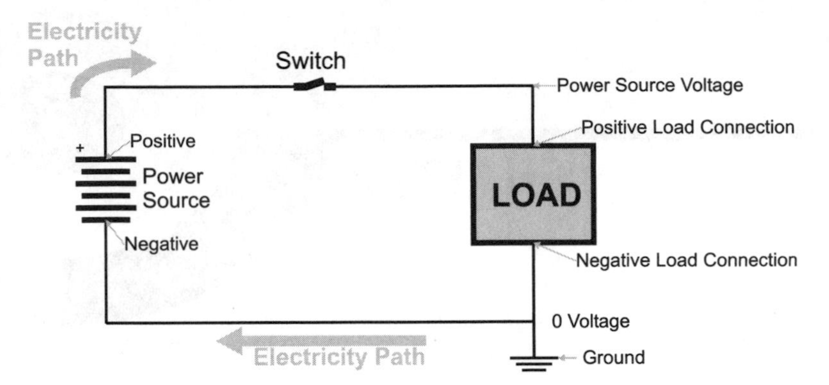

Fig. 3-1. Basic circuit diagram.

The most basic rule of electricity is that it can only move in a "closed circuit" (Fig. 3-1) in which a "power source" passes electricity to and then pulls it from a *load*. The power source has two connections that are marked with a "+" ("positive") and "−" ("negative") markings to indicate the "polarity" of the power source and the power source symbol consists of a number of pairs of parallel lines with the longer line in each pair representing the positive connection. The black lines connecting the power source to the load represent wires. When basic electricity is presented, this "load" is most often a lightbulb, because it turns on when electricity passes through it. As well as being a lightbulb, the load can be electrical motors, heater elements or digital electronic chips or any combination of these devices.

In the "electrical circuit" (or "schematic diagram") shown in Fig. 3-1 you can see that I have included a *switch,* which will *open* or *close* the circuit. When the switch is closed, electricity will flow through from the power source, to the load and back. If the switch is open or the wires connecting the power source to the load are broken, then electricity will not flow through the load.

As you are probably aware, electricity consists of electrons moving from the power source through the wires to the load and back to the power source. There are actually two properties of electricity that you should be aware of and they are analogous to the two properties of water flowing through a pipe. *Voltage* is the term given to the *pressure* placed on the electrons to move and *current* is the number of electrons passing by a point at a given time.

In the early days of electrical experimentation, it was Benjamin Franklin who postulated that electricity was a fluid, similar to water. As part of this supposition, he suggested that the electrical current flowed from the positive power supply connection to the negative. By suggesting that electrical current flowed from positive to negative, he started drawing electrical wiring diagrams or *schematics* (like the one in Fig. 3-1) with the electrical energy at the positive power supply connection being at the highest state. As the electrical current "flowed down" the page to the negative connection of the power supply, the energy of the electricity decreased. This method of drawing electrical circuits is clever and intuitive and caught on because it described what was happening in it.

Unfortunately, Franklin's suggestion that electrical current flowed from the positive to negative connections of the power source through the load was wrong. As we now know, electrons that make up electricity flow from the negative to positive connections of the power supply. This discovery was made about 150 years after his kite in a lightning storm experiment, so the notion that electrical current flowed from positive to negative was widely accepted and was never really challenged. For this reason, you should keep in mind that "electrical current flow" takes place in the opposite direction to "electron flow" in electrical circuits. This point trips many people new to electronics and I should state emphatically that the direction of current flow follows Franklin's convention.

Looking at the bottom right hand corner of Fig. 3-1, you will see a funny set of lines attached to the wiring lines – this is the circuit's "ground" connection. The circuit ground is another invention of Benjamin Franklin. If there is ever a large amount of electricity that finds its way into the circuit, it will have an "escape route" to prevent damage to the circuit's components or hurting anybody working with the circuit. The ground connection was literally a metal spike driven into the ground and connected to a home or barn's lightning rod. In modern structures, the "ground" is a connection to the metal pipe bringing in water.

Another term commonly used for a circuit's wire connections or wiring lines is "nets". The term originated when circuit analysis was first done on complex *networks* of wiring. It is used to describe the individual wiring connections in a circuit. I will use this term along with "wiring" and "lines" in this book interchangeably.

Like power supplies, many load devices also have connections that are marked with a positive ("+") and negative ("−") connections. When discussing the positive and negative connections of a basic two-wire load device, I like to use the terms, *anode* and *cathode* to describe the positive and negative connections of the load, respectively. The load's anode must always be connected to the positive terminal of the power supply and the load's cathode must always be connected to the negative terminal of the power supply. Reversing these connections may result in the device not working or even going so far as literally "burning out". To keep the terms anode and cathode straight, I remember that a "cathode ray tube" (i.e. your TV set) involves firing electrons, which are negative, at a phosphorus screen.

More complex load devices, like logic chips, also have positive and negative connections, but these connections are normally called *Vcc* or *Vdd* for the positive connection or *Gnd* and *Vss* for the negative (ground) connections.

When working with most basic digital electronic circuits, the binary value "1" is applied to a high, positive voltage (usually close to the voltage applied to the Vcc or Vdd pin of the chip). The binary value "0" is applied to low voltage (very close to the ground voltage level of the chip). This is generally considered intuitively obvious and can be easily remembered that a "1" input is the same as connecting an input to the power supply and a "0" input is the same as connecting an input to ground (resulting in "0" voltage). Similarly for outputs, when a "1" is output, you can assume that the chip can turn on a light. These conventions are true for virtually all basic electronic logic technologies; when you get into some advanced, very high speed logic, you may find that chips are designed with different operating conditions.

To simplify wiring diagrams, you will see many cases where the positive power connection and negative power connection are passed to terminal symbols to simplify the diagram and avoid the additional complexity of power and ground lines passing over the page and getting confused with the circuit "signal" lines.

When you are wondering how to connect an electronic device to its power supply, you can use Table 3-1 as a quick reference.

Table 3-1 Power wiring reference.

Positive ("+") connection	Negative ("−") connection	Comments
Red wire	Black wire	Wires connected to and between devices
Anode	Cathode	Diodes and capacitors
Vcc	Gnd	TTL
Vdd	Vss	CMOS

Basic Electronic Laws

Before starting to build your own digital electronics circuits, you should make sure that you are very familiar with the basic direct current electricity laws that govern how electricity flows through them. Don't worry if you have not yet been exposed to any direct current electrical theory, it's actually pretty simple and in the introduction to this chapter, I gave you a quick run down of how direct current circuits operate. I'm sure you were able to get through that without too many problems.

To make sure that you are clear on what *direct current* (also known as "DC") is, it consists of electricity running in a single direction without any changes. *Alternating current* ("AC") continuously changes from positive to negative (as shown in Fig. 3-2). AC is primarily used for high-power circuitry and not for any kind of digital electronics, cxccpt as something that is controlled by it. Digital electronics is powered by direct current, which consists of a fixed voltage which does not change level or polarity, as AC does.

As I indicated in the introduction, there are two components to electricity: *voltage* is the "pressure" applied to the electrons and *current* is the number of electrons that flow past a point or a set amount of time. I use the terms "pressure" and "flow" to help you visualize electricity moving in a wire as being the same as water flowing through a pipe. Using a water/pipe analogy can help you visualize how electricity moves and changes according to the conditions it is subjected to.

It should be obvious that the more pressure you apply to water in a pipe, the more water will pass through it. You can demonstrate this with a garden hose and a tap. By partially closing the tap, you are restricting the flow of the water coming from it, and the stream will not go very far from the end of the hose and very little water will flow out. When you completely open the tap, the water will spray out considerably further and a lot more water will be passing out the end of the hose. Instead of saying that you are closing the tap,

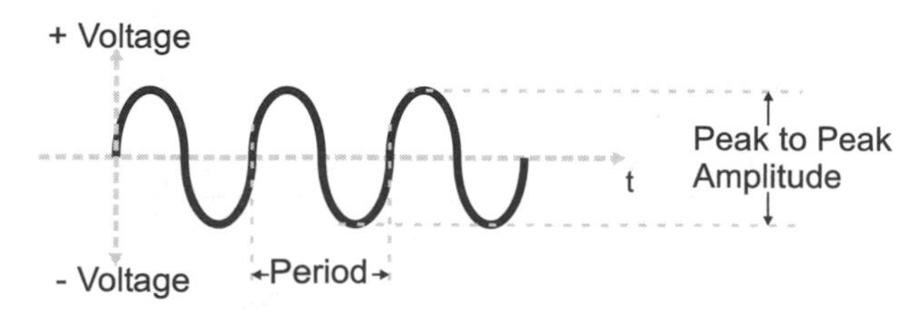

Fig. 3-2. Alternating current.

why don't you think of the closing tap as resisting the flow of water through the pipe and into the hose? This is exactly analogous to the *load* in a circuit converting electrical energy into something else. Electricity coming out of the load will be at a lower pressure (or voltage) than the electricity going into the load and the amount of current will be reduced as well.

When you visualized the pipe/tap/hose analogy, you probably considered that all the resistance in the circuit was provided by the tap – the pipe and the hose did not impede the water's flow in any way. This is also how we model how electricity flows in wires; the wires do not cause a drop in voltage and do not restrict the amount of current that is flowing in them. If you think about it for a moment, you will probably realize that this assumption means that the wires are "superconductors"; any amount of electricity and at any voltage could be carried in the wires without any loss.

The wires that you use are certainly not superconductors, but the assumption that the wires do not impede the flow of electricity is a good one as their resistance in most circuits is usually negligible. By assuming that the wires are superconductors, you can apply some simple rules to understand the behavior of electricity in a circuit.

Going back to the original schematic diagram in this chapter (see Fig. 3-1), we can relate it to the pipe/tap/hose example of this section. The circuit's power supply is analogous to the pipe supplying water to the tap (which itself is analogous to the electrical circuit's load). The hose provides the same function as the wires bringing the electrical current back to the power supply.

In the pipe/tap/hose example, you should be able to visualize that the amount of water coming through the hose is dependent on how much the tap impedes the water flow from the pipe. It should be obvious that the less the tap impedes the water flow, the more water will come out the hose. Exactly the same thing happens in an electrical circuit; the "load" will impede or "resist" the flow of electricity through it and, in the process, take energy from the electricity to do something with it.

The most basic load that can be present in a circuit is known as the "resistor" (Fig. 3-3), which provides a specified amount of resistance,

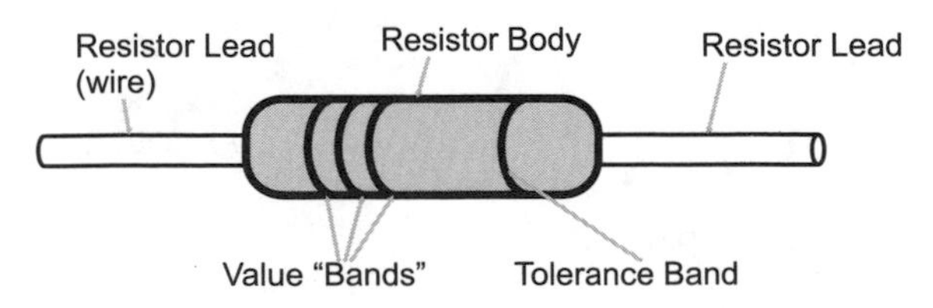

Fig. 3-3. Basic resistor.

measured in "ohms", to electricity. The "schematic symbol" is the jagged line you will see in various schematic diagrams in this book and in other sources. The schematic symbol is the graphic representation of the component and can be used along with the graphic symbol for a gate in a schematic diagram.

In traditional resistors, the amount of resistance is specified by a number of colored bands that are painted on its sides – the values specified by these bands are calculated using the formula below and the values for each of the colors listed in Table 3-2.

$$\texttt{Resistance} = ((\texttt{Band 1 Color Value} \times 10) + (\texttt{Band 2 Color Value})) \times 10^{\texttt{Band 3 Color Value}}\ \texttt{Ohms}$$

In the introduction to the chapter, I stated that power supplies provide electrons with a specific "pressure" called voltage. Knowing the voltage applied

Table 3-2 Resistor color code values.

Color	Band Color Value	Tolerance
Black	0	N/A
Brown	1	1%
Red	2	2%
Orange	3	N/A
Yellow	4	N/A
Green	5	0.5%
Blue	6	0.25%
Violet	7	0.1%
Gray	8	0.05%
White	9	N/A
Gold	N/A	5%
Silver	N/A	10%

to a load (or resistor), you can calculate the electrical current using Ohm's law which states:

> The voltage applied to a load is equal to the product of its resistance and the current passing through it.

This can be expressed mathematically as:

$$V = i \times R$$

where "V" is voltage, "R" is resistance and "i" is current. The letter "i" is used to represent current instead of the more obvious "C" because this character was already for specifying capacitance, as I will explain below. Voltage is measured in "volts", resistance in "ohms" and current in "amperes". For the work done in this book, you can assume that ohms have the units of volts/amperes and is given the symbol Ω; you can look up how these values are derived, but for now just take them for what I've presented here. With a bit of basic algebra, once you know two of the values used in Ohm's law, you can calculate the third.

Voltage, current, resistance, and, indeed, all the electrical values that you will see are part of the "SI" (Système Internationale), and its values are governed by SI standards. Each time a unit deviates by three orders of magnitude from the base value, the units are given a prefix that indicates the magnitude multiplier and these multipliers are listed in Table 3-3. For example, one thousandth of a volt is known as a "millivolt". The actual component values are normally given a single letter symbol that indicates its value. Most electronic devices, like resistors are given a two digit value that is multiplied by the power of ten which the symbol indicates. For example,

Table 3-3 Système Internationale magnitude of prefixes and symbols.

Power multiplier	Prefix	Symbol	Power multiplier	Prefix	Symbol
10^3	kilo	k	10^{-3}	milli	m
10^6	mega	M	10^{-6}	micro	µ
10^9	Giga	G	10^{-9}	nano	n
10^{12}	tera	T	10^{-12}	pico	p
10^{15}	peta	P	10^{-15}	femto	f

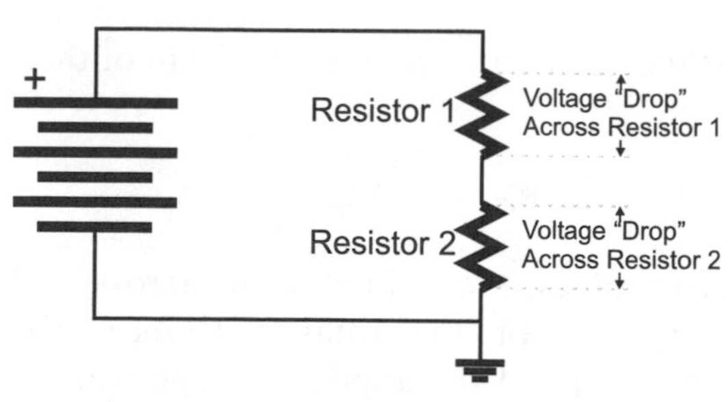

Fig. 3-4. Electrical circuit with two resistors in series.

thousands of units are given the prefix "k", so a resistor having a value of 10,000 ohms is usually referred to as having a value of "10 kohms", or most popularly "10 k".

Looking at more complex circuits, such as the two resistor "series" circuit shown in Fig. 3-4, you must remember that individual measurements must be taken across each resistor's two terminals; you do *NOT* make measurements relative to a common point. The reason for making this statement is to point out that the voltage across a resistor, which is also known as the "voltage drop", is dependent on the current flowing through it.

Using this knowledge, you can understand how electricity flows through the two series resistors in Fig. 3-4. The voltage applied to the circuit causes current to flow through both of the resistors and the amount of current is equal to the current passing through a single resistor value which is the sum of the two resistors. Knowing this current, and an individual resistor's value, you can calculate the voltage drop across each one. If you do the calculations, you will discover that the voltage drop across each resistor is equal to the applied voltage.

This may be a bit hard to understand, but go back to the pipe/tap/hose example and think about the situation where you had a pipe/tape/pipe/tap/hose. In this case, there would be a pressure drop across the first tap and then another pressure drop across the second tap. This is exactly what happens in Fig. 3-4: some voltage "drops" across Resistor 1 and the rest drops across Resistor 2. The amount of the drop across each resistor is proportional to its value relative to the total resistance in the circuit.

To demonstrate this, consider the case where Resistor 1 in Fig. 3-4 is 5 ohms and Resistor 2 is 8 ohms. Current has to flow through Resistor 1 followed by Resistor 2, which means that the total resistance it experiences is equivalent to the sum of the two resistances (13 ohms). The current through the two resistors could be calculated using Ohm's law, as voltage applied divided by Resistor 1 plus Resistor 2. The general formula for calculating equivalent the resistance of a series circuit is the sum of the resistances, which is written out as:

$$R_e = R_1 + R_2 + \ldots$$

Knowing the resistor values, the voltage drop across each resistor can be calculated as its fraction of the total resistance; the voltage across Resistor 1 would be 5/13ths of the applied voltage while the voltage across Resistor 2 would be 8/13ths of the applied voltage. Dividing the resistor values into the individual resistor voltage drops will yield the same current as dividing the applied voltage by the total resistance of the circuit.

Adding the two resistor voltage drops together, you will see that they total the applied voltage. This is a useful test to remember when you are checking your calculations, to make sure they are correct.

The properties of series resistance circuits are summed up quite well as Kirchoff's voltage law, which states that "the sum of the voltage drops in a series circuit is equivalent to the applied voltage and current is the same at all points in the circuit."

Along with being able to calculate the amount of current passing through a series resistor circuit and the voltage drop across each resistor, you can also calculate the voltage across each resistor in a parallel resistor circuit like Fig. 3-5 as well as the current through all the resistors. To do this, you have to remember Kirchoff's current law, which states that "the sum of the currents through each resistance is equivalent to the total current drawn by the circuit and the voltage drops across each resistor is the same as the applied voltage."

With each resistor in parallel, it should be fairly obvious that the voltage drop across each one is the same as the applied voltage, and the current flowing through each one can be calculated using Ohm's law. It should also

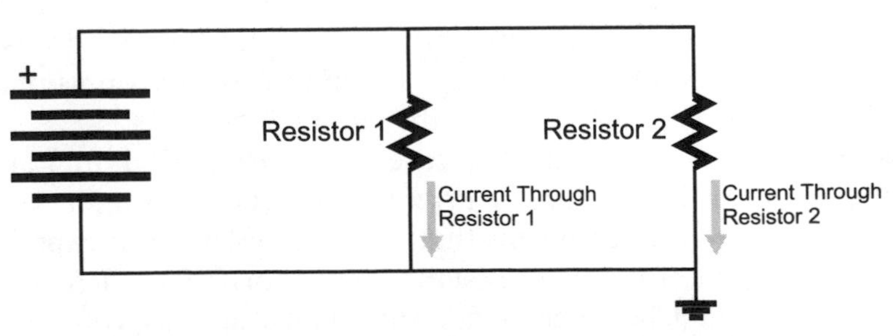

Fig. 3-5. Electrical circuit with two resistors in parallel.

be obvious that the current drawn from the power source is equivalent to the sum of the currents passing through each resistor.

If you were to calculate some different current values for different resistances, you would discover that the general formula for the equivalent resistance was:

$$R_e = 1/(1/R_1) + (1/R_2 + \ldots)$$

For the simple case of two resistors in parallel, the equivalent resistance can be expressed using the formula:

$$R_e = (R_1 \times R_2)/(R_1 + R_2)$$

Complex resistor circuits, made up of resistors wired in both series and parallel, like the one shown in Fig. 3-6, can be simplified to a single equivalent resistor by applying the series and resistor formulas that I have presented so far in this section. When doing this, I rccommend first finding the equivalent to the series resistances and then the equivalent to the parallel resistances until you are left with one single equivalent resistance.

The last piece of basic electrical theory that I would like to leave you with is how to calculate the power dissipated by a resistor. When you took Newtonian physics, you were told that power was the product of the rate at which something was moving and the force applied to it. In electrical circuits, we have both these quantities, voltage being the force applied to the electrons and current being the rate of movement. To find the power being dissipated (in watts), you can use the simple formula:

$$P = V \times i$$

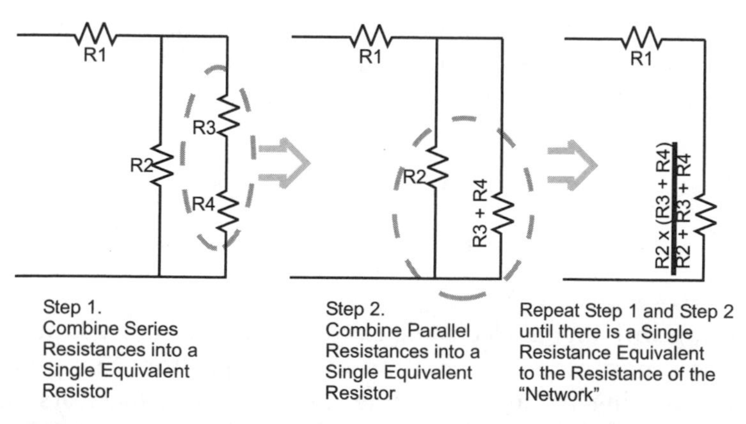

Fig. 3-6. Reducing the multiple resistor "network" into a single equivalent resistor.

or, if you don't know one of the two input quantities, you can apply Ohm's law and the formula becomes:

$$P = V^2/R$$
$$= i^2 \times R$$

I must point out that when you are working with digital electronics, most currents in the circuits are measured somewhere between 100 μA to 20 mA. This seemingly small amount of current minimizes the amount of power that is dissipated (or used) in the digital electronic circuits. I'm pointing this out because if you were to get a book on basic electronics you would discover that the examples and questions will usually involve full amperes of current – not thousands or tens of thousands as I have noted here. The reason why basic electronics books work with full amps is because it is easier for students to do the calculations and they don't have to worry about working with different orders of magnitude.

So far in these few initial pages of this chapter, I have gone through the same amount of material that is presented in multiple courses in electrical theory. Much of the background material has been left out as well as derivations of the various formulas. For the purposes of working with digital electronics, you should be familiar with the following concepts:

1. Electricity flows like water in a closed circuit.
2. The amount of current flow in a circuit is proportional to the amount of resistance it encounters.
3. Voltage across a load or resistance is measured at its two terminals.
4. Voltage is current times resistance (Ohm's law).
5. Power is simply voltage times current in a DC circuit.

The other rules are derivations of these basic concepts and while I don't recommend trying to work them out in an exam, what you do remember can be checked against the basic concepts listed above.

Capacitors

When working with digital electronic circuits, it is very important for you to understand the purpose and operation of the capacitor. Many people shy away from working at understanding the role of capacitors in digital electronics because the formulas that define their response to an applied voltage do not seem to be intuitive and many of them are quite complex. Further reducing the attractiveness of understanding capacitors is that they do not seem to be a basic component of digital electronics, and when they are used their value and wiring seems to be simply specified by a datasheet or an application note. I must confess that these criteria used to apply to me and I never understood the importance of capacitors in digital electronics until I was reviewing failure analysis of a 4 MB memory chip. As I will show, a dynamic RAM memory element (along with a MOSFET transistor) is essentially a capacitor, and the failure analysis of the chips showed how the differences in these capacitors affected their operation. One of the major conclusions of the failure analysis was that the memory chip wasn't so much a digital electronic device as a massive array of four million capacitors. This example is meant to show the importance of understanding the operation of capacitors and how they influence digital electronic circuits – being comfortable with the information in this section is more than good enough to use and successfully specify capacitors in digital electronic circuits.

The capacitor itself is a very simple energy storage device; two metal plates (as shown in the leftmost capacitor symbol in Fig. 3-7) are physically separated by a "dielectric" which prevents current from flowing between them. The dielectric is an insulator ("dielectric" is a synonym for "insulator") material which enhances the metal plates' ability to store an electric charge.

The capacitor is specified by the amount of charge it is able to store. The amount of charge stored in a capacitor (which has the symbol "C") is measured in "farads" which are "coulombs" per volt. One coulomb of electrons is a very large number (roughly 6.2×10^{18}) and you will find that for the most part you will only be working with capacitors that can store a very small fraction of a coulomb of electrons.

Knowing that farads are in the units of coulombs per volt, you can find the amount of charge (which has the symbol "Q") in a capacitor by using the formula:

$$Q = C \times V$$

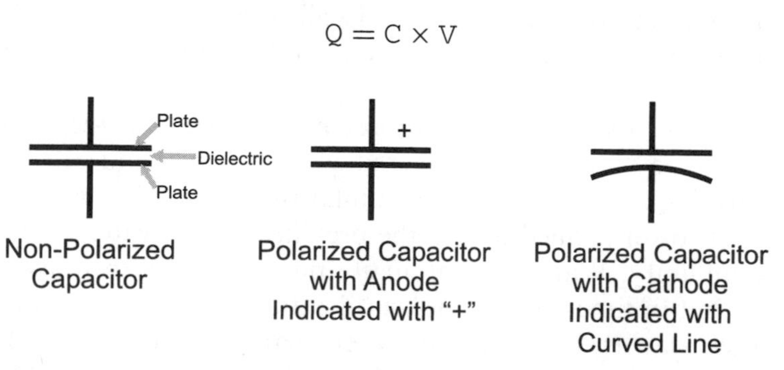

Fig. 3-7. Capacitor symbol.

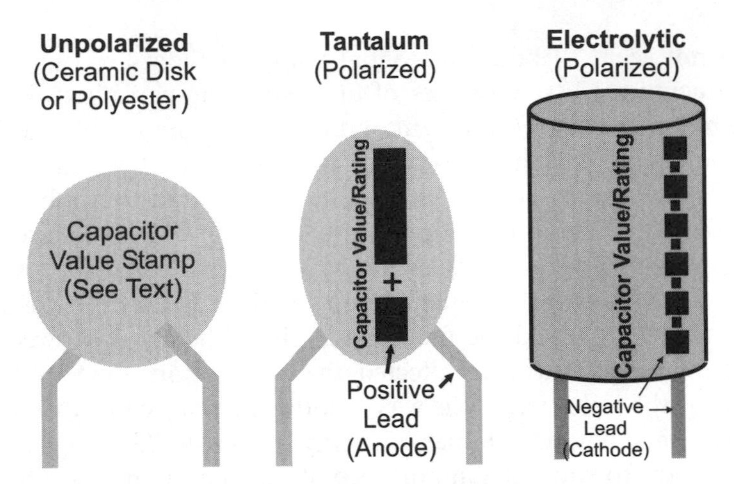

Fig. 3-8. Capacitor appearance and markings.

The fraction of a coulomb that is stored in a capacitor is so small, that the most popularly used capacitors are rated in millionth's ("microfarads" or "μF") or trillionth's ("picofarads" or "pF") of farads. Microfarads are commonly referred to as "mikes" and picofarads are often known by the term "puffs". Using standard materials (such as mica, polyester and ceramics), it is possible to build capacitors of a reasonable size of 1 microfarad (one millionth of a farad) but more exotic materials are required for larger value capacitors. For larger capacitors, the dielectric is often a liquid and the capacitor must be wired according to parameter markings stamped on it, as I have indicated in Fig. 3-8. These are known as "polarized" capacitors and either a "+" marking or a curved plate (as shown in Fig. 3-7) is used to indicate how the capacitor is wired in the schematic. Like other polarized components, the positive connection is called an "anode" and the negative a "cathode". Along with the markings, you should remember that the anode of a polarized two lead component is always longer than the cathode. The different lead lengths allow automated assembly equipment to distinguish between the two leads and determine the component's polarity.

Capacitors have two primary purposes in digital electronic circuits. The first is as a voltage "filter" (Fig. 3-9), reducing "spikes" and other problems on a wire carrying current. This use is similar to the use of a water tower in a city; the water tower is filled due to the pressure of the water being pumped into the community. Water is continually pumped to both houses and the water tower, but in times of high usage (like during the day when people are watering their lawns and washing their cars), water from the tower supplements the pumped water to keep the pressure constant. During the

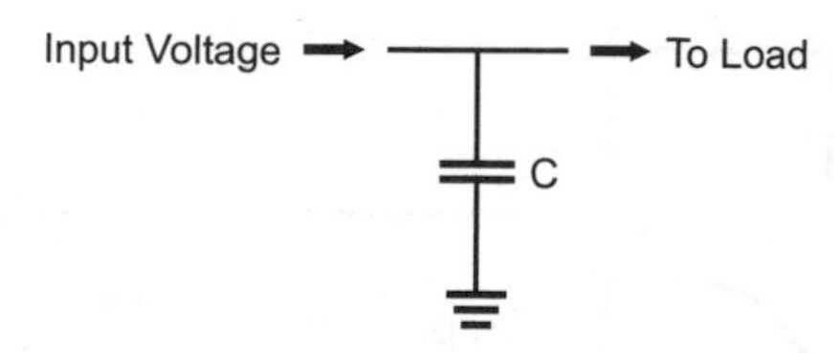

Fig. 3-9. Power supply filter using capacitor.

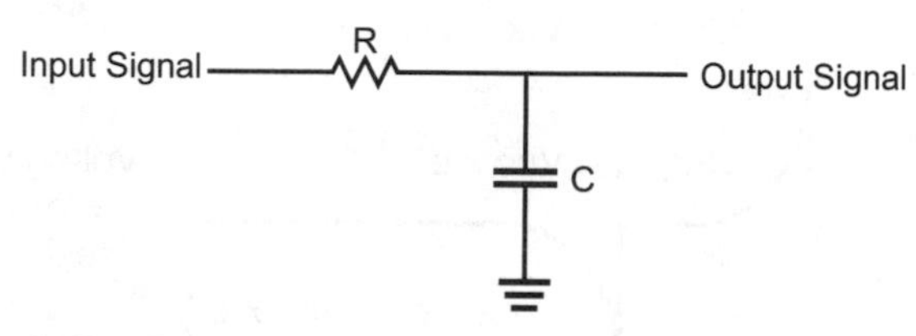

Fig. 3-10. Low-pass filter built from resistor and capacitor.

night, when few people are using water, the pumped water is stored in the water tower, in preparation for the next day's requirements.

When you look at digital electronic circuits, you will see two types of capacitors used for power filtering. At the connectors to the power supply, you will see a high value capacitor (10 μF or more) filtering out any "ripples" or "spikes" from the incoming power. "Decoupling" capacitors of 0.047 μF to 0.1 μF are placed close to the digital electronic chips to eliminate small spikes caused when the gates within the chips change state.

Large capacitors will filter out low-frequency (long-duration) problems on the power line while the small capacitors will filter out high-frequency (short-duration) spikes on the power line. The combination of the two will keep the power line "clean" and constant, regardless of the changes in current demand from the chips in the circuit.

The capacitor's ability to filter signals is based on its ability to accept or lose charge when the voltage across it changes. This capability allows voltage signals to be transformed using nothing more than a resistor and a capacitor, as in the "low-pass filter" shown in Fig. 3-10. This circuit is known as a low-pass filter because it will pass low-frequency alternating current signals more readily than high-frequency alternating current signals.

In digital electronics, we are not so much concerned with how a capacitor affects an alternating current as how it affects a changing direct current. Figure 3-11 shows the response, across Fig. 3-10's low-pass filter's capacitor and resistor, to a digital signal that starts off with a low voltage "steps" up to "V" and then has a falling step back to 0 V.

In Fig. 3-11, I have listed formulas defining the voltage response across the resistor and capacitor to the rising and falling step inputs. These formulas are

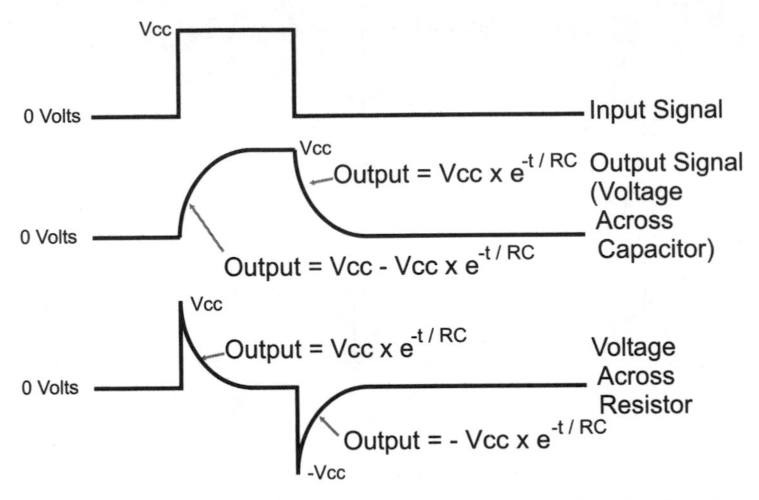

Fig. 3-11. Resistor/capacitor circuit response to changing input.

found within introductory college electricity courses by knowing that the voltage across the capacitor can be defined by using the formula:

$$V_C(t) = Q_C(t)/C$$

which simply states that the voltage across a capacitor at some point in time is a function of the charge within the capacitor at that point of time. The charge within the capacitor is supplied by the current passing through the resistor and the resistor limits the amount of current that can pass through it. As the voltage in the capacitor increases, the voltage across the resistor falls and as the voltage across the resistor falls, the amount of current that is available to charge the capacitor falls. It is a good exercise in calculus to derive these formulas, but understanding how this derivation works is not necessary for working with digital electronics.

There are two things I want to bring out from the discussion of low-pass filters. The first is that the response of the low-pass filter is a function of the product of the resistance and capacitance in the circuit. This product is known as the "RC time constant" and is given the Greek letter "tau" (τ) as its symbol. Looking at the formulas, you should see that by increasing the value of τ (either by using a larger value resistor or capacitor) the response time of the low-pass filter is increased.

This has two ramifications for digital electronics. The first should be obvious: to minimize the time signals take to pass between gates, the resistance and capacitance of the connection should be minimized. The second is more subtle: the resistor–capacitor response can be used to delay a signal in a circuit. This second issue with resistor–capacitor circuits is actually very useful in digital electronics for a number of different applications that I will discuss later in the book.

This is a very short introduction to capacitors and their operation in (digital) electronic circuits. Before going on, I would like to reinforce what I've said about their importance and recommend that you follow up this section's material by working through a book devoted to analog electronics.

Semiconductor Operation

Over the past 100 years, we have refined our ability to control the electrical properties of materials in ways that have made radios, TVs and, of course, digital electronic circuits possible. These materials have the ability to change their conductance, allowing current to pass through them under varying conditions. This ability to change from being an insulator to a conductor has resulted in these materials being called "semiconductors", and without them many of the basic devices we take for granted would be impossible.

The most basic electronic semiconductor device is the "diode". The electrical symbol and a sketch of the actual part is shown in Fig. 3-12. Diodes are a "one-way" switch for electricity; current will pass easily in one direction and not in the other. If you were to cut a silicon diode in half and look at its operation at a molecular level, you would see that one-half of the silicon was "doped" (infused with atoms) with an element which can easily give up electrons, which is known as an "N-type" semiconductor. On the other side of the diode, the silicon has been doped with an element that can easily accept electrons, a "P-type" semiconductor.

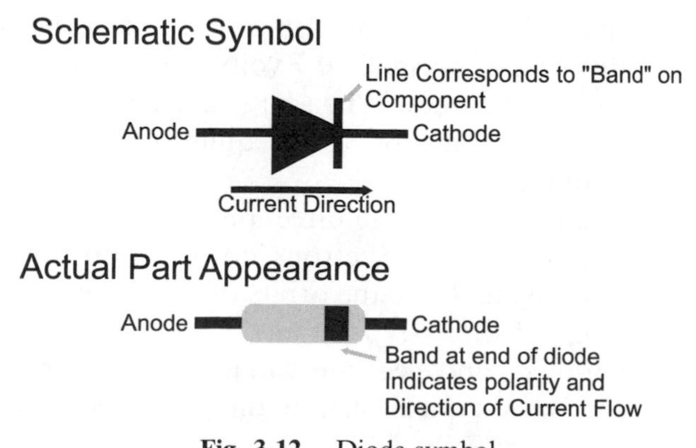

Fig. 3-12. Diode symbol.

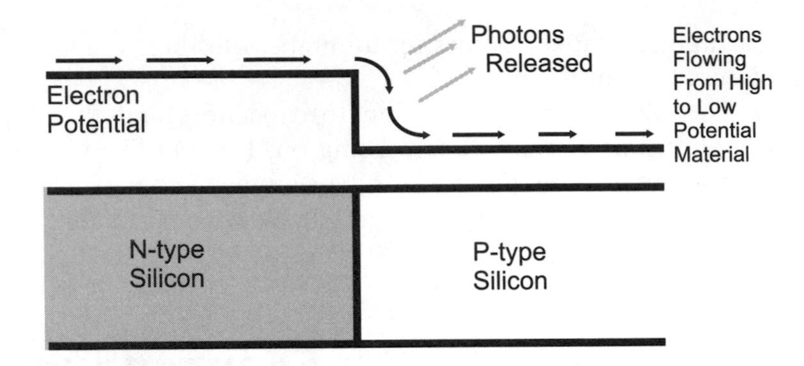

Fig. 3-13. Diode operation.

When a voltage is applied to the diode, causing electrons to travel from the atoms of the N-type semiconductor to the atoms of the P-type, the electrons "fall" in energy from their orbits in the N-type to the accepting orbit spaces in the P-type, as shown in Fig. 3-13. This drop in energy by the electron is accompanied by a release in energy by the atoms in the form of photons. The "quanta" of photon energy released is specific to the materials used in the diode – for silicon diodes, the photons are in the far infrared.

The voltage polarity applied to the diode is known as "bias". When the voltage is applied in the direction the diode conducts in, it is known as "forward biased". As you might expect, when the voltage is applied in the direction the diode blocks current flow, it is known as "reverse biased". This is an important point to remember, both for communicating with others about your designs and for understanding the operation of transistors, as explained below.

To keep the thermodynamic books balanced, the release in energy in terms of photons is accompanied by a corresponding voltage drop across the diode. For silicon diodes, this drop is normally 0.7 volts. The power equation I gave earlier ($P = V \times i$) applies to diodes. When large currents are passed through the diode and this is multiplied by 0.7 V, quite a bit of power can be dissipated within the diode.

If voltage is applied in the opposite direction (i.e. injecting electrons into the P-type side of the diode), the electrons normally do not have enough energy to rise up the slope and leave the orbits of the P-type atoms and enter the electron-filled orbits of the N-type atoms. If enough voltage is applied, the diode will "break down" and electrons will jump up the energy slope. The break down voltage for a typical silicon diode is 100 V or more – it is quite substantial.

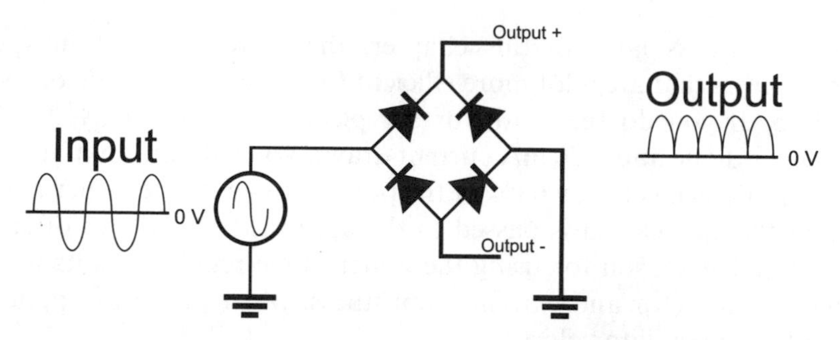

Fig. 3-14. "Full wave rectifier" using four diodes.

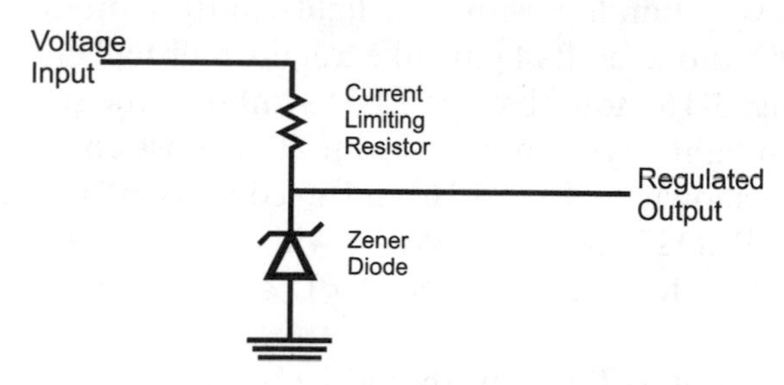

Fig. 3-15. Zener diode voltage regulator.

A typical use for a diode is to "rectify" AC to DC, as shown in Fig. 3-14, in which a positive and negative alternating current is converted using the four diodes to a "lobed" positive voltage signal, which can be filtered using capacitors, as discussed in the previous section.

Along with the simple silicon diode discussed above, there are two other types of diodes that you should be aware of. The first is the "Zener" diode which will break down at a low, predetermined voltage. The typical uses for the Zener diode is for accurate voltage references (Zener diodes are typically built with 1% tolerances) or for low-current power supplies like the one shown in Fig. 3-15. The symbol for the Zener diode is the diode symbol with the bent current bar shown in Fig. 3-15.

Building a power supply using this circuit is actually quite simple: the Zener diode's break down voltage rating will be the "regulated output" and the "voltage input" should be something greater than it. The value of the current limiting resistor is specified by the formula

$$R_{iLimit} = (V_{in} - V_{Zener})/i_{app}$$

where "i_{app}" is the current expected to be drawn (plus a couple of tens of percent margin). The power rating of the Zener diode should take into account the power dissipated if i_{app} was passing through it.

As I will discuss later in this chapter, there are a lot of inexpensive power regulators that are a lot more efficient than the Zener diode one shown in Fig. 3-15. If you do the math for a typical application (say 9 volts in, 5.1 volt Zener diode and a 20 mA current draw), you will find that at best it is 60% efficient (which is to say 60% of the power drawn by the Zener regulator circuit and the application is passed to the application, and can often be as low as 25%). The reason for using the Zener diode regulator is its low cost, very small form factor and extreme robustness. Most practical applications will use a linear regulator chip.

The other type of diode that I want to mention in this section is one that you are already very familiar with – the light-emitting diode or LED. As its name implies, this diode emits light (like a light bulb) when a current passes through it. In Fig. 3-16, note that the LED symbol is the same as the diode's symbol, but with light rays coming from it. The most common package for the LED is also shown in Fig. 3-16 and it consists of a rounded cylinder (somewhat like "R2D2" from Star Wars) with a raised edge at its base with one side flattened to indicate the LED's cathode (negative voltage connection).

There are a few points that you should be aware of with regard to LEDs. In the past few years, LEDs producing virtually every color of the rainbow (including white) have become available. I must point out that LEDs can only produce one color because of the chemistry of the semiconductors used to build them. You may see advertisements for two or three color LEDs, but these devices consist of two or three LEDs placed in the same plastic package and wired so that when current passes through its pins in a certain direction, a specific LED turns on.

Schematic Symbol

"Light Rays"
Indicating LED

Line Corresponds to "Flat" on
component's base

Current Direction

Actual Part Appearance

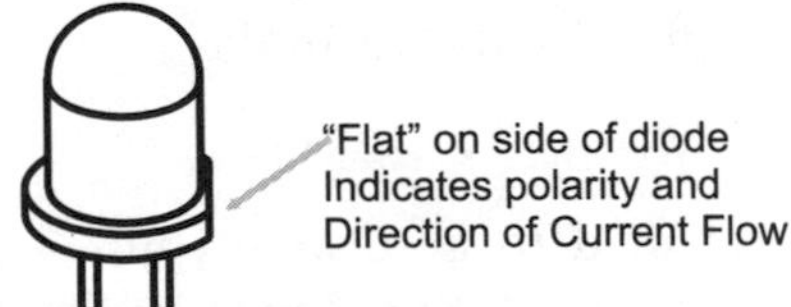

Fig. 3-16. LED symbol.

The brightness of a LED cannot be controlled reliably by varying the current passing through it, as you would with a light bulb. LEDs are designed to provide a set amount of light with current usually in the range of 5 to 10 mA. Reducing the current below 5 mA may dim its output or it may turn it off completely. A much better way to dim a LED is to use "pulse wave modulation" (PWM), in which the current being passed to the LED is turned on, and faster than the human eye can perceive, with varying amounts of on and off time to set the LED's brightness. I will discuss PWMs later in the book.

Finally, when I first introduced diodes, I noted that silicon diodes output photons of light in the far infrared and have a 0.7 volt drop when current passes through them. To produce visible light, LEDs are not made out of silicon, they are made from other semiconductor materials in which the energy drop from the N-type semiconductor to the P-type semiconductor produces light in the visible spectrum. This change in material means that LEDs do not have silicon's 0.7 V drop; instead, they typically have a 2.0 V drop. This is an important point because it will affect the value of the current limiting resistor that you put in series to make sure the LED's current limit rating is not exceeded or that it does not allow too much current in the circuit to pass through it, resulting in an unnecessary current drain.

It is always a source of amazement to me how many people do not understand how transistors work. For the rest of this section, I will introduce you to the two most common types of transistors and explain how they work as well as what applications they are best suited for. Understanding the characteristics of the two types of transistors is critical to understanding how digital logic is implemented and how you can interface it to different technologies.

As I explain the operation of the "bipolar" transistor, I will endeavor to keep to the "high level" and avoid trying to explain transistor operation using tools like the "small signal model", which is intimidating and obfuscates the actual operation of the device. Instead, I want to introduce you straight to the "NPN bipolar transistor" by its symbol and typical package and pinout for a small scale (low-power) device in Fig. 3-17.

As you have probably heard, a bipolar transistor can be considered a simple switch or a voltage amplifier, but you are probably mistaken on how it is controlled and how it actually works. The transistor is not voltage controlled (as you may have been led to expect); it is actually current controlled. The amount of current passing through the "base" to the "emitter" controls the amount of current that can pass from the "collector" to the emitter. The amount of current that can be passed through the collector is a multiple (called "beta" and given the symbol "β" or h_{FE}) of the

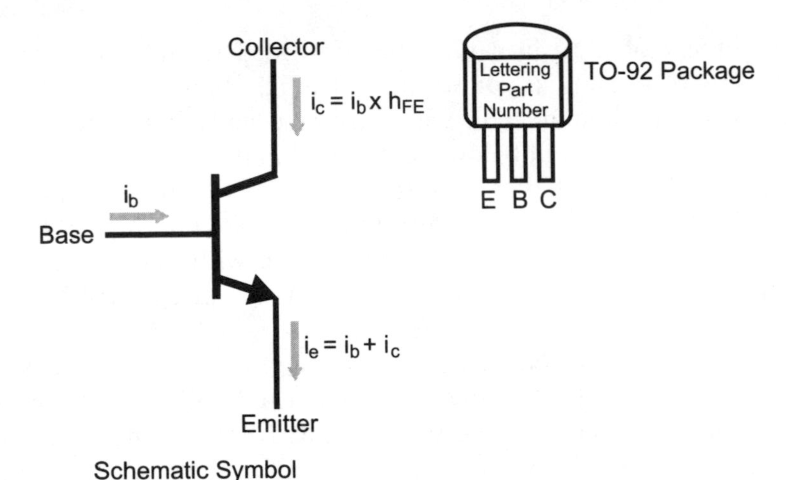

Fig. 3-17. NPN transistor symbol with parameters.

current flowing through the base; the bipolar transistor is actually an amplifier – a small amount of current allows a greater amount to flow. The simple formulas for the relationship between the base and collector currents are listed in Fig. 3-17.

I must point out that these formulas apply while the maximum collector current is in the "small signal" or "linear" operating range. As a physical device, a transistor can only allow so much current to flow through it; as it reaches this limit, increases in the transistor's base current will not result in a proportional increase in collector current. This operating region is known as the "non-linear" or "saturation" region and what happens in this situation can be easily understood by looking at what happens in a cross section of a transistor (Fig. 3-18).

A bipolar transistor consists of a P-type semiconductor sandwiched between two N-type semiconductors. This structure forms a reverse biased diode and no current can flow through it. With no current being injected into the NPN bipolar transistor, the P-type semiconductor is known as the "depletion region" because it does not have any electrons. When current is passed to the device, electrons are drawn through the P-type semiconductor via the emitter N-type semiconductor. As electrons are drawn into the P-type semiconductor, the properties of the P-type semiconductor change and take on the characteristics of the N-type semiconductors surrounding it and becomes known as the "conduction region". The more electrons that are drawn from the P-type semiconductor, the larger the conduction region bridging the two pieces of N-type semiconductor and the greater amount of current that can pass from the collector to the emitter. As more electrons are

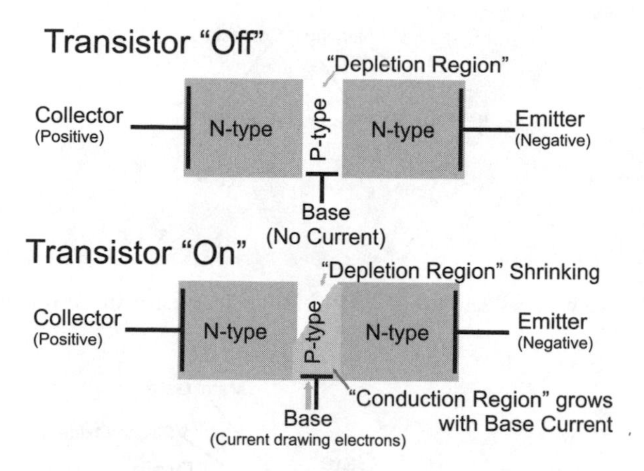

Fig. 3-18. NPN transistor operation.

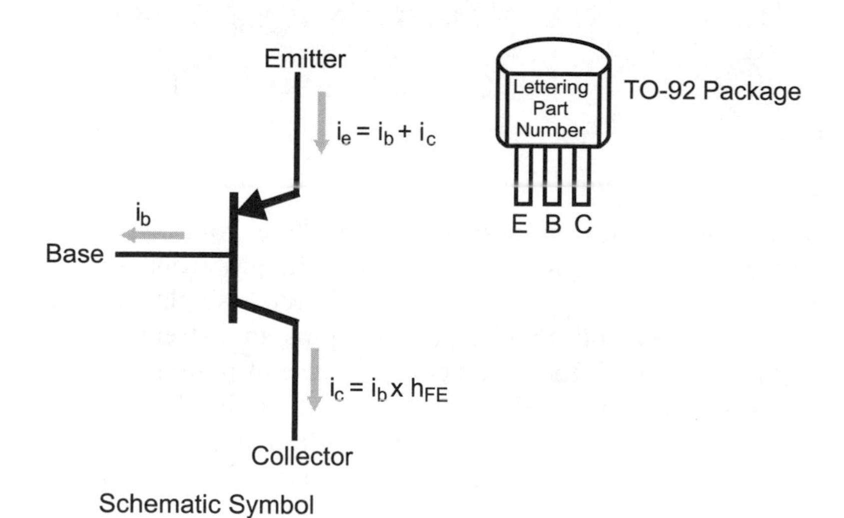

Fig. 3-19. PNP transistor symbol with parameters.

drawn from the P-type semiconductor, the conduction region grows until the entire P-type semiconductor of the transistor becomes "saturated".

The PNP bipolar transistor (Fig. 3-19) operates in the complete opposite way to the NPN transistor. It is built from an N-type semiconductor between two P-type semiconductors and to create a conduction region, electrons are injected into the base instead of being withdrawn, as in the case of the NPN bipolar transistor. As in the NPN bipolar transistor, the amount of collector current is a multiple of the base current (and that multiple is also called β or h_{FE}).

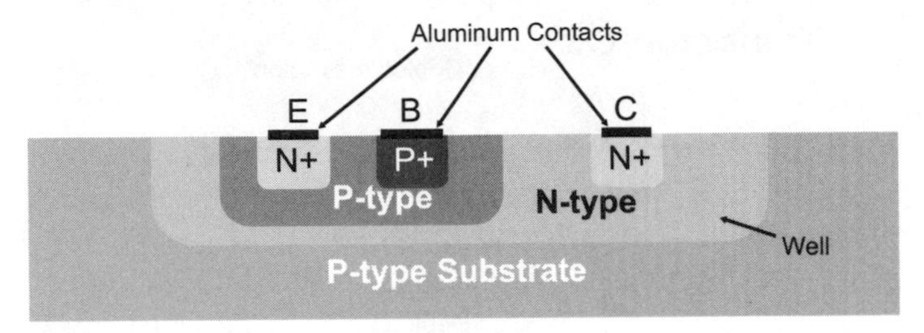

Fig. 3-20. Side view cross section of NPN bipolar transistor in an integrated circuit.

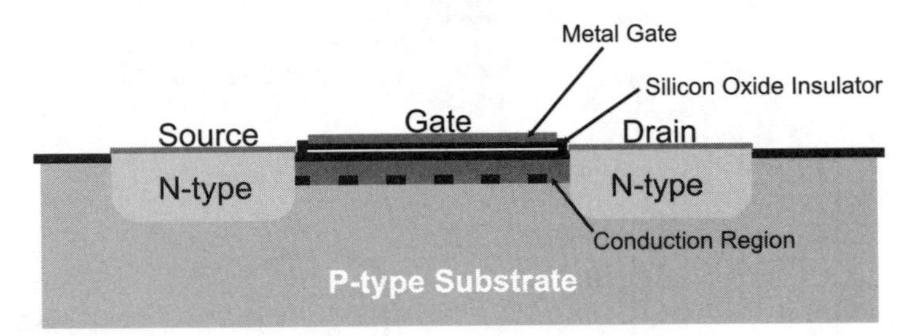

Fig. 3-21. N-channel MOSFET side view showing features during operation.

Bipolar transistor h_{FE} values can range anywhere from 50 to 500 and the amount of collector current they can handle ranges from a few tens of milliamps to tens of amps. As well as discrete (single) devices being inexpensive, they respond to changes in inputs in extremely short time intervals. You may think they are perfect for use in digital electronics, but they have two faults that make them less than desirable. First, the base current is actually a source of power dissipation in the device, which is usually not an issue when single transistors are used, but is of major concern when thousands or millions are used together in a highly complex digital electronic system.

Secondly, they take up a lot of chip "real estate" and are very expensive to manufacture. Figure 3.20 shows the side view of an NPN bipolar transistor built on a silicon chip. Instead of butting together different types of semiconductor, it is manufactured as a series of "wells", which are doped with the chemicals to produce the desired type of semiconductor by repeated operations. As many as 35 process steps are required to produce a bipolar transistor.

The N-channel enhancement "metal oxide silicon field effect transistor" (MOSFET) does not have these faults–it is built using a much simpler process (the side view of the transistor is shown in Fig. 3-21) that only requires one doping of the base silicon along with the same bonding of

aluminum contacts as the bipolar transistor. N-channel MOSFETs (as they are most popularly known) require nine manufacturing processes and take a fraction of the chip real estate used by bipolar transistors.

The N-channel MOSFET is not a current-controlled device, like the bipolar transistor, but a voltage-controlled one. To "turn on" the MOSFET (allow current to flow from the "source" to the "drain" pins), a voltage is applied to the "gate". The gate is a metal plate separated from the P-type silicon semiconductor substrate by a layer of silicon dioxide (most popularly known as "glass"). When there is no voltage applied to the gate, the P-type silicon substrate forms a reverse biased diode and does not allow current flow from the source to the drain. When a positive voltage is applied to the gate of the N-channel MOSFET, electrons are drawn to the substrate immediately beneath it, forming a temporary N-type semiconductor "conduction region", which provides a low-resistance path from the source to the drain. MOSFET transistors are normally characterized by the amount of current that can pass from the source to the drain along with the resistance of the source/drain current path.

The symbol for the N-channel MOSFET, along with its complementary device, the P-channel MOSFET are shown in Fig. 3-22. The P-channel MOSFET creates a conduction region when a negative voltage is applied to its gate. MOSFET transistors come in a variety of packages and some can handle tens of amps of current, but they tend to be very expensive.

MOSFETs do not have the issues of bipolar transistors; their gate widths (the measurement used to characterize the size of MOSFET devices) are, at the time of this writing, as small as 57 nm in high-performance microprocessors and memory chips. The voltage-controlled operation of MOSFETs eliminates the wasted current and power of the bipolar transistor's base, but while MOSFETs do not have the disadvantages of bipolar transistors, they do not have their advantages.

MOSFET transistors do not have a small signal/linear operating region; they tend to change from completely off to completely on (conducting) with a

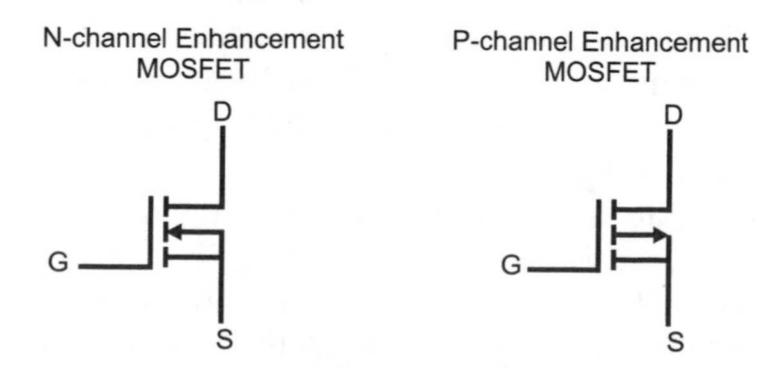

Fig. 3-22. MOSFET schematic symbols.

very small intermediate range. MOSFETs also tend to operate at slower speeds than bipolar devices because the gates become capacitors and "slow down" the signals, as I showed in the previous section. This point has become somewhat moot as modern MOSFET designs are continually increasing in speed, providing us with extremely high-speed PCs and other electronic devices. Finally, it is difficult to manufacture MOSFETs with high current capabilities; while high current MOSFETs are available, they are surprisingly expensive.

The characteristics of the two types of transistors give way to the conclusion that bipolar transistors are best suited to situations where a few high current devices are required. MOSFET transistors are best suited for applications where large numbers of transistors are placed on a single chip.

Today, for the most part, digital electronic designs follow these guidelines, but we are left with an interesting legacy. Despite being much simpler structurally and cheaper to manufacture, MOSFET transistors were only perfected in the late 1960s, whereas bipolar technology had already been around for 20 years and it was able to become entrenched as the basis for many digital electronic devices and chips. For this reason, you must be cognizant of the operating characteristics of bipolar transistors as well as those of MOSFET transistors. In the next section, many of these differences will become apparent.

Logic Gate Input and Output

If you have worked with digital electronics before, you probably have made a few assumptions about how the circuitry works and how you can demonstrate how digital electronic devices work. Chances are many of these assumptions are with regard to how gate and chip inputs and outputs work as well as how to properly interface them together and to different electronic devices. These assumptions are generally made on the evidence of by what somebody has seen with a voltmeter or logic probe and do not look at the underlying circuitry and how it works. In this section, I will give you a detailed introduction to the input and output pins on digital electronics and how they should be wired.

When we talk about digital electronics, we should identify the different technologies used. "Transistor to transistor logic" (TTL) is based on NPN bipolar transistors. TTL chips have the part number prefix "74" (i.e. a chip with four, two input NAND gates known as the "7400"). There are actually quite a few different technology chip families based on the 74xx "standard" pinout and operation and the technology is indicated by letter codes following the "74"; a chip marked with "74LS00" is a low-power, Shotkey four two-input NAND gate chips. Many of these technologies used with the 7400 series of chips are based on bipolar transistors, but some are based on MOSFET technology. These MOSFET technology based chips have the 74 prefix and a technology letter code containing a "C" (i.e. "C", "HC", "HCT"). Along with being used in 7400 series form factors, MOSFET devices are used in the "4000" series of logic chips. Understanding which type of transistor is used in a logic chip is critical to being able to successfully interface it to other chips or input/output devices.

When the term "TTL" is used, it is referring to bipolar transistor logic in the 7400 series. "CMOS" indicates MOSFET transistor logic used in the 74C00 and 4000 chip logic series.

Probably the biggest erroneous assumption that people have about digital logic is that TTL circuitry is voltage controlled. In the previous section, I emphasized the notion that bipolar transistors are *current* controlled and not voltage controlled. I'm sure that many people will argue with me and say that when they put a voltage meter to the input of a TTL gate, they saw a high voltage when a "1" was being input and a low voltage when a "0" was input. I won't argue with what they have seen; although I will state that the conclusion that TTL logic is voltage controlled made from these observations is incorrect.

The standard TTL input consists of an NPN bipolar transistor wired in the unusual configuration shown in Fig. 3-23. On the left side of this diagram, I have drawn a two input TTL gate which is implemented with a two emitter

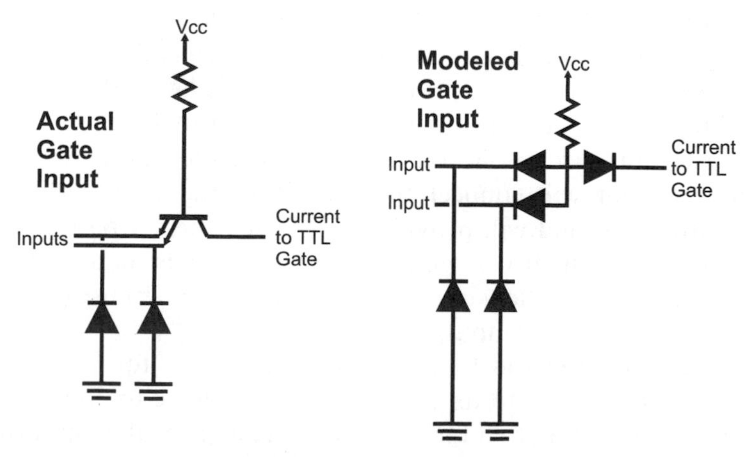

Fig. 3-23. Actual and model TTL input circuits.

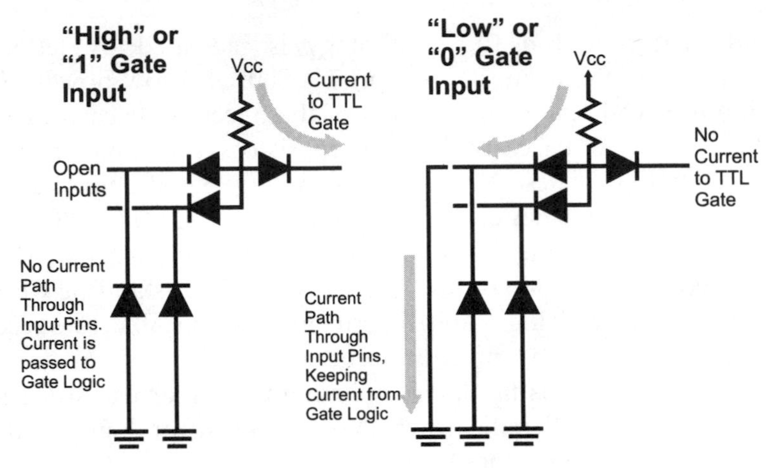

Fig. 3-24. Current control operation of TTL inputs.

NPN transistor – as unusual as this type of transistor sounds, they really do exist. To understand how the input works, I replaced the two emitter NPN transistor with the three diode equivalent "model" on the right side of Fig. 3-23.

Normally, an NPN transistor passes current from its base to the emitter, but when wired in the TTL input configuration, the base current does not have a path through the transistor's emitters and passes through the transistor's collector to the gate logic. Figure 3.24 shows this situation along the other case where one of the input transistor's emitter's is tied to ground and the base current passes through the emitter and not the collector. The logic connected to the input NPN transistor's collector responds depending on whether or not current is available from the collector.

Obviously a simple switch, connected to ground, will allow current to pass through the emitter but you are probably wondering how other logic devices can control this device. A typical logic device output looks like Fig. 3-25 and consists of two transistors: one that will connect the output to the device power and one that will connect the output to the device ground. This transistor path to ground will provide the emitter current path of the chip. When the output is a high voltage (the top transistor is on and the bottom one is off), no current will flow into the TTL input gate because of the reverse diode nature of the emitter input pin.

The TTL output shown in Fig. 3-25 is known as a "totem pole" output because of its resemblance to its namesake. If you were to connect a totem pole output to a TTL input and measured the voltage at the input or output pins, you would see a high voltage, which the gate connected to the input

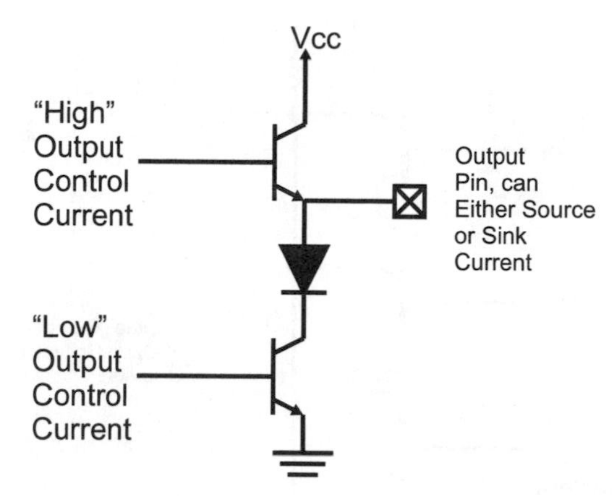

Fig. 3-25. TTL "totem pole" output.

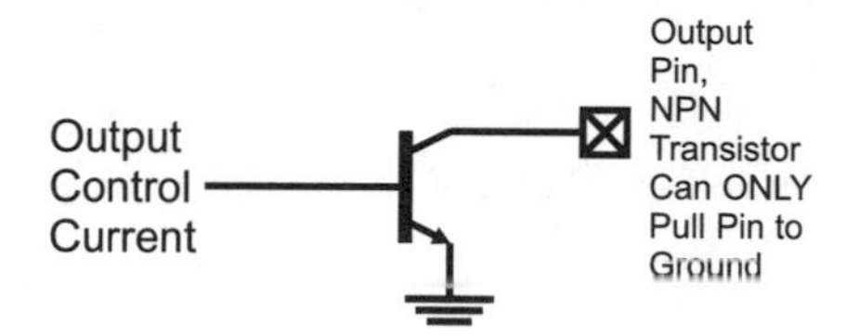

Fig. 3-26. TTL open collector output.

would respond to as a "1". When a low voltage is output, the TTL gate will respond as if a "0" was input. What you are not measuring is the current flow between the two pins.

There are two terms used in Fig. 3-25 that I should explain. When a transistor is connected to the power supply of a chip and is turned on, it is said to be "sourcing" current. When a transistor is connected to ground and is turned on, the transistor is said to be a current "sink". I will use these terms throughout the book and you will see in other books and references any time a device is either supplying ("sourcing") or taking away ("sinking") current.

There is another type of output which does not source any current and is known as the open collector output (Fig. 3-26). This output typically has two uses. The first is it can pull down voltages which are greater than the positive voltage applied to the chip. Normally these voltages are less than 15 V and can only source 10 to 20 mA. For higher currents and voltages, discrete transistors must be used.

By not sourcing any current, these outputs can be "ganged" together in parallel, as I have shown in Fig. 3-27. This circuit is known as a "dotted AND" because it only outputs a 1 if all the outputs are "high" and each

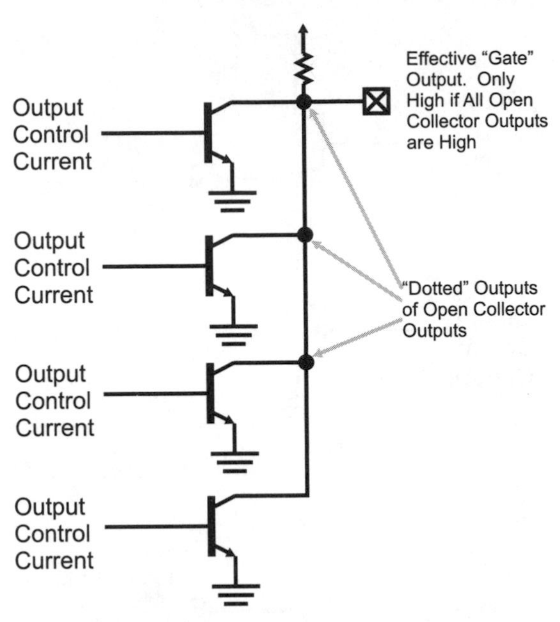

Fig. 3-27. Multiple "open collector" outputs combined to form a "dotted AND" gate.

transistor is "off" and not pulling the common output line to ground. Note that there must be a pull up resistor connected to the output to provide a high-voltage, low-current source. Dotted AND gates are useful in a variety of different situations ranging from circuits where an arbitrary number of outputs can control one line or where digital outputs and buttons are combined. (I will discuss this in more detail later in the book.)

Totem pole outputs are the recommended default gate output because you can easily check voltage levels between intermediate gates in a logic string. As I will show later in this chapter, you cannot use a voltmeter or logic probe to check the logic levels if a TTL gate is driven by an open collector output. Along with this, a CMOS input is connected to an open collector (or open drain, as I will discuss below) output. Then there will be no high voltage for the gate to operate. The only cases where an open collector/open drain output should be used is when you are wiring a dotted AND gate or are switching an input that is operating at a voltage different from the gate's power.

TTL output pins are internally limited to only sink or source around 20 mA of current, which limits the number of inputs that it can drive. If you were to do the math, you would discover that when a TTL input is pulled low, 1.075 mA of current is passed through the output pin (this was found by assuming the base/emitter voltage of a transistor is 0.7 volts and the current limiting resistor connected to the input transistor's base is 4 k, which is typical for TTL inputs.

Along with the totem pole and the open collector outputs, there is also the "tri-state driver" output, which cannot only source or sink current but can be turned "off" to electrically isolate itself from the circuit that it is connected to. I will discuss tri-state drivers later in the book, when I present busses and multiple devices on the same line.

Knowing that each TTL input requires a current sink of just over 1 mA and most TTL outputs can sink up 20 mA, you might expect the maximum number of TTL inputs driven by a single output (which is called "fanout") to be 18 or 19. The actual maximum fanout is 8 to ensure that there is a comfortable margin in the output to be able to pull down each output in a timely manner. Practically, I would recommend that you try to keep the number of inputs driven by an output to two and never exceed four. Some different technologies that you work with, do not have the same electrical drive characteristics and may not be designed to pull down eight inputs of another technology; so, to be on the safe side, always be very conservative with the number of inputs you drive with a single output.

Re-reading the last sentence of the previous paragraph, you might wonder if any potential low-drive situations could be improved by wiring multiple outputs together. This must be avoided because of the danger that the gates will switch at different times, resulting in large currents passing through the gate output circuitry, and not through the net the outputs are connected to.

The CMOS logic gate input (Fig. 3-28) is quite a bit simpler than the TTL gate input and much easier to understand. The CMOS input and, as I will explain, the output, consist of a balanced P-channel MOSFET and an N-channel MOSFET wired as a very high gain amplifier. The slightest positive or negative voltage applied to this input circuit will cause the

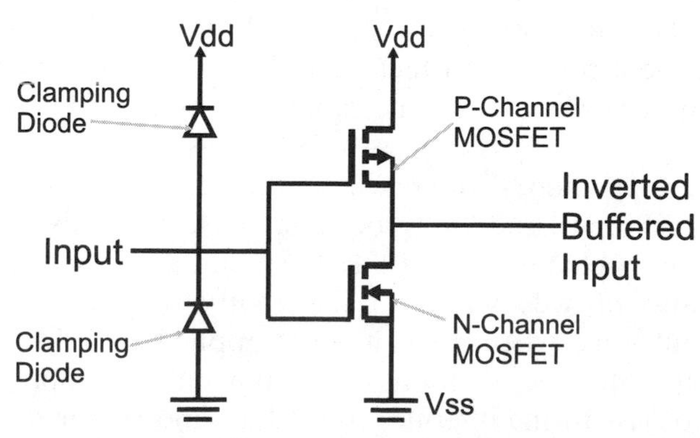

Fig. 3-28. Basic CMOS input/output circuit.

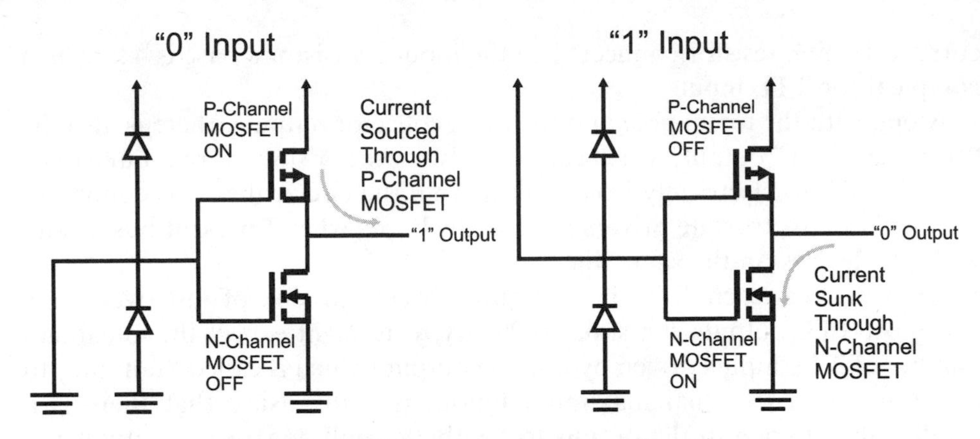

Fig. 3-29. CMOS gate response for different inputs.

appropriate transistor to turn on and either source current (in the case where a negative voltage is applied and the P-channel MOSFET turns on) or sink current (a positive voltage will turn on the N-channel MOSFET). This operation can be seen in Fig. 3-29.

One interesting aspect of the two MOSFET transistors that I have shown wired as an inverter is that they not only provide the ability to sense and respond to voltage inputs, but as the voltage controls transistor switches, they are also effective totem pole output circuits as well. Not only are MOSFET transistors much easier to place on a piece of silicon semiconductor and can be placed in a smaller amount of surface area but also gates built from them are also much simpler than their TTL counterparts.

When the P-channel MOSFET is removed from the output of a CMOS gate, its output is said to be "open drain". This term refers to the drain of the N-channel MOSFET that is not connected to a transistor which can source current in just the same way as an "open collector" TTL output transistor and does not have a transistor which can source current. The CMOS logic open drain output works exactly the same way as the TTL open collector output.

The two "clamping diodes" are placed in the circuit to hold the voltages to within Vdd (power input) and Vss (ground) and are primarily there to protect the P-channel and N-channel MOSFETs from damage from static electricity. These diodes also provide you with the ability to power a CMOS chip through its input pins; when no voltage is applied to Vdd but there is a high-voltage input to one or more input pins, the clamping diodes will allow current to pass to the internal MOSFETs and power the circuit. This is usually an undesirable side effect and one that you should watch for.

The clamping diode function is provided in TTL by the diode and the bipolar transistor emitter that makes up a TTL gate input. Whereas CMOS logic requires additional diodes built into the circuitry, TTL has this function built in.

Unlike TTL, CMOS logic is voltage controlled; there is no path for current to enter or leave the MOSFET's gate circuitry. This has some interesting side effects that you should be aware of. The first is that while at first glance of the inverter operation in Fig. 3-29 it appears that there is no current flow if the output of the CMOS input transistors was another CMOS gate, there actually is a very small amount of change passed to the gates of the transistor from Vdd when the P-channel MOSFET is turned on and this charge is sunk to Vss when the N-channel MOSFET is turned on. This transfer of charge grows with the number of CMOS gates as well as the speed that the gates switch; the faster they switch the more charge that is transferred over time. As I discussed at the start of this chapter, the measurement of charge movement over time is current.

Earlier in the book, I said that the basic gate used in CMOS logic circuits is the NOR gate (just as the NAND gate is the basic gate used in TTL). Before leaving this chapter, I would like to show you the circuit used by a CMOS NOR gate (Fig. 3-30). If you trace through the operation of the four MOSFETs that make up this circuit, you will discover that the only time both P-channel MOSFETs are on (and voltage/current from Vdd is passed to the "Output") is when the two inputs are low, which matches the expected operation of the NOR gate.

The reason why the NOR gate was selected for use as the basic CMOS logic gate has to do with how MOSFETs and other circuits are put

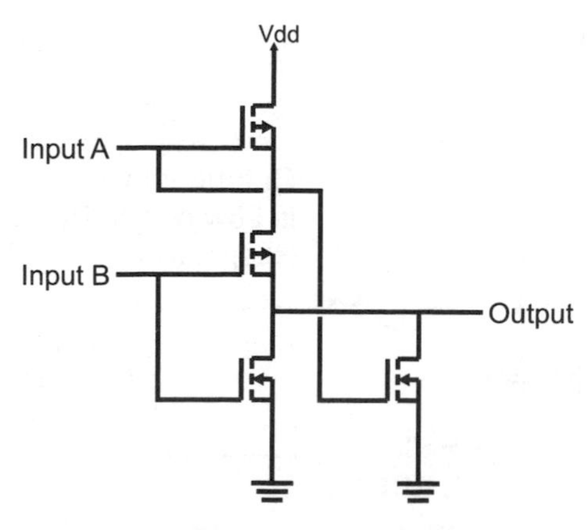

Fig. 3-30. CMOS NOR gate.

down on a silicon semiconductor. The NOR gate is the most efficient while the NAND (which would make the basic building blocks of TTL and CMOS logic the same) cannot be accomplished as easily and in as small amount of space.

The last point I want to make about inputs and outputs is how to wire them when you want to hold them at a specific state (high/"1" or low/"0"). While you could connect the pins directly to power (for a high input) and ground (for a low input), I want to show you the recommended way of doing this and explain why you should go through the extra effort. Connecting the input to high value is accomplished using a 10 k resistor (called a "pull up"), as I show in Fig. 3-31. This circuit will allow input to be temporarily wired to ground (for testing or circuit debug), without causing a short circuit (a low-resistance path between positive and negative power voltage).

Providing a "pull down" (connection to ground) is not quite so simple; the single resistor pull up of Fig. 3-31 is input into an inverter, as shown in Fig. 3-32. This circuit allows the pull up to be connected to ground for testing and debug (changing the input of the gate to a high from a solid low) just as in the pull up case.

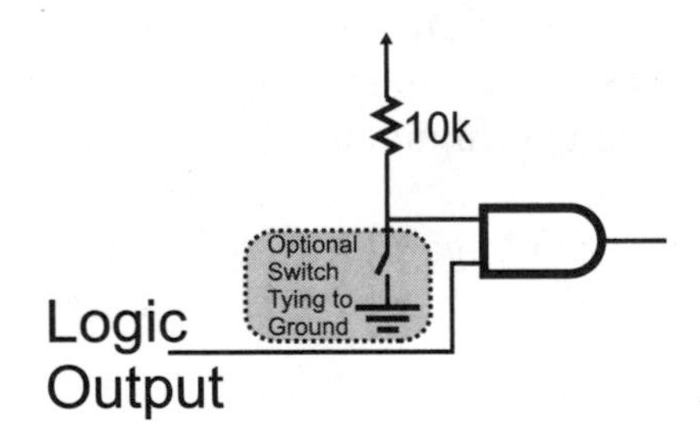

Fig. 3-31. The best method of implementing a "pull up".

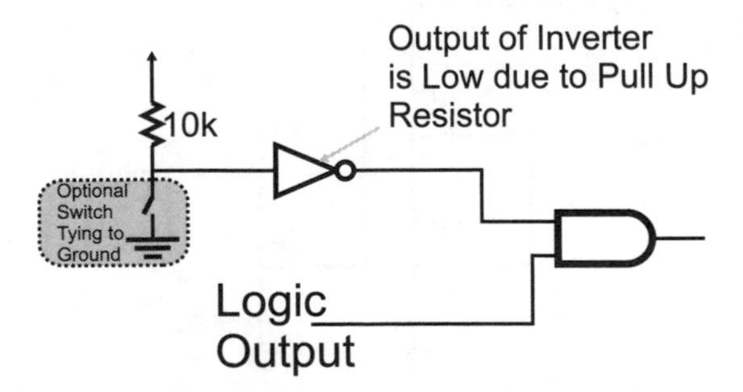

Fig. 3-32. Recommended way to "pull down" a logic input.

If you have followed the gate explanations up to this point, you might be feeling like these methods of tying the gates to pull ups and pull downs is "overkill". I admit that these methods may seem more complex than just wiring the inputs to positive or negative power, but there are a number of reasons for specifying that pull ups and pull downs are wired in this way. For TTL, to make an input high all the time it can be simply left unconnected and to pull it down it can be pulled directly to ground; the 1 mA of current that will flow through the gate to ground should not be an excessive amount of current. For CMOS logic, the input pin can be tied directly to Vdd (positive power) for a high input and Vss (negative power) for a low input – there will be no current flow in either case. It is important to understand the three reasons why I recommend using the pull up resistor or the pull up resistor and inverter.

First, as I said above, it allows you to temporarily change the input value by connecting the resistor voltage to negative voltage without worrying about damaging any part of the circuit. Secondly, it allows simple test equipment to change the state of the input pin for testing without potentially overloading the circuit or the tester. This is a very important consideration when you are designing a product for mass production. Finally, this method can be used for both TTL and CMOS logic without regard to what type of logic is being used. I realize that going through the rigor of following these recommendations increases the complexity of a circuit as well as increasing the number of gates required, its cost and power consumption. In many cases, you will not feel that it is necessary, but if you decide to forgo using pull ups and inverted pull ups, make sure you understand what are the tradeoffs and the risks of the decision.

Simple Digital Logic Circuit Development

Many people do not realize that it is quite easy to build sample digital electronic logic circuits that demonstrate the concepts that have been presented to you as well as let you try out your own simple experiments. If you have, or are taking, a course in digital electronics, it probably includes a well-equipped laboratory in which you worked through a number of experiments. You do not need to replicate this laboratory at home if you wish to experiment with digital electronics. As I will show in this chapter, you can come up with a very capable digital logic circuit test kit for less than $20 and use parts available in modest electronics stores (like "Radio Shack").

Chances are, you are familiar with a variety of different electrical power sources: the ones that comes to mind first are batteries. There are a confusing number of different batteries that you can choose from, ranging from simple "AA" batteries that cost a few cents to the batteries used in the International Space Station that weigh (on Earth) 1200 pounds and cost over $200,000 each. Along with batteries, electricity can also be produced by generators, solar cells and fuel cells. Within your home you can access electrical power very conveniently through outlets in the walls, although this power is *alternating current* ("AC") and not the *direct current* ("DC") required for digital logic. AC power coming from the sockets in your home will have to be reduced and rectified into DC.

When you are experimenting with simple electronics, I think it's best to use a power source that is definitely "low end"; "alkaline" and rechargeable nickel–metal hydride ("NiMH") batteries are widely available to power your experiments. TTL digital electronic chips generally operate between 4.5 and 5.5 volts – you could come up with a combination of batteries that will provide 5 volts to your circuit, or convert a 9 volt radio battery output to 5 volts using a "regulator". Rather than going through this effort and potential expense for TTL, I am going to recommend that you use CMOS digital logic chips that can be powered by 9 volts directly.

A 9 volt battery "clip" (Fig. 3-33) will cost you just a few cents and a bag of them can be bought for a dollar or so. For the purposes of the digital logic circuit test kit, you should look for a 9 volt battery clip that either has wire's individual strands soldered together (the ends of the wires will look silver, shiny and attached together) or has a single strand. The wires will be covered

Fig. 3-33. 9 volt battery clip with red (positive) and black (negative) wires attached to it.

in a red and black plastic insulation and the strands will poke out the ends for a 1/4 inch or so.

Make sure the strands of the 9 volt battery clip wires are either soldered together or the wires consist of a single strand, because the wires from the battery clip will be pushed into holes and clamped by copper springs to provide power for the test circuits. Loose, individual strands break easily, can short with other loose wires or become a tangled mess, none of which are good things.

The battery clip is only one part of the wiring that will be used with the digital logic circuit test kit. By itself, the battery clip brings power out of the 9 volt battery conveniently, but is difficult to work with when you are working with chips and even moderately complex circuitry. The "breadboard" and wiring kit (Fig. 3-34) provide a customizable platform in which chips and other electronic components can be inserted into and easily wired together.

"Breadboards" allow you to simply and quickly wire up your own prototyping circuits. From the top, a breadboard looks like a sea of holes, but if you were to "peel back" the top (Fig. 3-35), you would see that the holes are actually interconnected, with the central groups of holes connected outwards and the outermost two sets of holes connected along the length of the breadboard.

The central holes are spaced so that DIP chips can be placed in the breadboard and wired into the circuit easily. The outside two rows

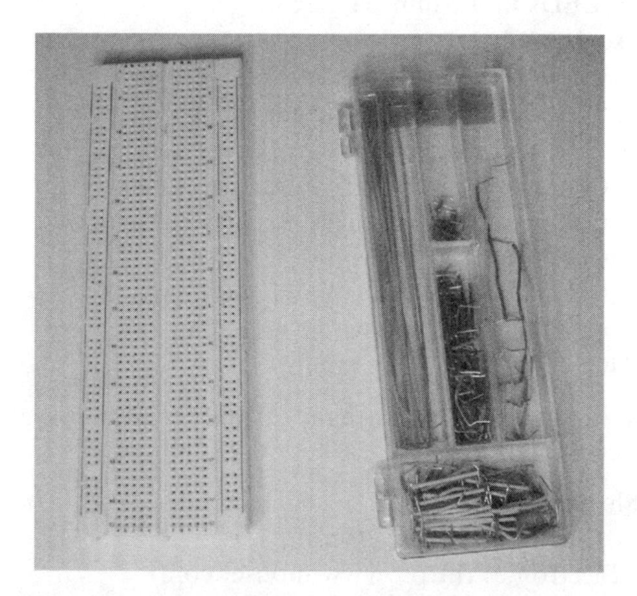

Fig. 3-34. Breadboard with a wiring kit.

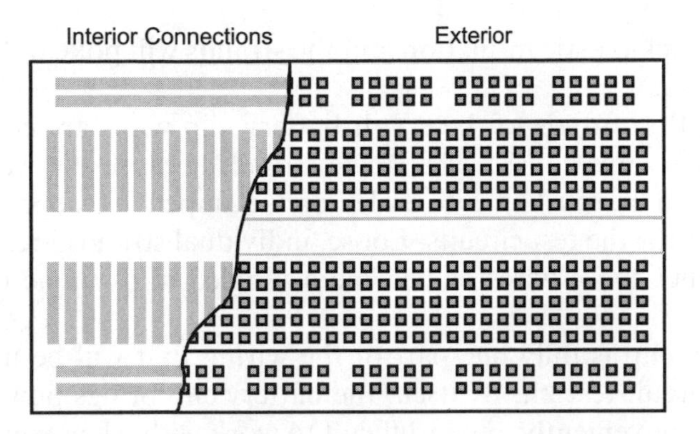

Fig. 3-35. "Breadboard" with interior connections shown.

of holes, I use as power "buss bars" and connect the power source to them directly.

Along with the breadboard, you can either buy a pre-cut and stripped wiring kit (shown in Fig. 3-34) or a roll of 24-gauge solid core wire and some needle nose pliers, wire clippers and maybe some wire strippers. For convenience, I usually go with the wiring kit as it costs just a few dollars.

Along with buying the battery clip, breadboard and wiring kit, you should also buy:

1. 5 or so LEDs in a 5 mm package
2. 10 or so 1k, 1/4 watt resistors
3. 10 or so 0.01 μF ceramic capacitors
4. One 555 oscillator/monostable chip
5. 5 or so SPDT switches, that can be inserted into the breadboard
6. One 74C00 quad two-input NAND gates chips
7. One 74C02 quad two-input NOR gates chips
8. One 74C04 hex inverter chip
9. One 74C08 quad two-input AND gates chip
10. One 74C32 quad two-input OR gates chip
11. One 74C74 dual D-flip flop chip.

All these parts should cost you less than $20 and are available at a fairly wide variety of sources including:

- Radio Shack (http://www.radioshack.com)
- Digi-Key (http://www.digikey.com)
- Mouser Electronics (http://www.mouser.com)
- Active Components (http://www.active-electronics.com).

You will not require any test equipment (such as a Digital Multi-Meter) for this kit and the sample circuits that I will present in this book.

Testing a Simple TTL Inverter

So far I have used the term "load" when I've described the electronic devices that are to be used in a circuit, but before going on, I want to familiarize you with the basic, "dual in-line package" "chip" (Fig. 3-36). The "chip" consists of a rectangular plastic box which has a series of metal pins (or connections) coming out from the two long sides. These pins are the electrical connections that are to be made to make up the digital logic circuits as well as provide power to the chip. As I have shown in Fig. 3-36, there can be one or two "pin 1" indicators on each chip (not all chips have both indicators) and the pins are numbered by going counterclockwise around the top of the chip.

Before leaving this chapter, I would like to show both how easy it is to create a simple circuit to test out ideas and parts of applications as well as demonstrate how the TTL gate works. You should have a pretty good idea of how to wire in the chip, but you probably have some questions on how to create useful inputs and outputs to see what's happening. The output will simply consist of a resistor and a LED – when the chip's output is high, the LED will be on. Providing the same function for the input, a LED that is on

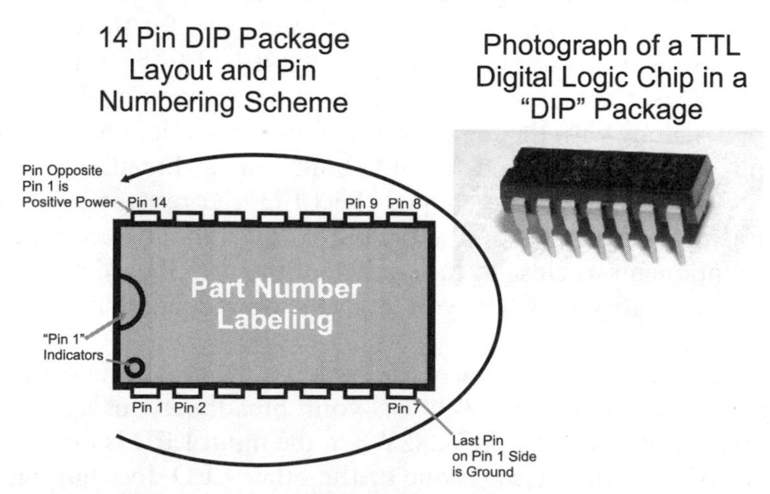

Fig. 3-36. Integrated circuit "dual in-line package" ("DIP") – aka a "chip."

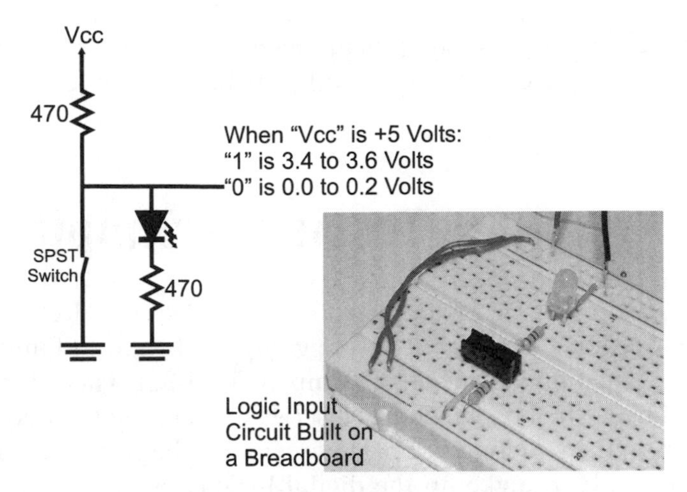

Fig. 3-37. Switch input circuit with LED.

when the input is high is a bit more difficult and uses the circuit shown in Fig. 3-37.

This input circuit probably seems to be much more complex than I have led you to believe is necessary, but there are some requirements that were important for this circuit to meet so that it could be used in a variety of different situations. The first requirement was that it had to work for both TTL (using 5 volt power) as well as CMOS logic (powered from 5 to 9 volts). By providing a direct path to ground, the low voltage requirement of CMOS logic and the current path to ground for TTL was provided. Next, it had to light a LED when the input was high and turn it off when the input was low; the switch will provide a zero impedance current path for the current from the positive power to bypass the LED. Finally, it had to be easy for you to wire and check over in case it doesn't seem to be working properly.

In Fig. 3-37, along with the logic input circuit schematic, I have included a photograph of the completed circuit built on a breadboard. In the photograph, notice that I have clipped the LED and resistor leads to keep the circuit as neat as possible on the breadboard. I strongly recommend that you keep components as close to the surface of the breadboard as possible to minimize your confusion when you are starting to build more complex circuits.

To demonstrate the operation of the inverter, you can build the circuit shown in the left side of Fig. 3-38 on your breadboard using the wiring diagram on the right side of Fig. 3-38. When the input LED is on, the output LED will be off and vice versa. If one or the other LED does not light, then first check your wiring followed by the polarity of the LEDs – the flat side of

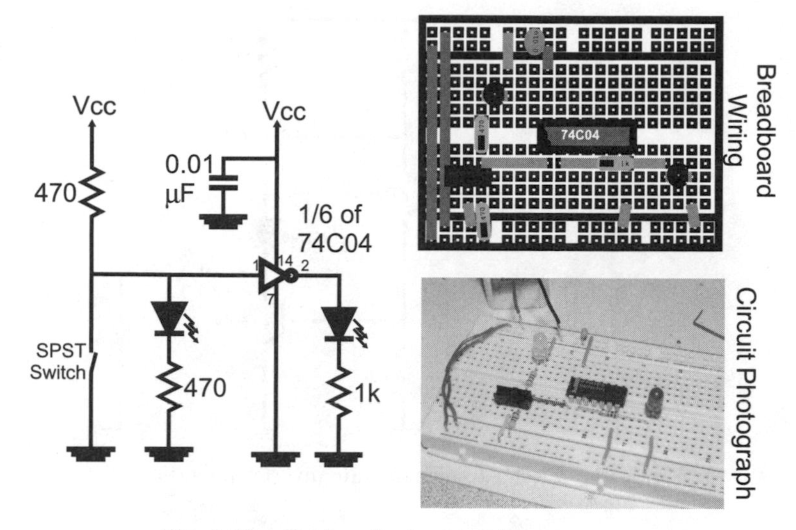

Fig. 3-38. Putting the input with an inverter.

the LED must be connected to the negative voltage (Vss) connection of your circuit.

To build the inverter test circuit, you will need the following parts:

- Breadboard
- 9 volt battery
- 9 volt battery clip
- 74C04 CMOS hex inverter chip
- Two 5 mm LEDs
- Two 470 Ω 1/4 watt resistors
- 1 k 1/4 watt resistor
- 0.01 µF capacitor (any type)
- Breadboard mountable switch (Digi-Key EG1903 suggested).

The only part that you might have some problems finding is the breadboard mountable switch (the EG1903 is a single-pole, double-throw switch with three posts 0.100 inch apart). This part is fairly unique and if you don't want to go through the trouble of ordering the part from Digi-Key, you can either add wires to another switch or simply connect the circuit to the Vss connection to simulate the switch closing (in this case, the LED will go off indicating a low input, just as if a switch were in circuit).

The 74Cxx family of chips are CMOS logic that are pin and output current compatible with 74LSxx TTL chips. The 74C04 used in the circuit shown in Fig. 3-38 demonstrates the operation of the NOT gate (or inverter) to quite good effect. The 74C04 does not demonstrate the operation of a TTL gate all

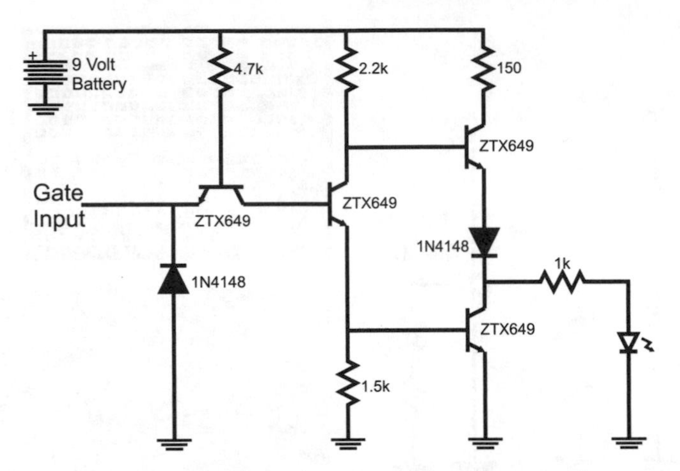

Fig. 3-39. Circuit to demonstrate inverter operation.

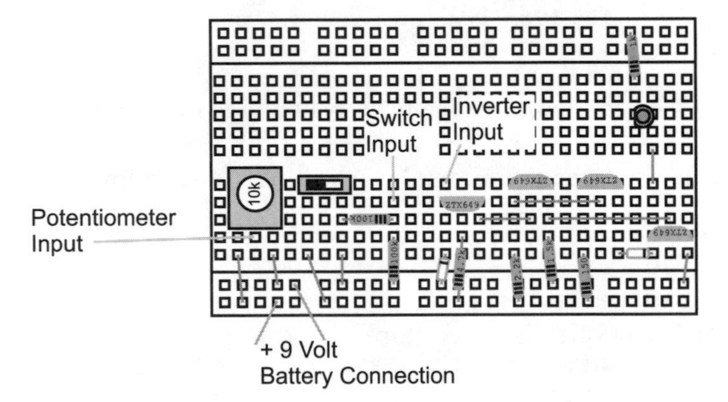

Fig. 3-40. Breadboard wiring diagram for the TTL inverter circuit.

that well, so if you have a few moments, I suggest that you build the circuit shown in Fig. 3-39 (wired according to Fig. 3-40) and test it out – externally, it will seem to work identically to the 74C04 circuit shown in Fig. 3-38, but there are a few differences that you can experiment with.

The parts that you will need for this circuit are:

- Breadboard
- 9 volt battery
- 9 volt battery clip
- Four 2N3904 NPN bipolar transistors
- Two 1N914 (or equivalent) silicon diodes
- Two 5 mm LEDs
- 150 Ω 1/4 watt resistor

- Two 470 Ω 1/4 watt resistors
- 1 k 1/4 watt resistor
- 1.5 k 1/4 watt resistor
- 2.2 k 1/4 watt resistor
- 4.7 k 1/4 watt resistor
- 100 k 1/4 watt resistor
- 10 k potentiometer
- Breadboard mountable switch (Digi-Key EG1903 suggested).

Going through the circuit, you can see that current flows through the circuit in two different directions, as shown in Figs. 3-41 and 3-42. When the input is

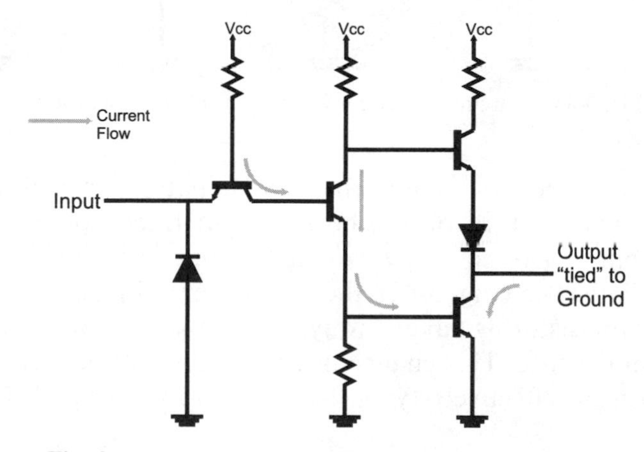

Fig. 3-41. TTL inverter with a "1" or floating input.

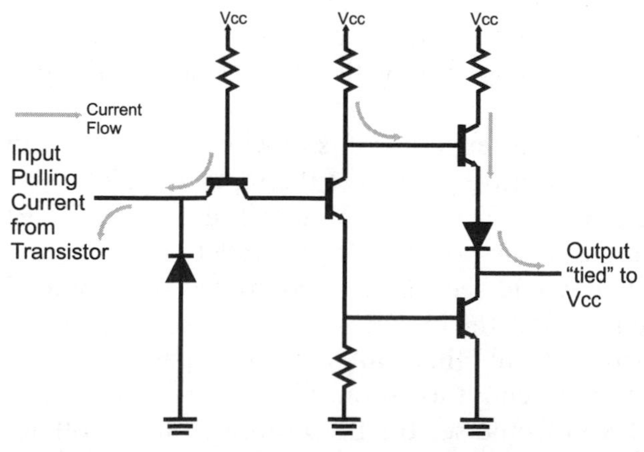

Fig. 3-42. TTL inverter with a "0" input.

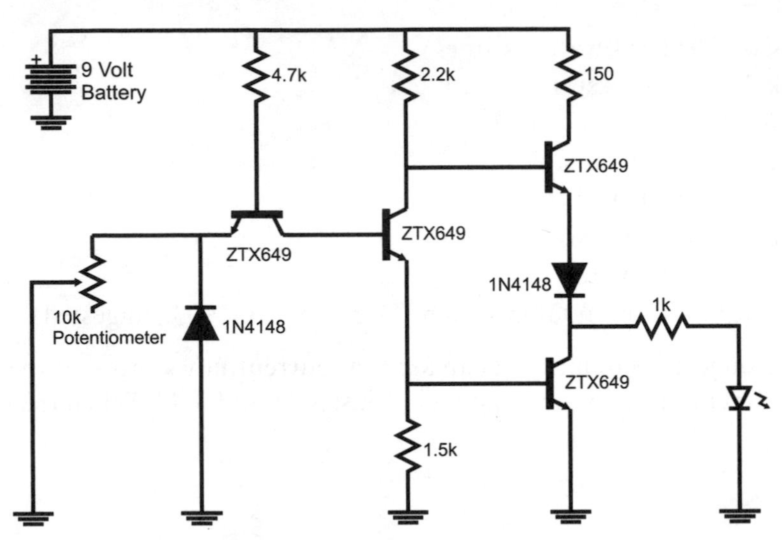

Fig. 3-43. Circuit to test current draw for TTL operation.

"high" (LED on) and you follow the current path, you will see that the current will ultimately turn on the bottom right transistor, connecting the gate's output pin to ground ("low" voltage output). When current is drawn from the TTL input pin (Fig. 3-42), the current that ultimately turned on the bottom right transistor is taken away, resulting in a different path for currents within the gate. This change in current flow ultimately turns on the top right transistor, effectively tying the output to power and driving out a "high" voltage.

Once you have built the circuit and tested it, you can now look at the operating aspects of it by putting a potentiometer in the circuit, as I have shown in Fig. 3-43, and adjust it until the LED either flashes on and off or dims. If you have a digital multi-meter (DMM), you will find that the *threshold* current is about 1 mA, with a voltage across the potentiometer of around 0.5 volts.

The final aspect of this experiment is to wire the inverter's input as shown in Fig. 3-44 and alternatively connect the input (passing through the 100 k resistor) to the power in or ground. You will find that the LED never turns on regardless of the switch position. If you were to measure the voltage at the 100 k resistor, you would see that it is connected directly to the power and ground connections, but the circuit seems to ignore the ground connection. The 100 k resistor prevents the 1 mA of current passing through to ground, resulting in the LED being turned on. If you were to repeat this experiment with the 74C04, you would see the LED turning on and off according to the voltage at the 100 k resistor.

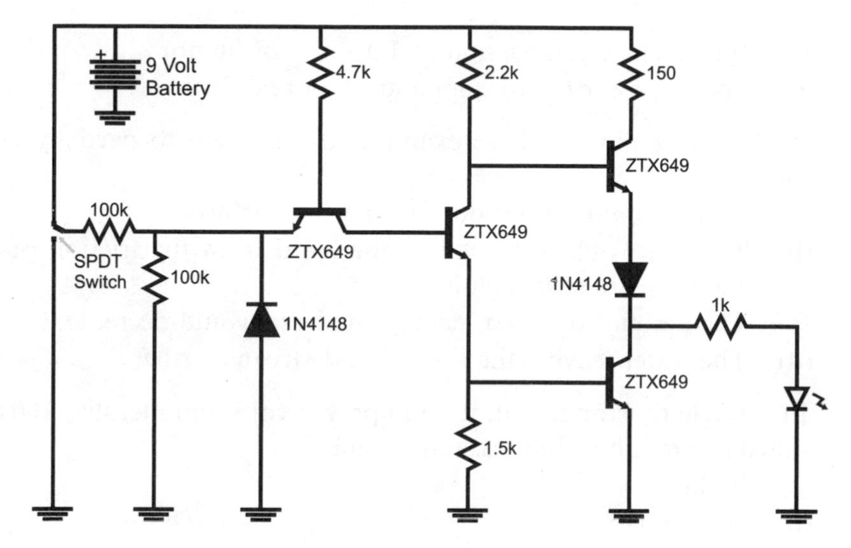

Fig. 3-44. Circuit to test voltage control of TTL operation.

In this chapter, I have given you a brief tutorial in basic electronics, an introduction to semiconductors and a method that you can use to build test circuits to experiment with digital electronics. In these few pages, I have covered the material included in several high school and college courses. It was not my intention to overwhelm you, but provide you with enough information to understand what is happening in a digital electronic circuit as well as give you a few basic rules to help you avoid problems, or if things aren't working as you would expect, to have some ideas on where to look for the problems.

Quiz

1. Electricity must:
   (a) Change polarity 60 times a second
   (b) Flow between the planets
   (c) Be equal in all parts of a circuit
   (d) Flow in a closed, continuous loop
2. Every electrical circuit has three parts:
   (a) Breadboards, batteries and electronic parts
   (b) Power source, load and conductors

(c) Intelligence, compassion and a sense of humor
(d) Speed, power (or torque) and corporeal form

3. In the water pipe/tap/hose example, if you were to partially close the tap:
(a) Water would stream out faster from the hose
(b) The tap would get hot in your hand from the friction of the water passing through it
(c) The amount of water leaving the hose would decrease
(d) The water leaving the hose would stream further

4. In a single resistor circuit, if you apply 9 volts and measure 100 mA flowing through it, the resistance value is:
(a) 9 ohms
(b) 900 ohms
(c) 90 ohms
(d) 1,111 ohms

5. The equivalent resistance of a 10 ohm and 20 ohm resistor in parallel:
(a) Is always zero
(b) 30 ohms
(c) 7.5 ohms
(d) 6.7 ohms

6. A diode is said to be "forward biased" when:
(a) A positive voltage is applied to the "bar" painted on the side of the diode
(b) Electrons are injected into the P-type semiconductor of the diode
(c) Current flows into the diode through the end which doesn't have a band painted on it
(d) More than 0.7 volts is applied to it

7. If a bipolar transistor with an h_{FE} of 150 had a "small signal operating region" base current of 1 μA to 1 mA, what base current would be required to allow 10 mA collector current?
(a) This is impossible to answer because 10 mA collector current is greater than 1 mA.
(b) 1 mA
(c) 67 μA
(d) 667 μA

8. The basic TTL gate is:
   (a) The NOT gate
   (b) The AND gate
   (c) The NOR gate
   (d) The NAND gate
9. Totem pole outputs are best used:
   (a) When there are multiple outputs tied together as a "dotted AND"
   (b) To drive electric motors
   (c) As the default output type used in digital electronic circuits
   (d) When high-speed operation of the digital electronic circuit is required
10. The dual in-line package:
    (a) Is a standard method for packaging digital electronic chips
    (b) Is used because part numbers cannot be stamped on bare chips
    (c) Allows for an easy visual check to see whether or not the part was damaged by heat
    (d) Facilitates effective cooling to the chip inside

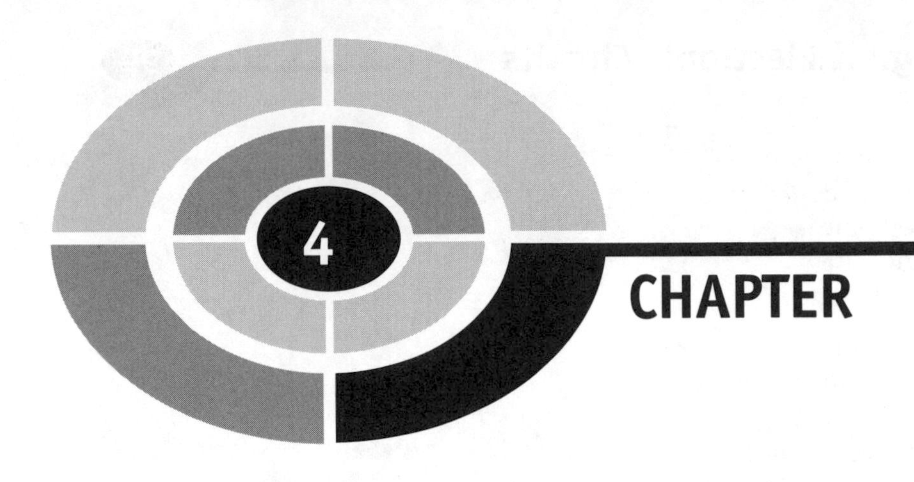

Number Systems

Working through the book to this point, you should be comfortable with combining multiple single bit values together in a variety of different ways to perform different combinatorial circuit functions. Along with being able to meet the basic requirements, you should be able to optimize the circuit to the fewest number of gates that is available within the technology that you are going to use. This skill is very useful in itself, but it is only scratching the surface of what can be done with digital electronics; most data consists of more than a single bit (which can have only two values) to process, and working with multiple single bits of data can be cumbersome. What is needed is a methodology for combining bits together so they can represent larger values that can be simply expressed.

The solution to this issue is to combine bits in exactly the same way as a 10-value character is combined to produce the decimal numbers that you are familiar with. While on the surface, combining bits does not seem to be directly analogous to decimal numbers, by using the same method that decimal numbers are produced, multi-bit numbers (which are most often described as "binary") numbers can be produced.

In primary school, you learned that the four-digit number "1,234" was built out of four digits, any of which could have the 10 values "0", "1", "2",

"3", "4", "5", "6", "7", "8" and "9". When listing the different values for a digit, zero is stated because the number "10" is actually a two digit number. The number of different values for each digit is referred to as its "base" or "radix". It is important to note that the first value is always zero and the last value is the base minus one.

When expressing each digit, its value was stated by the "column" it was in ("ones", "tens", "hundreds", "thousands", etc.). For example, the second column of "1,234" is the "hundreds" column and in 1234, there are two hundreds.

In high school, you would have been introduced to the concept of exponents and instead of expressing each digit in the number by the column, you would express it by the digit multiplier. So, 1,234 could now be written out as:

$$1 \text{ Thousand} = 1 \times 1{,}000 = 1 \times 10^3$$

$$2 \text{ Hundreds} = 2 \times 100 = 2 \times 10^2$$

$$3 \text{ Tens} = 3 \times 10 = 3 \times 10^1$$

$$4 \text{ Ones} = 4 \times 1 = 4 \times 10^0$$

The beauty of expressing a number in this way is that each digit's multiplier is mathematically defined as a power of the base. Using this format, it is possible to create a numbering system using single bits to represent "binary" numbers.

For example, four bits could be put together with the bit containing the least significant digit labelled "Bit_0", the second least significant as "Bit_1", the second most significant as "Bit_2" and the most significant as "Bit_3". The term significance when applied to bits is used to express the magnitude of the bit's multiplier. For example, Bit_0, which is multiplied by 2^0 or 1, has less significance than Bit_3 which is multiplied by 2^3 or 8.

Using the same exponent format as was used to define the decimal number 1,234, the four-bit binary number could be defined as:

$$Bit_3 \times 2^3$$

$$Bit_2 \times 2^2$$

$$Bit_1 \times 2^1$$

$$Bit_0 \times 2^0$$

and written out in a similar format to a decimal number. Collectively, the number is written out as a series of ones and zeros, in a similar manner to that of a decimal number.

Many books go into great length trying to explain how to convert a decimal number to a binary number. I won't go into the same amount of detail because the algorithm to do this is really quite simple: you simply start at some most power of two and work your way down, writing out a "1" for each time the subtraction the power of two results in a positive number or zero and a "0" when the difference is negative.

Written out as part of a "C" program, converting a decimal number to a character four-bit binary number is accomplished by the following statements:

```
for (i = 4; i! = 0; i - -)
    if ((DecVal - (2 * *(i - 1)))> = 0)
    {                          //Can Take Away Digit Value
        DecVal = DecVal - (2 * *(i - 1));
        Bit[i - 1] = `1´;
    } else                     //Result of subtraction is negative
        Bit[i - 1] = `0´;   //Can't take away value
```

Note that I start at "4" and subtract one for the actual bit value in the example code above.

Demonstrating the algorithm, consider the case where you wanted to express the decimal number "11" as a four-bit decimal. In Table 4-1, I have listed each step of the program with the variable values at each step.

Converting binary numbers to decimal is very easy because the power of two of each digit that has a value of "1" are summed together. The "C" code

Table 4-1 Converting decimal 11 to binary 1011.

'Bit' variable	i	DecVal	DecVal – ((2 ** (i – 1))
xxxx	4	11	3 (Positive)
1xxx	3	3	–1
10xx	2	3	1 (Positive)
101x	1	1	0 (Positive)
1011	0	0	N/A

Table 4-2 Converting binary 0110 to decimal 6.

'Bit' variable	i	Bit[i – 1] != 0?	DecVal
0110	4	No	0
0**1**10	3	Yes	4 (DecValue + (2 ** (3 – 1)))
01**1**0	2	Yes	6 (DecValue + (2 ** (2 – 1)))
011**0**	1	No	6
0110	0	N/A	6

to convert a value in "Bit" to a decimal value is:

```
DecVal = 0;                    // Initialize the Decimal
                                  Value Variable
for (I = 4; i> = 0; i – –)     // Repeat for four bits
  if(Bit[i – 1] == "1")        // Add Digit Value if Digit Not 0
    DecValue = DecValue + (2 * *(i – 1));
```

In Table 4-2, I have listed the process of converting the binary number 0110 to decimal and you should note that I have highlighted the bit that is being tested.

Before going on, I would like to point out that there can be a lot of confusion with regards to using binary numbers with decimal numbers or numbers of different bases. To eliminate the confusion, you should always identify the binary numbers by placing a percentage sign ("%") or surrounding it with the letter "B" and two single quotes ("' "). Using these conventions, the bit pattern converted in Table 4-1 would be written out as %0110 or B'0110'. The % character put before a binary number is a common assembly language programming convention. The letter "B" and the single quotes around the number is the format used in "C" programming and will be the convention that I use in this book.

Another area of confusion with regards to binary numbers is how they are broken up for easier reading. Each group of three digits in a decimal number is usually separated from other groups of digits by use of a comma ("," in North America and a period or dot (".") in Europe and other parts of the world). When working with binary numbers, instead of separating each three digit group with a punctuation character, it is customary to use a blank to separate four digit groups. Using the conventions outlined here, the eight bit number 10111101 would be written out as:

B′1011 1101′

This is the binary number format convention that I will use for the rest of the book.

Base 16 or Hexadecimal Numbers

As I will show in this and the next section, having programming experience is a two-edged sword – it will help you understand certain concepts (such as the "bit" and some data structures like the ones presented in this and the next section), but it will blind you to other opportunities. The goal of these sections is to illustrate how bits can be grouped together to make your design efforts more efficient as well as making it easier for you to both see possibilities for the design and articulate them to other people.

Creating binary numbers from groups of bits, as I demonstrated in the introduction to this chapter, is quite easy to do, but can be very cumbersome to write out as well as transfer correctly. You may also have difficulty in figuring out exactly how to express the number, asking should it be passed along starting from the most significant or least significant bit. At the end of this chapter's introduction, I left you with the number B'1011 1101' and you should agree that telling somebody its value is quite cumbersome; for example, you might say something like, "The eight bit, binary number, starting with the most significant bit is one, zero, one, one, one, one, zero and one."

It is much more efficient to combine multiple bits together into a single entity or digit.

The most popular way of doing this is to combine four bits together as a "hexadecimal" digit which has 16 different values. This numbering system has a base of 16. If you are familiar with programming, chances are you are familiar with hexadecimal digits (which is often contracted to the term "hex"), which I have listed out with their decimal and binary equivalents in Table 4-3.

To create a way of expressing the 16 values, the first 10 hexadecimal values are the same as the 10 decimal number values, with the following six being given letter codes. This is why I included the "phonetic" values for the hexadecimal values greater than 9; the letter names "B", "C" and "D" can be easily confused, but their phonetic representations are much clearer.

Table 4-3 Hexadecimal digits with binary, decimal equivalents and phonetic values.

Decimal	Binary	Hex	Phonetic	Decimal	Binary	Hex	Phonetic
0	B'0000'	0	Zero	8	B'1000'	8	Eight
1	B'0001'	1	One	9	B'1001'	9	Nine
2	B'0010'	2	Two	10	B'1010'	A	Able
3	B'0011'	3	Three	11	B'1011'	B	Baker
4	B'0100'	4	Four	12	B'1100'	C	Charlie
5	B'0101'	5	Five	13	B'1101'	D	Dog
6	B'0110'	6	Six	14	B'1110'	E	Easy
7	B'0111'	7	Seven	15	B'1111'	F	Fox

I tend to place a lot of importance to using conventions when expressing letters. You may be tempted to make up your own letter codes or use the aviation phonetic alphabet (Table 4-4) when communicating hexadecimal values to other people ("AF" could be "Apple-Frank" or "Alpha-Foxtrot" instead of "Able-Fox"). I would like to discourage this for two reasons: the first is that the person you are talking to will have to mentally convert your words into letters and then hex digits – this process is complicated when unexpected words are used. Secondly, I prefer using the phonetic codes in Table 4-3 for hex values and the aviation phonetic codes for letter codes.

Multi-digit hexadecimal numbers are written out in a similar way as decimal or binary numbers with each digit multiplied by 16 to the power of the number of value's position. For a 16 bit number (four hexadecimal digits), the digit multipliers are listed below:

$$\texttt{HexDigit}_3 \times 16^3 = \texttt{HexDigit}_3 \times 4{,}096$$

$$\texttt{HexDigit}_2 \times 16^2 = \texttt{HexDigit}_2 \times 256$$

$$\texttt{HexDigit}_1 \times 16^1 = \texttt{HexDigit}_1 \times 16$$

$$\texttt{HexDigit}_0 \times 16^0 = \texttt{HexDigit}_0 \times 1$$

To indicate a hex number, you should use one of the programming conventions, such as putting the prefix "0x0" or "$" at the start of the hexadecimal

Table 4-4 Aviation phonetic codes.

Letter	Phonetic	Letter	Phonetic	Letter	Phonetic
A	Alpha	J	Juliet	S	Sierra
B	Beta	K	Kilo	T	Tango
C	Charlie	L	Lima	U	Uniform
D	Delta	M	Mike	V	Victor
E	Echo	N	November	W	Whiskey
F	Foxtrot	O	Oscar	X	X-Ray
G	Gulf	P	Papa	Y	Yankee
H	Hotel	Q	Quebec	Z	Zulu
I	India	R	Romeo		

value. The same formatting convention used with binary numbers (X'##', where "##" are the hex digits) could also be used. For this book, I will be expressing hexadecimal numbers in the format 0x0## which is visually very different from binary numbers, which should help to immediately differentiate them.

To convert a decimal number to a character 16 bit hexadecimal number, you can use the "C" algorithm shown below. Note that I have used the C modulo ("%") operation which returns the remainder from an integer division operation and not its dividend.

```
for (i = 16; i! = 0; i = i - 4)
{
  if ((DecVal/(16 * (1<<(i - 8)))) > 9)
    Hex[(i/4) - 1] = (DecVal/(16 * (1<<(i - 8)))) - 10 + `A´;
  else
    Hex[(i/4) - 1] = (DecVal/(16 * (1<<(i - 8)))) + `0´;
  DecValue = DecValue%(16 * (1<<(i - 8));
} //  rof
```

Going the other way, to convert a four hexadecimal digit number to decimal you can use the algorithm:

```
DecVal = 0;
for(i = 4; i! = 0; i − −)
   if((Hex[i − 1] >= "A")&&(Hex[i − 1] <= "F"))
      DecVal = (DecValue * 16) + (Hex[i − 1] − `A´ + 10);
   else
      DecVal = (DecValue * 16) + (Hex[I − 1] − `0´);
```

Many books provide a conversion table between binary, hexadecimal and decimal numbers, but I would like you to be familiar with the conversion algorithms written out above as well as buy yourself an inexpensive scientific calculator which has ability to convert between base systems. The ability to convert between the base systems is actually quite simple and available in many basic scientific calculators which cost $10 or less. Understanding how to convert between base systems and having an inexpensive calculator will enable you to perform the conversions faster and with more flexibility than using a table, which is limited in the number of different values it can present.

If you are familiar with numbers in different languages, then you will know that the prefix "hex" actually refers to the number "six" and not "16". The actual prefix for 16 is the term "sex" and in the early days of computers, this was (obviously) a source of some amusement. When IBM introduced the System/360, in the early 1960s, the company was uncomfortable with releasing something that was programmed in "sexadecimal", fearing that it might upset some users. To avoid any controversy, all documentation for the System/360 was written using the 16 bit "hexadecimal" numbering system presented here. The System/360 was a wild success, becoming the first "computer for the masses" and many people's first experience in programming and electronics. The term "hexadecimal" became the popular term for 16 bit numbers and displaced the more correct "sexadecimal."

Binary Coded Decimal

In the early days of programming, data structures were often the result of a curious blend of trying to come up with a data format that best suited the programmer and what best suited the current hardware. One of the more

Table 4-5 Decimal digits with binary and BCD decimal equivalents.

Decimal	Binary	BCD	Decimal	Binary	BCD
0	B'0000'	0	8	B'1000'	8
1	B'0001'	1	9	B'1001'	9
2	B'0010'	2	10	B'1010'	Invalid
3	B'0011'	3	11	B'1011'	Invalid
4	B'0100'	4	12	B'1100'	Invalid
5	B'0101'	5	13	B'1101'	Invalid
6	B'0110'	6	14	B'1110'	Invalid
7	B'0111'	7	15	B'1111'	Invalid

enduring structures that came from this time is the "binary coded decimal" (most often referred to by its acronym "BCD") which used four bits, like hexadecimal values, but only allowed the values of zero through nine rather than the full 16 values that were possible (as shown in Table 4-5). The reason for using this data structure has largely disappeared in computer systems, but it is still a viable and useful method of handling data in digital electronics and one that you should keep in your "hip pocket" when you design circuits.

The original reason for using the BCD data format in computer programming was its elimination of the need to add code to the program to convert a binary or hex number into decimal. The code storage required for the conversion was expensive and the processors were nowhere near as possible as what is available today. Using decimal values was actually an optimal way of processing data in these old systems.

The lasting legacy of this is the number of standard chips that can process BCD values just as easily as other standard chips can process hexadecimal values and will allow you to design circuitry that works with decimal values just as easily as if you were working with hex values.

While this is getting a bit ahead of things, I want to give the example of designing a delay that holds back a signal for 100 seconds. Using traditional binary logic, which only works with bits that are a power of two, you would have to design a circuit that compares a counter value and indicates when the value "100" was reached and reset itself. When using digital electronic chips that are designed for BCD values, the comparator function is not required, as each BCD digit cannot be greater than "9" and, cascaded together, they can only count to a maximum value of "99" to "00".

This may seem like a trivial example, but you will find a number of cases like this one where you will have to create circuits that work on base 10 data and by using chips which are designed for BCD values, the complexity of your work will be greatly reduced.

Going back to Table 4-5, the production of the "invalid" indication is worthy of some discussion as it provides a good example of how gate optimization is not always as straightforward as you might expect.

In most BCD chips, if the value of 10 or more is passed in the binary bits, then the value is converted to zero and a carry indication is output. Using the tools presented in Chapter 2, you should be able to derive the sum of products formula for the positive active "invalid" indicator as:

$$\texttt{Invalid} = (\texttt{A3} \cdot \texttt{A2}) + (\texttt{A3} \cdot \texttt{A2})$$

and using the conversion formulas of Chapter 2, you would simplify the "invalid" formula above to:

$$\texttt{Invalid} = \texttt{A3} \cdot (\texttt{A2} + \texttt{A1})$$

Figure 4-1 shows the AND/OR gates for this function along with the "NAND equivalent" function beneath it. The NAND equivalent was chosen by assuming that the function would be implemented in TTL. While this circuit looks a bit complex, if you follow it through, you will find that it provides the same function as the AND/OR combination above it.

It will probably surprise you to find out that this circuit is not optimal by any measurement: you can do better in terms of the number of gates, the time

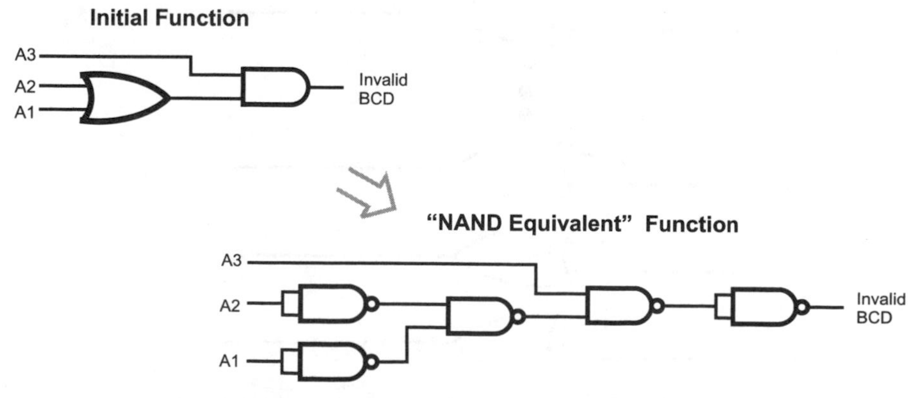

Fig. 4-1. Basic circuit for detecting values over "9" with reduction.

it takes a signal to pass through the gates and in providing a constantly timed output. The circuit at the bottom half of Fig. 4-1 will respond in two gate delays if A3 changes and in four gate delays if A2 changes. For many circuits, this is not a problem, but when you are working with high-performance designs, a variable output delay can result in the application not working correctly and being almost impossible to debug.

A much better approach to optimizing the circuit is to work at converting it to the basic gate used by the technology that you are working with and then optimizing this. Going back to the original "Invalid" equation:

$$\texttt{Invalid} = (\texttt{A3} \cdot \texttt{A2}) + (\texttt{A3} \cdot \texttt{A2})$$

I can convert the OR to a NAND, by inverting its two parameters (according to De Morgan's theorem), ending up with:

$$\texttt{Invalid} = \texttt{!}(\texttt{!}(\texttt{A3} \cdot \texttt{A2}) \cdot \texttt{!}(\texttt{A3} \cdot \texttt{A2}))$$

It is probably astounding to see that the function provided by the mess of NAND gates in Fig. 4-1 can be reduced to the three simple gates required by the formula above. Along with reducing the number of gates, you should also notice that the maximum number of gate delays is two, regardless of which bit changes.

Looking at the NAND circuits in both diagrams, you are probably at a loss as to how you could reduce the NAND circuit in Fig. 4-1 to the three

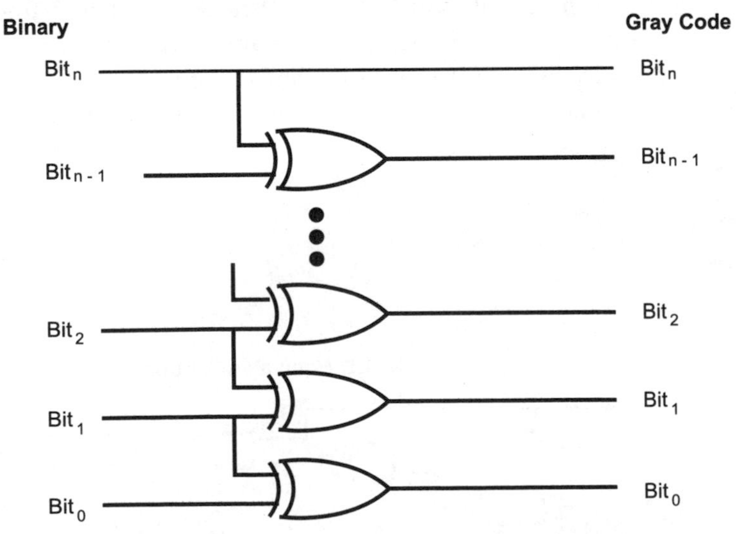

Fig. 4-2. Simple digital electronic circuit to convert a binary number to a Gray code.

gates of the optimized circuit. Personally, I would be surprised if you could; when I look at the two circuits, they look like they provide completely different functions.

What I want to leave you with is an example of how looking at a logic function from different perspectives can result in radically different circuits with surprisingly different parameters. In the first case, I reduced three gates to two, to end up with six NAND gates, while in the second, I avoided reducing the basic function and converted it directly to a much more efficient three NAND gate circuit.

In going through this exercise to produce the "invalid" output for BCD, I hope that you can apply this knowledge for creating circuits that work with different base systems than just a power of two. In some cases, you may have to work with numbers that are base 9 or 13 and using the example here, you should have some idea of how to keep the values within certain "bounds".

Gray Codes

I hope I have convinced you of the usefulness of using Gray codes for inputs when you are illustrating how digital electronic logic functions respond to inputs. I must point out, however, that Gray codes were originally created for a much different function – they were designed for use in position sensors as the single changing bit allowed hardware to be designed to respond to a single changing bit and not the potentially several bits of a binary sequence. By only changing one bit at a time, absolutely precise positioning of marking sensors (causing all changing bits being sensed at the exact same instant) was not required.

Gray codes were invented by Frank Gray of Bell Labs in the mid 1950s and has a "hamming value" of 1. The hamming value is the number of bits that change between one value and the next. A four bit binary number can have all four bits change as it increments or decrements; a Gray code never has more than one bit change during incrementing or decrementing operations.

Chances are, you would not have any trouble coming up with a two bit Gray code (b'00', b'01', b'11' and b'10') and in a pinch, you would be able to come up with a three bit Gray code (b'000', b'001', b'011', b'010', b'110', b'111', b'101' and b'100'). I suspect that if you were given the task of coming up with any more bits than this, you would be stumped.

In trying to come up with a way of explaining how Gray codes worked, I noticed that when a new most significant bit was set, the previous values were ORed with this bit, but written out in reverse order. In some texts, this property is recognized by calling Gray codes a *binary reflected code*. Looking at Table 4-6, you can see that I created a four bit

Table 4-6 Building Gray codes from previous value.

Binary value	Gray code	Comments
B'0000'	B'0000'	1 bit Gray code, Gray code = binary value
B'0001'	B'0001'	1 bit Gray code, Gray code = binary value
B'0010'	B'0011'	2 bit Gray code, last value of bit 1 (B'1') list ORed
B'0011'	B'0010'	2 bit Gray code, first value of bit 1 (B'0') list Used
B'0100'	B'0110'	3 bit Gray code, last value of bit 2 (B'10') list ORed
B'0101'	B'0111'	3 bit Gray code, next last value of bit 2 bit (B'11') list ORed
B'0110'	B'0101'	3 bit Gray code, next last value of bit 2 (B'01') list ORed
B'0111'	B'0100'	3 bit Gray code, first value of bit 2 (B'00') list ORed
B'1000'	B'1100'	4 bit Gray code, last value of bit 3 (B'100') list ORed
B'1001'	B'1101'	4 bit Gray code, next last value of bit 3 (B'101') list ORed
B'1010'	B'1111'	4 bit Gray code, next last value of bit 3 (B'111') list ORed
B'1011'	B'1110'	4 bit Gray code, next last value of bit 3 (B'110') list ORed
B'1100'	B'1010'	4 bit Gray code, next last value of bit 3 (B'010') list ORed
B'1101'	B'1011'	4 bit Gray code, next last value of bit 3 (B'011') list ORed
B'1110'	B'1001'	4 bit Gray code, next last value of bit 3 (B'001') list ORed
B'1111'	B'1000'	4 bit Gray code, first value of bit 3 (B'000') list ORed

Gray code by taking the eight values of the three bit code, reversing them and setting bit 3.

This could be written out as a computer program algorithm as:

```
for  (i = 1; i < #Bits; i++)  //Loop Around for Each Code
    if(1 == i)                //-Trivial Initial Case
    {
     GrayCode[0] = 0;  GrayCode[1] = 1;
    } else {                  //Copy Previous Codes with
                              //New MSB Set
       for(j = (1 << (i - 1)),k = ((1 << (i - 1)) - 1);
        j < (1 << i); j++,k--)
          GrayCode[j] = (1 << (i - 1)) + GrayCode[k];
    }        //fi
```

This code demonstrates how Gray codes are produced, but is not the optimal method for producing Gray codes (it is actually an "order n_2" algorithm, which means that every time the number of bits is doubled, the amount of time required to produce the values is quadrupled). Along with this, it is not easy to create digital logic hardware that will create these codes.

Fortunately, individual binary codes can be converted to Gray codes using the circuit shown in Fig. 4-2, which simply implements the formula:

```
Gray Code = Binary ^ (Binary >> 1)
```

Going the other way (from Gray code to binary) is a bit more complex and while it uses n – 1 (where "n" is the number of bits) XOR gates, like converting binary codes to Gray codes, the output of each XOR gate is required as an input to the next least significant bit, as shown in Fig. 4-3. The output of the circuit is not correct until the most significant bit has passed through each of the XOR gates to the least significant bit.

To perform the data conversion a simple formula cannot be used. Instead the following algorithm is required:

```
GrayCode = Binary; Shift = 1;
While((GrayCode >> Shift)! = 0)
```

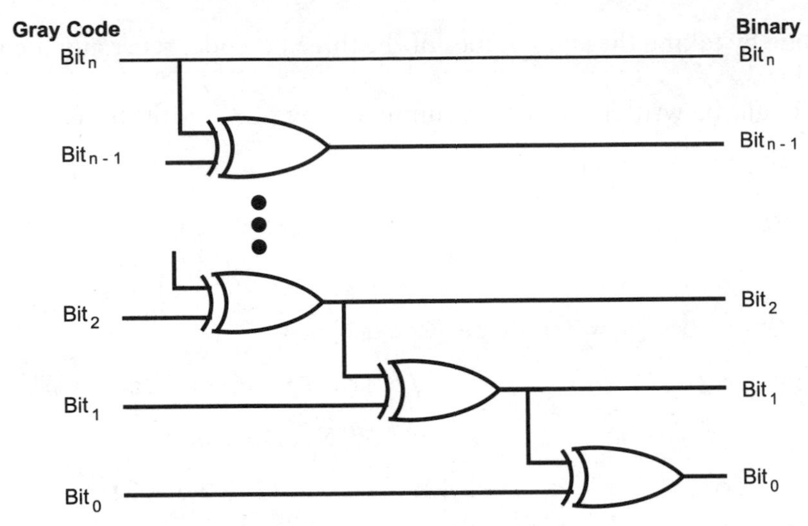

Fig. 4-3. Digital electronic circuit to convert a Gray code to a binary number.

```
    {
        GrayCode = GrayCode ^ (GrayCode >> Shift);
        Shift = Shift * 2;
    } // elihw
```

I find it very difficult to explain exactly how this code works, except to say that with each iteration of the while loop, the "Gray code" value gets shifted down more and more to move the most significant bits into position for XORing with the less significant bits. To convince yourself that the algorithm works, you might want to perform a "thought experiment" on it and list the changing value of "Gray code" as I have done in Table 4-7.

In this chapter, more than anywhere else in the book, I have used sample computer programs to show how different values can be produced. This is a somewhat different approach to explaining how multi-bit binary data conversions are implemented and one that takes advantage of the ubiquity of the personal computer and the ability of most technical students to perform even rudimentary programming.

Using computer code to help demonstrate how the conversions are done should also give you another method for processing binary values as well as of testing formulas and optimizations. I always find it useful to have a number of different ways to solve a problem, or test a potential solution,

Table 4-7 Working through the shifting values of the Gray code convention algorithm.

Initial bit values	Shift = 1 bit values	Shift = 2 bit values	Shift = 4 bit values
B7	B7	B7	B7
B6	B6 ^ B7	B6 ^ B7	B6 ^ B7
B5	B5 ^ B6	B5 ^ B6 ^ B7	B5 ^ B6 ^ B7
B4	B4 ^ B5	B4 ^ B5 ^ B6 ^ B7	B4 ^ B5 ^ B6 ^ B7
B3	B3 ^ B4	B3 ^ B4 ^ B5 ^ B6	B3 ^ B4 ^ B5 ^ B6 ^ B7
B2	B2 ^ B3	B2 ^ B3 ^ B4 ^ B5	B2 ^ B3 ^ B4 ^ B5 ^ B6 ^ B7
B1	B1 ^ B2	B1 ^ B2 ^ B3 ^ B4	B1 ^ B2 ^ B3 ^ B4 ^ B5 ^ B6 ^ B7
B0	B0 ^ B1	B0 ^ B1 ^ B2 ^ B3	B0 ^ B1 ^ B2 ^ B3 ^ B4 ^ B5 ^ B6 ^ B7

and I suggest that along with the various tools and computer algorithms presented in this book that you try to come up with methods for yourself that will help you design and test digital electronic circuits more efficiently.

Quiz

1. If you had a number system that was base 5, the most significant value in a digit would be:
   (a) 6
   (b) 10
   (c) 4
   (d) 5

2. The eight bit binary equivalent to decimal 47 is:
   (a) 0010 1111
   (b) B'0010 1111'

(c) 101111
(d) 1011 11

3. The third most significant digit in the decimal number "1234" is:
   (a) The hundreds column
   (b) 3
   (c) 1
   (d) No digit can be the third most significant
4. To verbally tell somebody the hex number value 0x04AC you would say:
   (a) "Four-Able-Charlie"
   (b) "Hexadecimal Four-Eh-See"
   (c) "Hexadecimal Four-Apple-Charlie"
   (d) "Hexadecimal Four-Able-Charlie"
5. The decimal number "123" in hexadecimal is:
   (a) 0x0123
   (b) B'0111 1011'
   (c) 7B
   (d) 0x07B
6. The four bit hexadecimal number 0x01234 expressed in decimal is:
   (a) 1,234
   (b) 4,660
   (c) B'0001 0010 0011 0100'
   (d) 0x04D2
7. Binary coded decimal is defined as:
   (a) Ten bits providing ten different values
   (b) Four bits providing ten numeric values and six control codes
   (c) Four bits providing ten numeric values
   (d) Five bits with each bit providing two values for a total of 10
8. BCD should:
   (a) Never be used
   (b) Used with circuits that operate with base 10 numbers
   (c) Only be used when you've run out of binary chips
   (d) Used when values are not expected to exceed 9
9. B'0110' in binary, using the formula Gray code = binary ^ (binary ≫ 1) can be converted to the Gray code:
   (a) B'1010'
   (b) B'0110'

(c) B'0101'
(d) B'0111'

10. The Gray code B'0010' corresponds to the binary value:
    (a) B'0011'
    (b) Unknown because more data is required
    (c) B'1101'
    (d) B'0010'

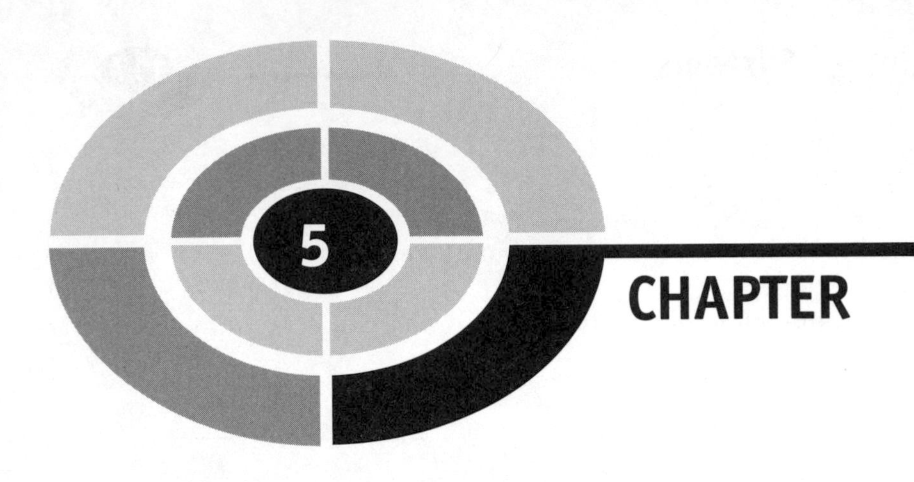

Binary Arithmetic Using Digital Electronics

Before going into showing how basic binary arithmetic operations are performed in digital electronic circuits, I thought it would be useful to review how you would perform basic arithmetic operations. Before discussing how many binary arithmetic operations there are, some different characteristics of binary numbers should be discussed. I realize that much of the material in this chapter introduction is a review of work that you first did in grade school, but often when confronted with situations that require you to develop binary arithmetic operations in digital electronics, this basic information can easily be forgotten and standard devices that provide this function are often overlooked.

Table 5-1 Necessary decimal addition pairs to memorize.

0+0									
0+1	1+1								
0+2	1+2	2+2							
0+3	1+3	2+3	3+3						
0+4	1+4	2+4	3+4	4+4					
0+5	1+5	2+5	3+5	4+5	5+5				
0+6	1+6	2+6	3+6	4+6	5+6	6+6			
0+7	1+7	2+7	3+7	4+7	5+7	6+7	7+7		
0+8	1+8	2+8	3+8	4+8	5+8	6+8	7+8	8+8	
0+9	1+9	2+9	3+9	4+9	5+9	6+9	7+9	8+9	9+9

When you first learned to add decimal numbers together, you probably were required to memorize all 100 different combinations of single digit parameters when only 55 are really required. In Table 5-1, I have listed the 55 pairs which have to be memorized; the remaining 45 pairs do not have to be memorized because of the commutative law which states:

$$A + B = B + A$$

and means that the number pairs like "4 + 7" and "7 + 4" are equivalent.

The result of adding each of these two parameters produces either a single digit or double digit sum. The double digit sum indicates that the value of the result is greater than could be represented in a single digit of the number base. For decimal numbers, the maximum value that can be represented by a single digit is "9". Looking at the general case, the maximum value that can be represented by a number system is the base minus one. So, for the binary number system (base 2), the maximum value is "1"; for hexadecimal (base 16), the maximum value is "15" (or "0x0F").

The leftmost digit of a double digit number is known as the "carry" digit.

In Chapter 4, I showed how multi-digit numbers are made up of single digit values multiplied by powers of their base. Knowing the sums of the 55 addition pairs of Table 5-1, multi-digit numbers can be added together

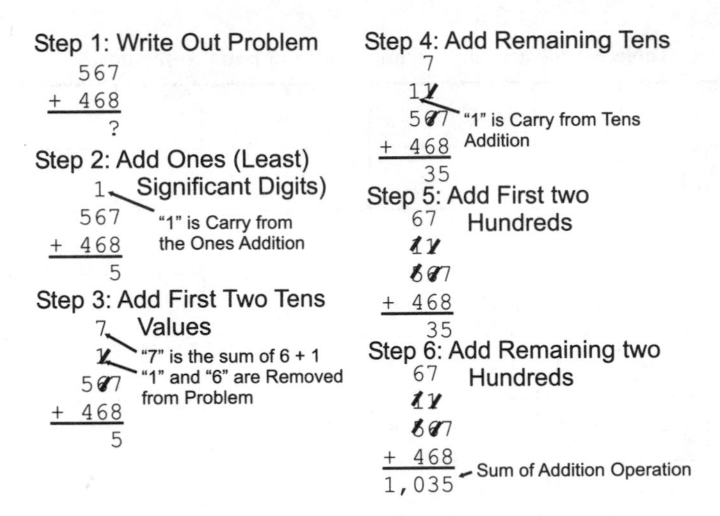

Fig. 5-1. The process steps used to add two three digit numbers together.

by working through pairs of numbers, as I show in Fig. 5-1. This is a rather pedantic way of showing addition and I'm sure that when you add two multi-digit numbers together, you are much more efficient, but when you were learning, this was probably the process that you went through.

While saying that you are much more efficient, it really comes down to the idea that you are able to recognize that one plus another number is the same as incrementing the other number. You are still only adding one digit at a time and the carry is "rippling" to the next significant digit. Carry "ripple" is an important concept that will be discussed in more detail in the next section.

Subtraction has many of the same issues as addition, but with some additional complexities. The first being that you cannot simplify your memorization of the 100 pairs of subtracted parameters as you did for addition; the commutative law does not apply to subtraction as it did for addition. For example,

$$6 - 4 \neq 4 - 6$$

Next, if the number being taken away is greater than the original number, the result (or "difference") could be less than 0 or "negative". There is a very big question on how to represent that negative number. Typically, it is represented as a value with a "minus" or "subtraction" sign in front of it, e.g. "–2".

The minus sign is only used when the digit cannot "borrow" from the next significant digit, as shown in Fig. 5-2. The result of 25 minus 9 is 16, with the ones borrowing 10 from the tens column (the next significant digit) to allow the operation to proceed without a negative result.

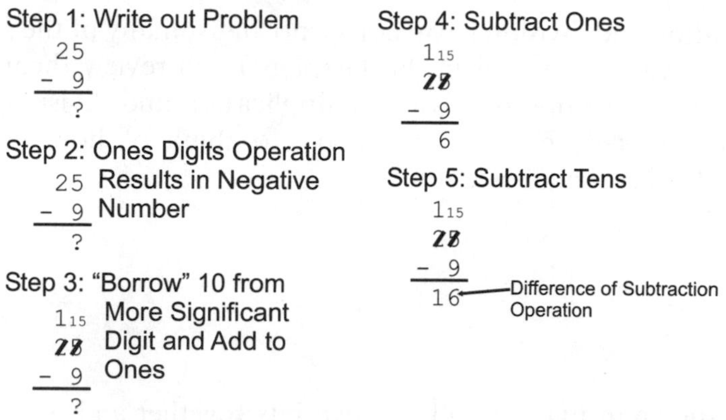

Step 1: Write out Problem
```
  25
-  9
----
   ?
```
Step 2: Ones Digits Operation Results in Negative Number
```
  25
-  9
----
   ?
```
Step 3: "Borrow" 10 from More Significant Digit and Add to Ones
```
  1 15
  2 5
-   9
----
    ?
```
Step 4: Subtract Ones
```
  1 15
  2 5
-   9
----
    6
```
Step 5: Subtract Tens
```
  1 15
  2 5
-   9
----
  1 6
```
Difference of Subtraction Operation

Fig. 5-2. Multi-digit subtraction with borrowing.

Subtraction can also be expressed as adding a negative value and can be written out as:

$$\mathrm{A} - \mathrm{B} = \mathrm{A} + (-\mathrm{B})$$

This should not be a surprise to you unless you consider the following philosophical question. What would happen if infinity was arbitrarily defined as one million (1,000,000)? Instead of adding a minus sign to our value to make it negative, we could subtract it from "infinity".

For example, if we had the problem:

$$8 - 5 = 8 + (-5)$$

we could define "–5" as one million subtract 5 or "999,995". Now, going back to the addition of the negative number and substituting in "999,995" for "–5" we get:

$$\begin{aligned} 8 - 5 &= 8 + (-5) \\ &= 8 + 999{,}995 \\ &= 1{,}000{,}003 \end{aligned}$$

Since a million is defined as infinity and has no meaning, it can be taken away from the result, leaving us with the difference of 8 minus 5 being "3". This method may seem to be overly complex, but I will show you how this applies to digital electronics later in the chapter.

Like addition, the method presented here for subtraction is carried out a single digit at a time with the need to borrow from the next more significant digit being similar to passing the carry digit in addition. Like the "ripple carry" in addition, the "borrow" in subtraction can also be thought of as a "ripple" operation.

Multiplication and division have, not surprisingly, many of the same issues and when I discuss them later in this chapter, I will review them with you. Before reading the section discussing multiplication and division, I suggest that you review these operations and try to think of how they can be accomplished using digital electronics.

Adders

The circuit shown in Fig. 5-3 will add two bits together and output the sum ("S") bit along with a carry ("C") bit, if both inputs are "1" and the sum is "2", which is greater than the maximum number that can be represented by the number base (which is 1 for binary). Table 5-2 is a truth table, showing the output of each bit for different input values. You should be able to see that the "sum" bit is 1 when one or the other (but not both) of the two input bits is 1 and the "carry" bit is 1 only when both input bits are 1.

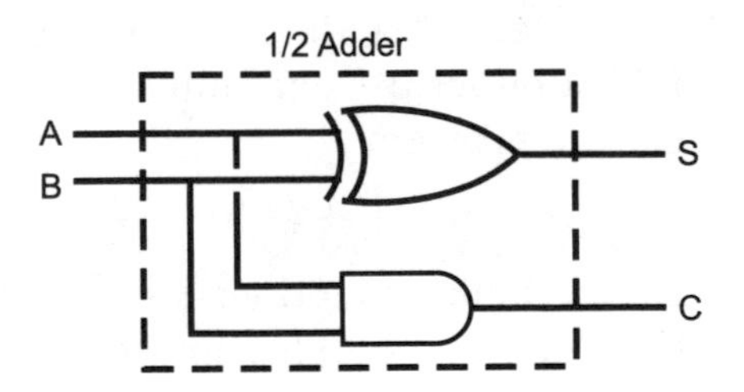

Fig. 5-3. Half adder circuit.

Table 5-2 Half adder operation truth table.

"A" input	"B" input	"Sum" bit	"Carry" bit
0	0	0	0
0	1	1	0
1	1	0	1
1	0	1	0

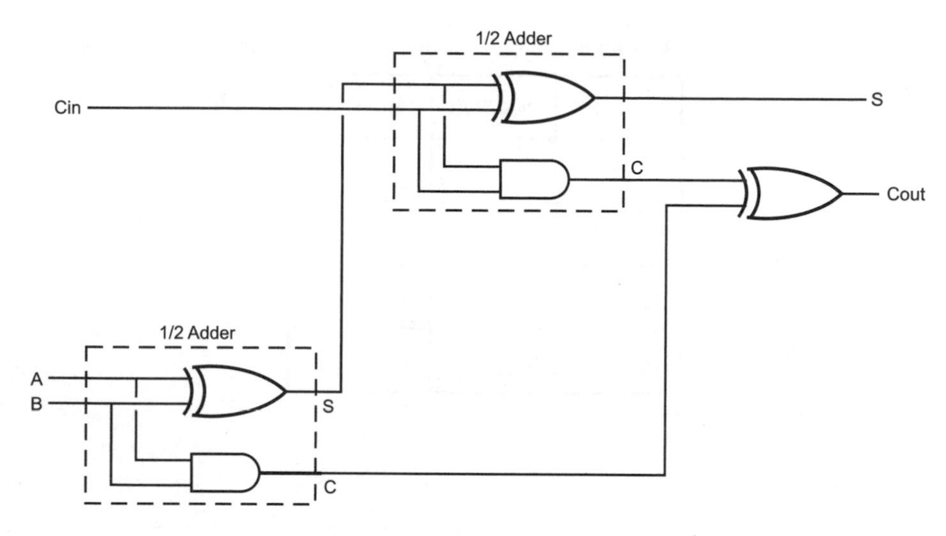

Fig. 5-4. Full adder circuit.

The adder is the first practical use most people have for the XOR gate and its function can be seen very clearly in Table 5-2 for the sum bit. Along with the XOR gate providing the function for the sum bit, you should also recognize that the carry bit is the output of a simple AND gate.

This simple digital electronic circuit is known as a "half adder" because it will handle half the operations required of the general case addition circuit. The "full adder" (Fig. 5-4) starts with a half adder and adds another bit (which is the less significant bit's "carry" output) to its sum. Put another way, the full adder adds three individual bits together (two bits being the digit inputs and the third bit assumed to be carry from the addition of the next least-significant bit addition operation, known as "C_{in}").

You can analyze the operation of the full adder to check on its operation. The sum bit is 1 only if one or three of the input bits is 1. In the half adder, I showed that the sum bit could be written out as:

$$\text{Sum} = A \otimes B$$

and should only be "1" if only one of the two inputs was 1. To understand the logic required to produce the sum bit for the three bit full adder, I created Table 5-3 in which the XOR output of the A and B inputs was given a single column entry. From the data presented in this table, you can see that the sum could be expressed as:

$$\text{Sum} = (A \otimes B) \otimes C$$

Table 5-3 Full adder sum bit operation truth table.

"A ⊗ B"	"C_{in}" bit	"Sum" bit
0	0	0
0	1	1
1	1	0
1	0	1

which, if you look back at Fig. 5-4, is exactly how it is implemented in the "full adder".

The carry output bit is 1 if two or three of the input bits are 1. As an exercise, you may wish to create a truth table and reduce it down to see if you can match the carry gate logic of Fig. 5-3, but you can write out and reduce a sum of products equation quite easily:

$$\begin{aligned}\mathtt{Carry} &= (\mathtt{A} \cdot \mathtt{B} \cdot !\mathtt{C}_{\mathtt{in}}) + (!\mathtt{A} \cdot \mathtt{B} \cdot \mathtt{C}_{\mathtt{in}}) \\ &\quad + (\mathtt{A} \cdot !\mathtt{B} \cdot \mathtt{C}_{\mathtt{in}}) + (\mathtt{A} \cdot \mathtt{B} \cdot \mathtt{C}_{\mathtt{in}}) \\ &= (!\mathtt{C}_{\mathtt{in}} \cdot (\mathtt{A} \cdot \mathtt{B})) + (\mathtt{C}_{\mathtt{in}} \cdot (\mathtt{A} \cdot \mathtt{B})) \\ &\quad + (\mathtt{C}_{\mathtt{in}} \cdot (\mathtt{A} \cdot !\mathtt{B}) + (!\mathtt{C}_{\mathtt{in}} \cdot (!\mathtt{A} \cdot \mathtt{B}) \\ &= ((!\mathtt{C}_{\mathtt{in}} + \mathtt{C}_{\mathtt{in}}) \cdot (\mathtt{A} \cdot \mathtt{B})) + (\mathtt{C}_{\mathtt{in}} \cdot (\mathtt{A} \cdot !\mathtt{B})) \\ &\quad + (\mathtt{C}_{\mathtt{in}} \cdot (!\mathtt{A} \cdot \mathtt{B})) \\ &= (\mathtt{A} \cdot \mathtt{B}) + (\mathtt{C}_{\mathtt{in}} \cdot ((!\mathtt{A} \cdot \mathtt{B}) + (\mathtt{A} \cdot !\mathtt{B}))\end{aligned}$$

Knowing that

$$((!\mathtt{A} \cdot \mathtt{B}) + (\mathtt{A} \cdot !\mathtt{B})) = \mathtt{A} \otimes \mathtt{B}$$

the equation for the carry output of the full adder can be written out as:

$$\mathtt{Carry} = (\mathtt{A} \cdot \mathtt{B}) + ((!\mathtt{A} \otimes \mathtt{B}) \cdot \mathtt{C}_{\mathtt{in}})$$

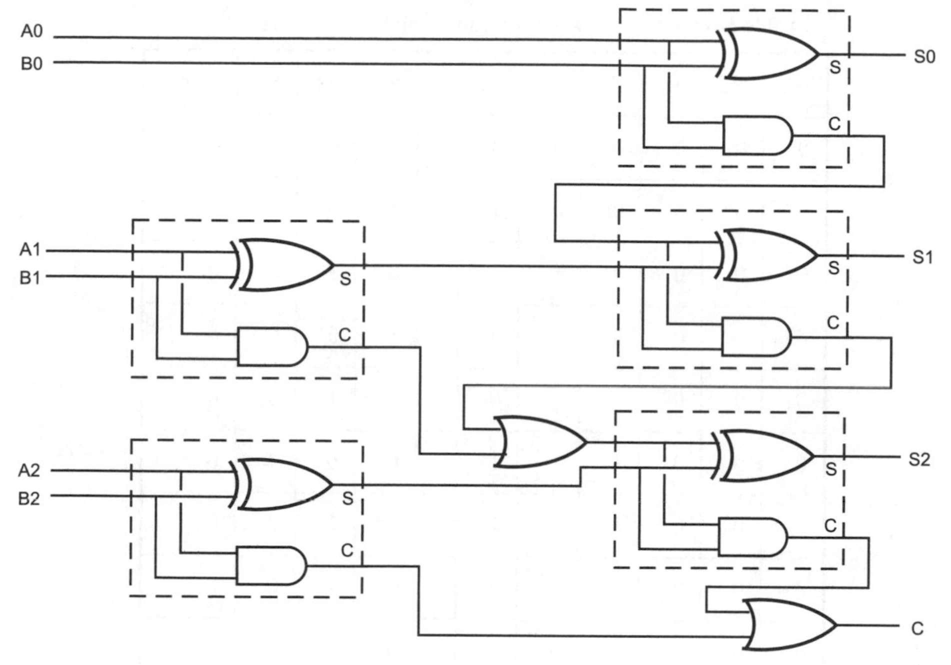

Fig. 5-5. Three bit "ripple adder" circuit.

which is *exactly* the carry logic circuit shown in Fig. 5-2. This type of analysis is useful to do when you are trying to puzzle out what a circuit is doing or to confirm that it is doing exactly what you expect it to do. It is also good practice of using the logic equation optimization skills first presented in Chapter 2.

Multiple full adders can be chained together (like in Fig. 5-5) to produce a multi-bit adder in which the carry results for each bit "ripples" through the various adder circuits. For most applications, this "ripple carry adder" can be used safely, but in something like your PC's processor, where quite a few bits are required and the adder is expected to execute quickly, the time required for the carry to ripple through the adders is prohibitive.

The solution to this problem is the "carry look-ahead" adder in which each bit takes not only the appropriate bits for input but also all the least significant bits that can affect the bit. The length of time the carry look-ahead adder needs to produce a sum is generally independent of the number of bits in the operation (unlike the time the ripple adder requires to produce a sum which is a function of the number of bits). Table 5-4 lists the different inputs and expected outputs for a three bit carry look-ahead adder. Reducing the

Table 5-4 Carry look-ahead adder input/output truth table.

A2	B2	A1	B1	A0	B0		S0	S1	S2	Carry
0	0	0	0	0	0		0	0	0	0
0	0	0	0	0	1		1	0	0	0
0	0	0	0	1	1		0	1	0	0
0	0	0	0	1	0		1	0	0	0
0	0	0	1	1	0		1	1	0	0
0	0	0	1	1	1		0	0	1	0
0	0	0	1	0	1		1	1	0	0
0	0	0	1	0	0		0	1	0	0
0	0	1	1	0	0		0	0	1	0
0	0	1	1	0	1		1	0	1	0
0	0	1	1	1	1		0	1	1	0
0	0	1	1	1	0		1	0	1	0
0	0	1	0	1	0		1	1	0	0
0	0	1	0	1	1		0	0	1	0
0	0	1	0	0	1		1	1	0	0
0	0	1	0	0	0		0	1	0	0
0	1	1	0	0	0		0	1	1	0
0	1	1	0	0	1		1	1	1	0
0	1	1	0	1	1		0	0	0	1
0	1	1	0	1	0		1	1	1	0
0	1	1	1	1	0		1	0	0	1

(*continued*)

Table 5-4 Continued.

A2	B2	A1	B1	A0	B0		S0	S1	S2	Carry
0	1	1	1	1	1		0	1	0	1
0	1	1	1	0	1		1	0	0	1
0	1	1	1	0	0		0	0	0	1
0	1	0	1	0	0		0	1	1	0
0	1	0	1	0	1		1	1	1	0
0	1	0	1	1	1		0	0	0	1
0	1	0	1	1	0		1	1	1	0
0	1	0	0	1	0		1	0	1	0
0	1	0	0	1	1		0	1	1	0
0	1	0	0	0	1		1	0	1	0
0	1	0	0	0	0		0	0	1	0
1	1	0	0	0	0		0	0	0	1
1	1	0	0	0	1		1	0	0	1
1	1	0	0	1	1		0	1	0	1
1	1	0	0	1	0		1	0	0	1
1	1	0	1	1	0		1	1	0	1
1	1	0	1	1	1		0	0	1	1
1	1	0	1	0	1		1	1	0	1
1	1	0	1	0	0		0	1	0	1
1	1	1	1	0	0		0	0	1	1
1	1	1	1	0	1		1	0	1	1

(*continued*)

Table 5-4 Continued.

A2	B2	A1	B1	A0	B0		S0	S1	S2	Carry
1	1	1	1	1	1		0	1	1	1
1	1	1	1	1	0		1	0	1	1
1	1	1	0	1	0		1	1	0	1
1	1	1	0	1	1		0	0	1	1
1	1	1	0	0	1		1	1	0	1
1	1	1	0	0	0		0	1	0	1
1	0	1	0	0	0		0	1	1	0
1	0	1	0	0	1		1	1	1	0
1	0	1	0	1	1		0	0	0	1
1	0	1	0	1	0		1	1	1	0
1	0	1	1	1	0		1	0	0	1
1	0	1	1	1	1		0	1	0	1
1	0	1	1	0	1		1	0	0	1
1	0	1	1	0	0		0	0	0	1
1	0	0	1	0	0		0	1	1	0
1	0	0	1	0	1		1	1	1	0
1	0	0	1	1	1		0	0	0	1
1	0	0	1	1	0		1	1	1	0
1	0	0	0	1	0		1	0	1	0
1	0	0	0	1	1		0	1	1	0
1	0	0	0	0	1		1	0	1	0
1	0	0	0	0	0		0	0	1	0

information from this table, I have listed the equations for the three sum bits and the carry bit below:

$$
\begin{aligned}
\mathtt{S0} &= (\mathtt{A0} \otimes \mathtt{B0}) \\
\mathtt{S1} &= ((\mathtt{A0} \otimes \mathtt{B0}) \cdot (\mathtt{A1} \otimes \mathtt{B1})) \\
&\quad + ((\mathtt{A1} \otimes \mathtt{B1}) \cdot !(\mathtt{A0} + \mathtt{B0})) \\
&\quad + (!(\mathtt{A1} \otimes \mathtt{B1}) \cdot (\mathtt{A0} \cdot \mathtt{B0})) \\
\mathtt{S2} &= ((\mathtt{A2} \otimes \mathtt{B2}) \cdot !(\mathtt{A2} \cdot \mathtt{B2})) \\
&\quad + (!(\mathtt{A2} \otimes \mathtt{B2}) + (\mathtt{A1} \cdot \mathtt{B1})) + (!(\mathtt{A2} \cdot \mathtt{B2}) \\
&\quad + (\mathtt{A1} \otimes \mathtt{B1}) \cdot (\mathtt{A0} \cdot \mathtt{B0})) \\
\mathtt{Carry} &= (\mathtt{A2} \cdot \mathtt{B2}) + ((\mathtt{A2} \otimes \mathtt{B2}) \cdot (\mathtt{A1} \cdot \mathtt{B1})) \\
&\quad + (\mathtt{A2} \otimes \mathtt{B2}) \cdot (\mathtt{A1} \otimes \mathtt{B1}) \cdot (\mathtt{A0} \cdot \mathtt{B0}))
\end{aligned}
$$

It was a major effort on my part to reduce the equations for each sum bit and the carry bit. To do this, I used the truth table reduction method discussed in Chapter 2. To reduce the number of terms in the resulting sum of products equations, I first deleted all the instances where the specific bit was not "1"–in every case, this reduced the number of instances by half. Next, I worked at combining instances that were similar and found that rather than combining "don't care" bits, I found a number of places where two bits were XORed together. In the resulting equations, I kept the "XOR" terms in, even though when the "technology optimization" stage of the development effort is completed, these gates will be reduced to the technology's basic gates.

If you read through the equations and try to understand them, you will find that they do make a kind of sense. Obviously, as more bits are added to the carry look-ahead adder, the circuit becomes much more complex. Despite this complexity, the carry look-ahead is the most efficient way to provide an adder circuit for large bit words in fast applications.

Subtraction and Negative Numbers

As you might expect, binary subtraction has many of the same issues as addition, along with a few complexities that can make it harder to work with. In this section, I will introduce some of the issues in implementing a practical "subtracter" as well as look at some ways in which subtraction can be implemented easily with an existing addition circuit.

To make sure we're talking the same language, I want to define the terms that I will be using to describe the different parts of the subtraction operation. The "horizontal" arithmetic equation:

$$\texttt{Result} = \texttt{A} - \texttt{B}$$

can be written out "vertically" as:

```
      A    <– Minuend
     –B    <– Subtrahend
 ------
 Result    <– Difference
```

The "minuend" and "subtrahend" terms are probably something that you forgot that you learned from grade school. I use them here because they are less awkward than referring to them as the "value to be subtracted from" and the "value subtracted". The term "difference" as being the result of a subtraction operation is generally well understood and accepted.

When you carry out subtraction operations, you do it in a manner that is very similar to how you carry out addition; each digit is done individually and if the digit result is less than zero, the base value is "borrowed" from the next significant digit. With the assumption that subtraction works the same way as addition, you could create a "half subtracter", which is analogous to the half adder and could be defined by the truth table shown in Table 5-5

The "difference" bit is simply the minuend and subtrahend XORed together, while the "borrow" bit (decrementing the next significant digit) is only true if the minuend is 0 and the subtrahend is 1. The borrow bit can be defined as the inverted minuend ANDed with the subtrahend. The equations

Table 5-5 "Half subtracter" defining truth table.

Minuend	Subtrahend	Difference	Borrow
0	0	0	0
0	1	1	1
1	1	0	0
1	0	1	0

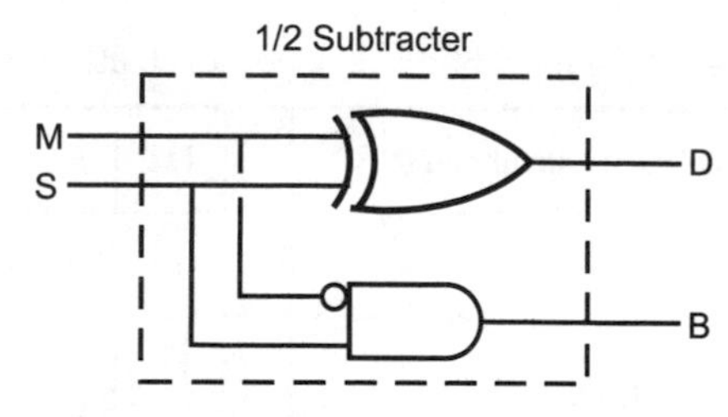

Fig. 5-6. Half subtracter circuit.

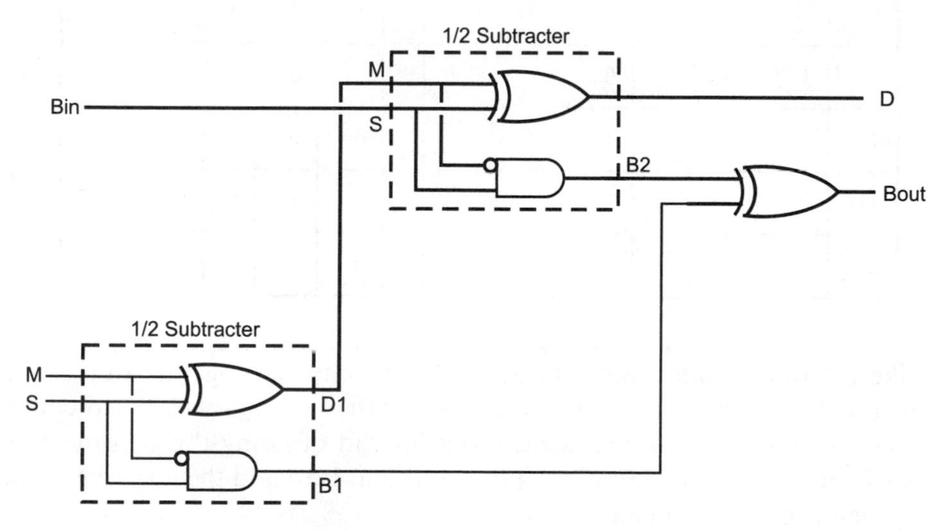

Fig. 5-7. Full subtracter circuit.

for the half subtracter are listed below and the subtracter building block is shown in Fig. 5-6.

$$\texttt{Difference} = \texttt{Minuend} \otimes \texttt{Subtrahend}$$

$$\texttt{Borrow} = !\texttt{Minuend} \cdot \texttt{Subtrahend}$$

The small circle on the single input indicates that the value is inverted before being passed into the gate. This convention avoids the need for putting a full inverter symbol in the wiring diagram of a digital circuit and is often used in chip datasheets to indicate inverted inputs to complex internal functions.

Two half subtracters can be combined into a "full subtracter", just as two half adders can be combined to form a full adder (Fig. 5-7). In Fig. 5-7, I have labeled the two half subtracters, so that their operation can be listed in Table 5-6, to test the operation of the full subtracter.

Table 5-6 Full subtracter operation truth table.

Bin	Minuend ("M")	Subtrahend ("S")		D1	B1	D	B2	Bout
0	0	0		0	0	0	0	0
0	0	1		1	1	1	0	1
0	1	1		0	0	0	0	0
0	1	0		1	0	1	0	0
1	1	0		1	0	0	0	0
1	1	1		0	0	1	1	1
1	0	1		1	1	0	0	1
1	0	0		0	0	1	1	1

Like the ripple adder, full subtracters can be chained together to create a multi-bit subtracter circuit (Fig. 5-8) and a "borrow look-ahead" (to coin a phrase) subtracter could be designed, but instead of going through the pain of designing one, there is another option and that is to add the negative of the subtrahend to the minuend.

In the introduction to this chapter, I introduced the idea of negative numbers as being the value being subtracted from an arbitrary large number and showed an example that produced "−5" in a universe where infinity was equal to one million. When you first went through this example, you might have thought that this was an interesting mathematical diversion and an illustration as to how negative and positive numbers converge when they approach infinity. This concept, while seemingly having little application in the "real world", is very useful in the digital domain.

In the digital domain, the term "infinity" can be replaced with "word size" and if the most significant bit of the word is considered to be the "sign" bit, positive and negative numbers can be expressed easily. In Table 5-7, I have listed the decimal, hex as well as the positive and negative values which take into account that a negative number can be written as:

$$-A = \text{Word Size} - A$$

This negative value is known as a "two's complement" negative number and is the most commonly used negative bit format used. There is a "one's

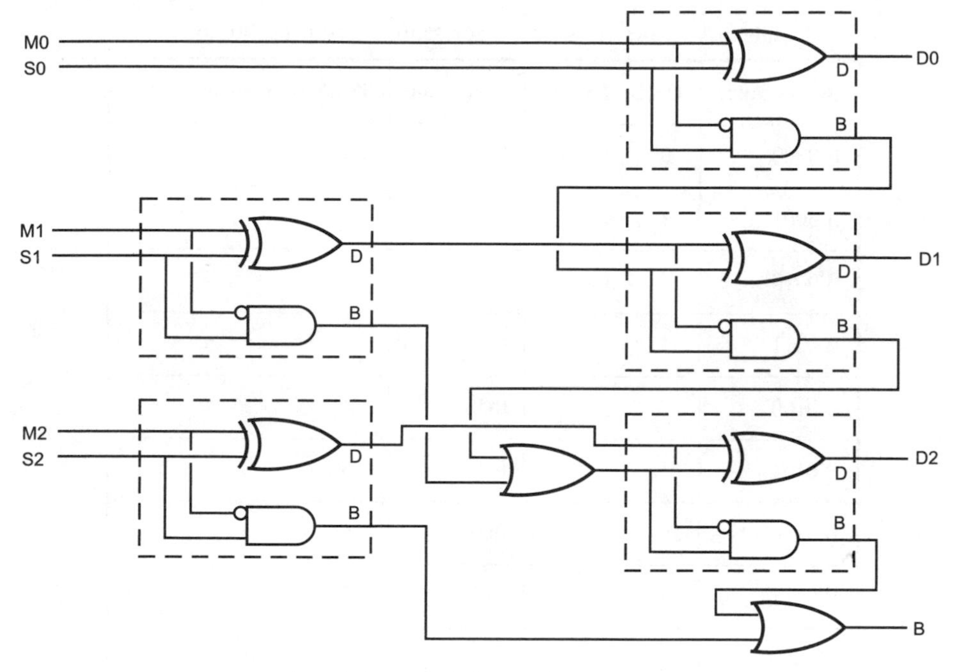

Fig. 5-8. Three bit "ripple subtracter" circuit.

complement" number format, but it does not avail itself as efficiently as two's complement to allow for easier subtraction and addition of negative numbers.

Looking at the formula above, you are probably confused as to why it would be used because it requires both a subtraction operation as well as an addition operation to carry out one subtraction operation. Negating a number in two's complement format does not actually require a subtraction operation; it can be done by inverting each bit (XORing each bit with 1) and then incrementing the result. Using the values of Table 5-7, you can demonstrate how a positive value is negated.

For example, to negate the value "5", the following steps are used:

1. Each bit of the number is XORed with "1". B'0101' becomes B'1010'.
2. The XORed result is incremented. B'1010' becomes B'1011', which is "−5".

The opposite is also true: the individual bits can be inverted and the result incremented to convert a negative two's complement value to a positive.

Once the value has been negated, it can be simply added to the other parameter, as I show in Fig. 5-9. There are three things that you should be

Table 5-7 Different ways of representing a four bit number.

Binary value	Decimal value	Hex value	Two's complement value
B'0000'	0	0x00	0
B'0001'	1	0x01	1
B'0010'	2	0x02	2
B'0011'	3	0x03	3
B'0100'	4	0x04	4
B'0101'	5	0x05	5
B'0110'	6	0x06	6
B'0111'	7	0x07	7
B'1000'	8	0x08	−8
B'1001'	9	0x09	−7
B'1010'	10	0x0A	−6
B'1011'	11	0x0B	−5
B'1100'	12	0x0C	−4
B'1101'	13	0x0D	−3
B'1110'	14	0x0E	−2
B'1111'	15	0x0F	−1

aware of before leaving this section. The first is the use of the "V" shaped mathematical function symbols in Fig. 5-9; these symbols indicate that two parameters are brought together to form one output. I use this symbol when a group of bits (not just one) are passing through the same operation.

You might be wondering why instead of simply inverting the individual bits of the value to be converted to a negative two's complement value, I XOR the bits with the value 1. The reason for doing this is in the interests

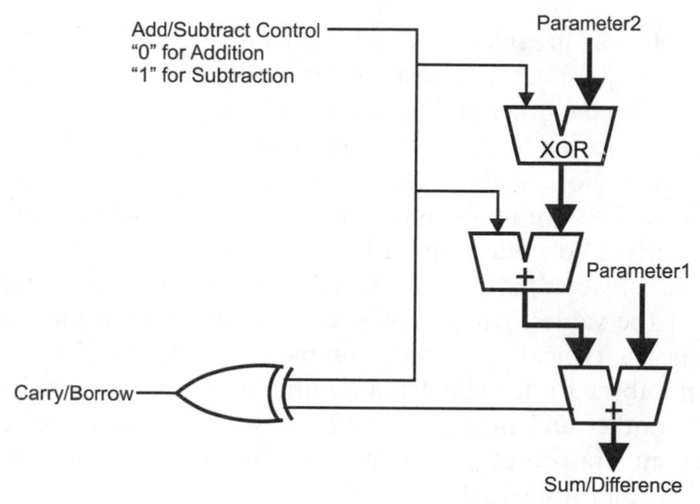

Fig. 5-9. Integrated adder/subtracter circuit.

of practicality and looking ahead. In Fig. 5-9, I show a circuit in which two parameters can be added together or one can be subtracted by the other – the "switch" control for which operation is selected. If a 1 is passed to the "Parameter2" circuitry, each bit of Parameter2 is XORed with 1, inverting each bit and a 1 is passed to the Parameter2 adder, which increments the value. If a zero is passed to the Parameter2 circuitry, the bits of Parameter2 are not inverted and zero is added to the output of the XOR function, resulting in an unchanged value of Parameter2 being passed to the adder with Parameter1. To net it out, if a "1" is passed to this circuit, Parameter2 is subtracted from Parameter1; if a "0" is passed to the circuit, the two parameters are added together.

The last point to note is that the "carry" output of the final adder is a negated "borrow" output when the subtraction operation is taking place. To integrate the operation of the "carry/borrow" bit with the add/subtract switch bit, this bit is set when a carry or borrow of the next significant word is required, regardless of the operation.

Magnitude Comparators and Bus Nomenclature

Along with being able to add and subtract binary values, you will find the need to compare binary values to determine whether or not a value is less than, equal to or greater than another value. Just as if this were a programming requirement, to test two binary values together, you would subtract one from the other and look at the results. An important issue when comparing a value made up of multiple bits is specifying how it is to be represented in logic drawings and schematic diagrams. In the previous section I touched on both these issues, in this section, I want to expand upon them and help you to understand a bit more about them.

When you are comparing two binary values, you are comparing the *magnitude* of the values, which is where the term "magnitude comparator" comes from. The typical magnitude comparator consists of two subtracters which either subtract one value from another and vice versa or subtract one value from another and then compare the result to zero. In either case, the magnitude comparator outputs values indicating which value is greater than the other or if they are equal.

Figure 5.10 shows a basic comparator, which consists of two subtracters utilizing the negative addition discussed in the previous section. The differences are discarded, but the !borrow outputs are used to determine if the negative value is greater. If the !borrow outputs from the two

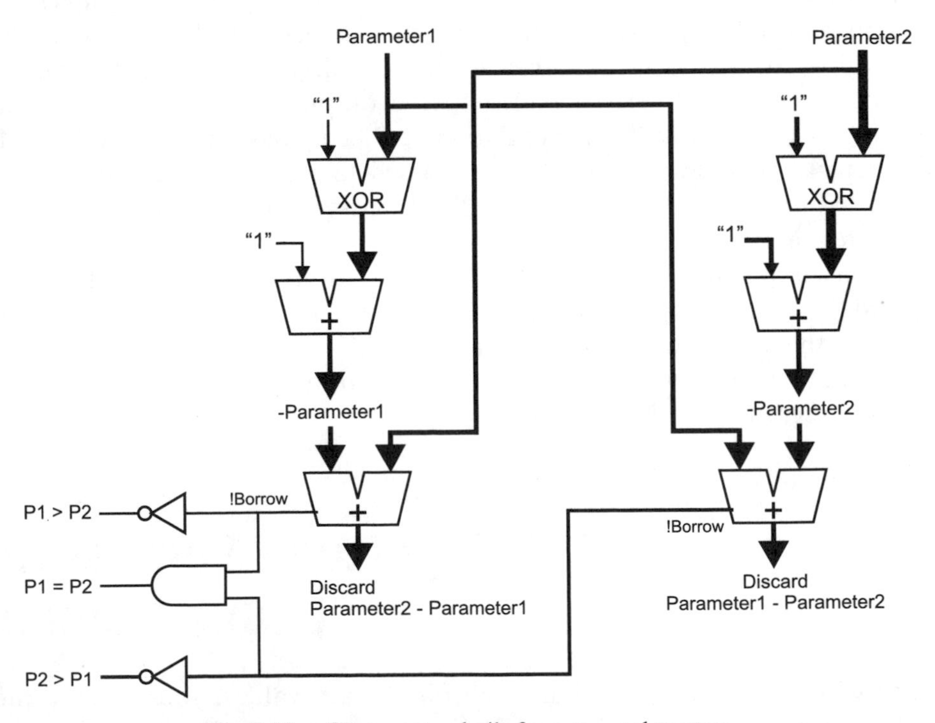

Fig. 5-10. Comparator built from two subtracters.

subtracters are both equal to "1", then it can be assumed that the two values are equal.

If one value is subtracted from the other to determine if one is lower than the other and if the value is not lower (i.e. !borrow is not zero), the result can then be compared to zero to see if the value is greater than or equal to the other. This method is probably less desirable because it tends to take longer to get a valid result and the result outputs will be valid at different times. Ideally, when multiple outputs are being produced by a circuit, they should all be available at approximately the same time (which is the advantage of the two subtracter circuit shown in Fig. 5-10 over this one).

If you are working with TTL and require a magnitude comparator, you will probably turn to the 7485, which is a four bit magnitude comparator consisting of two borrow look-ahead subtracters to ensure that the outputs are available in a minimum amount of time and are all valid at approximately the same time.

In Fig. 5-10 (as well as the multi-bit subtracter shown in the previous section), I contained related multiple bits in a single, thick line. This very common method of indicating multiple related bits is often known as a "bus". Other methods include using a line of a different color or style. The advantage of grouping multiple bits that function together like this should be obvious: the diagram is simpler and it is easier to see the path that related bits take.

When I use the term "related bits", I should point out that this does not only include the multiple bits of a binary value. You may have situations where busses are made up of bits which are not a binary value, but perform a similar function within the circuit. For example, the memory control lines for a microprocessor are often grouped together as a bus even though each function is provided by a single bit (memory read, memory write, etc.) and they are active at different times.

As well as indicating a complete set of related bits, a bus may be broken up into subsets, as shown in Fig. 5-11. In this diagram, I have shown how two four bit magnitude comparators can be "ganged" together to provide a comparison function on eight bits. The least significant four bits are passed to the first magnitude comparator and the most significant four bits are passed to the second magnitude comparator. The bits are typically listed as shown in Fig. 5-11, with the most significant bit listed first and separated from the least significant bit by a colon. In very few cases will you see the width of the bus reduced to indicate a subset of bits as I have done in Fig. 5-11; most design systems will keep the same width for a bus, even if one bit is being used in it.

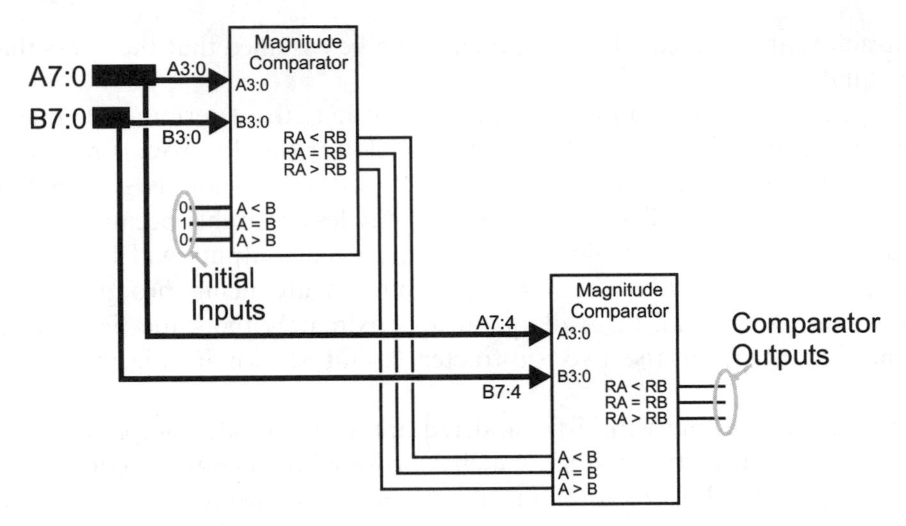

Fig. 5-11. Cascading two four bit comparators.

Before going on, I want to make some comments about Fig. 5-11 as it provides a function that is often required when more bits must be operated on than is available by basic TTL or CMOS logic chips. To carry out the magnitude comparison operation on eight bits, I used two four bit magnitude comparator chips (modeled on the 7485) with the initial state inputs (marked "Initial Inputs" on Fig. 5-11) to start the chip off in the "neutral" state as if everything "upstream" (before) was equal to each other and the chip's bits as well as any "downstream" (after) bits will determine which value is greater or if the two values are equal. This is a typical method for combining multiple chips to provide the capability to process more bits than one chip is able to.

Multiplication and Division

As you've read through this chapter, you should have noticed that there are usually many different ways of implementing different digital electronics functions. Each of the different implementation methods have their own advantages and tradeoffs – it is up to the application designer to understand what are the important ones. Nowhere is this more true than when you start discussing multiplication and division; there are a number of different methods of performing these arithmetic operations, each with their own characteristics.

Off the top of my head, I can come up with five different ways to multiply two binary numbers together. Before listing the different methods, I should make sure that I have defined the terms used in multiplication. The "multiplicand" is the value that is multiplied by the "multiplier" and typically remains static. The "multiplier" is the number of times that the multiplicand is added together to form the result or "product".

It should go without saying that if you had to multiply by a power of 2 (i.e. 1, 2, 4, 8, 16, etc.) a true multiplication operation is not required at all; the operation can be accomplished by shifting the multiplicand. For example to multiply a value by 4, you would simply shift the value to the left two times. Division is the same, except the value is shifted to the right.

Understanding the basic terms "multiplier" and "multiplicand" leads to a second method to implement a multiplication function in software – the multiplicand is added multiplier number of times to the product. It can be written out in "C" as:

```
Product = 0;                        //Clear result accumulator
for (i = 0; i < Multiplier; i++)
                                    //Repeatedly add Multiplicand
    Product = Product + Multiplicand;
```

This method is painfully slow (especially for large multiplier values) and is difficult to implement in combinatorial digital logic. It is also different from the method which you were taught in school in which the multiplicand is shifted up by the radix for each digit of the multiplier. Using this method, "123" decimal is multiplied by "24" decimal in the format:

```
   123
×   45
   ---
   615
+ 492   < -Sum Products of each digit
  ----
  5535
```

In the first line of the solution, I multiplied the multiplicand "123" by the least significant digit of the multiplier followed by multiplying the multiplicand by 10 (the radix) followed by the next significant digit of the multiplier. Once the multiplicand has been multiplied by each digit of the multiplier (along with the appropriate multiplication of the digit position), each product is added together to get the final result.

This method lends itself extremely well to working within binary systems. Rather than multiplying the multiplicand repeated by the radix for each digit, the multiplicand is simply shifted to the right (which is multiplying by two) for each bit of the multiplier. If the multiplier bit is zero, then the value added to the product is zero. The binary multiplication operation for 123 by 45 is:

```
          B'01111011'
        × B'00101101'
          B'01111011'     <– Bit 0 of Multiplier is not Zero
         B'00000000'      <–Bit 1 of Multiplier is Zero
        B'01111011'       <–Bit 2
       B'01111011'        <–Bit 3
      B'00000000'         <–Bit 4
     B'01111011'          <–Bit 5
    B'00000000'           <–Bit 6
 + B'00000000'            <–Bit 7
 B'0001010110011111'      <– Product(5,535 Decimal)
```

This is much more efficient than the previous version in terms of execution time and not significantly more complex than the other version. The "C" code that implements it is:

```
Product = 0;                          //Clear result accumulator
for (i = 0; i < log2 (Multiplier); i++)
{
    if ((Multiplier & 1) == 1)        //If LS Bit Set, Add Current
    Product = Product + Multiplicand;           //to Product
    Multiplier = Multiplier >>1; //Shift down Multiplier
    Multiplicand = Multiplicand << 1;    //Shift up
                                              //Multiplicand
} // rof
```

The first version is known as "Order n" because it loops once for each value of the multiplier. The shift and add version shown directly above is known as "Order log_2" because it executes the log_2 of the word size of the multiplier. For the eight bit multiplication example shown here, for the first method, up to 255 loops (and addition operations) may have to be executed. For the second example, a maximum of 8 loops (one for each bit, which is a simple way to calculate log_2 of a number) is required.

The final method of multiplying two numbers together is known as "Booth's algorithm" and looks for opportunities to reduce the number of addition operations that are required by rounding the multiplier up and then adjusting the product by subtracting the multiplier's zero bits that were ignored. For the example given in this section (123 multiplied by 45), Booth's algorithm would recognize that 45, B'00101101' rounds up to 64 (B'01000000'). Multiplying a binary number by 64 is the same as shifting left six times.

To adjust the product, the basic multiplicand (multiplied by 1) along with all the instances where the multiplier has bit value of zero (in this case, bits one and four) have to be taken away from the rounded up value. For this example, the multiplication operation would look like:

$$
\begin{aligned}
&(\mathtt{B'01111011'} << 6) - \mathtt{B'01111011'} - (\mathtt{B'01111011'} << 1) \\
&\quad - (\mathtt{B'01111011'} << 4) \\
&= \quad \mathtt{B'01111011000000'} \\
&- \qquad\qquad\quad \mathtt{B'01111011'} \\
&- \qquad\qquad \mathtt{B'011110110'} \\
&- \qquad \mathtt{B'011110110000'} \\
&\overline{\quad \mathtt{B'01010110011111'}}
\end{aligned}
$$

which is the same result as we got above for a bit less work. Booth's algorithm can produce a product in fewer, the same or more addition/subtraction operations as the previous method, so care must be taken to make sure that it is only used when it is going to provide an advantage.

Each of the three methods presented so far requires the ability to "loop" through multiple iterations. This is a problem for most digital electronic circuits, as it not only requires a "clock" to synchronize the operations but it will also most likely take up more time than is desired. When a digital logic multiplier is designed, it typically looks something like Fig. 5-12. This circuit is wired to add all the multiplicand bits together for each possible multiplier values.

The multiplier bits are taken into account by ANDing them with each of the multiplicand bits. If a multiplier bit is zero, then the shifted up multiplicand bits are not passed to the multi-bit adders.

There are a couple points about the multi-bit adders that you should be aware of. The first is that the maximum number of input bits for the adders used in the multiplier circuit is the number of bits in the multiplicand plus the $\log_2$ value of the multiplier. Secondly, as drawn, the adders are connected in a "ripple" configuration – a commercial circuit would probably wire the

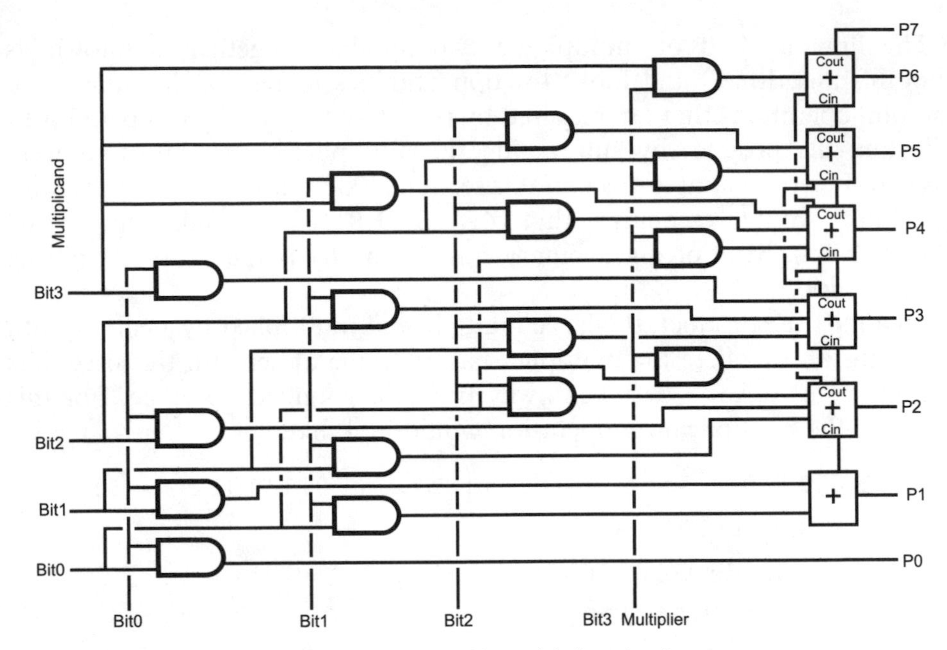

Fig. 5-12. Four bit multiplier circuit.

adders together as a carry look-ahead to minimize the time required for the multiplication operation to take place.

Before leaving the topic of multiplication, I should point out that all the methods presented here will handle multiplication of two's complement negative numbers "natively". This is to say that no additional gates must be added to support the multiplicand or multiplier being negative.

Division is significantly more difficult to implement and is very rarely implemented in low-cost devices. Handling negative values considerably complicates the division operation and in this section, as well as most commercial solutions, negative values cannot be used for the dividend or divisor. To avoid the hardware complexities of division, software intensive solutions are normally done such as a repeated subtraction:

```
Quotient = 0;                    //Clear result accumulator
while (Dividend >= Divisor)      //Repeat Subtraction
{
    Dividend = Dividend - Divisor;
    Quotient = Quotient + 1;
} //  elihw
```

The bit shifting method shown for multiplication can also be used, but before comparisons can start, the divisor should be shifted up the word size of the dividend. To follow the bit shifting division code listed below you might want to do a thought experiment and single step through it with arbitrary values to see exactly how it works:

```
Quotient = 0;                    //Clear result accumulator
Divisor = Divisor << DividendWordSize;
DivisorPos = 1<<DivendendWordSize;
                                 //Current Divisor Bit Shift
while (DivisorPos != 0)
                              //Repeat While Divisor Shifted
{
if (Dividend >= Divisor)   //Can you take away current
{
  Dividend = Dividend - Divisor;
  Quotient = Quotient + DivisorPos;
} // fi
Divisor = Divisor >> 1;          //Shift Down Divisor
DivisorPos = DivisorPos >> 1;
} // elihw
```

At the end of both these division algorithms, "Quotient" contains the quotient of the division operation and "Dividend" contains the remainder.

The bit shifting division algorithm could be implemented using digital electronic gates as I demonstrated for the bit shifting multiplication algorithm, but you will find that it is quite a bit more complex than the bit shifting multiplication application in Fig. 5-12. This does not mean that there are some tricks you cannot implement if a division operation is required.

For example, if you wanted to divide a value by a known, constant value, you could multiply it by its fraction of 256 (rather than the typical 1) and then divide by 256 (which is accomplished by shifting right by eight bits). For example, dividing 123 by 5 would be accomplished by multiplying 123 by 51 (256 divided by 5) and shifting the product (6,273) to the right by 8 bits to get the result 24. While this method seems like it's complex, it is actually quite easy to implement in digital electronics.

Quiz

1. 6 − 5 is the same as:
   (a) 6 + (−5)
   (b) 5 − 6
   (c) 999999
   (d) B'1111 1111'
2. In a universe where infinity (the highest possible number) is one million (1,000,000); "−11" could be represented as:
   (a) Only −11
   (b) 999,988
   (c) 999,989
   (d) 89
3. A "half adder":
   (a) Can perform an addition operation on two bits
   (b) Can add half the bits together of an addition operation
   (c) Combines the "carry" outputs of a "full adder" to produce the correct result
   (d) Is built from half the number of gates as a full adder
4. A "ripple adder" is not used in a PC or workstation processor because:
   (a) Its complexity can affect the operation of other arithmetic functions
   (b) The result is often wrong by one or two bits
   (c) The delay required for the signal to pass through the gates can be unacceptably long
   (d) It cannot handle the 32 or 64 bit addition operations required
5. B'10' − B'01' passed through two full subtracters produces the result:
   (a) Cannot be done because a borrow operation is required
   (b) B'01' with borrow = 1
   (c) B'10' with borrow = 0
   (d) B'01' with borrow = 0
6. Converting the four bit, two's complement value "−4" to a positive number by inverting each bit and incrementing the result produces the bit pattern:
   (a) B'0100'

(b) Which is five bits long and is invalid
(c) B'0011'
(d) B'1100'

7. Busses are made up of:
(a) Multiple bits of a single value
(b) Multiple bits passing to the same subsystem of the application
(c) The highest speed signals in the application
(d) Related bits

8. Multiplying two four bit numbers by repeated addition:
(a) Will require up to 4 addition operations
(b) Will require up to 15 addition operations
(c) Cannot be implemented in digital electronics
(d) Is the fastest way of performing binary multiplication

9. Multiplying a binary number by 16 can be accomplished by:
(a) Clearing the least significant four bits
(b) Shifting left four bits
(c) Shifting right four bits
(d) Setting the least significant four bits

10. Dividing an eight bit value by the constant value 6 is best accomplished by:
(a) Using the repeated subtraction method
(b) Using the bit shifting method
(c) Shifting the value to the right by two and then shifting the value to the right by 1 and adding the two values.
(d) Multiplying by 256/6 (42) and shifting the product to the right by 8 bits

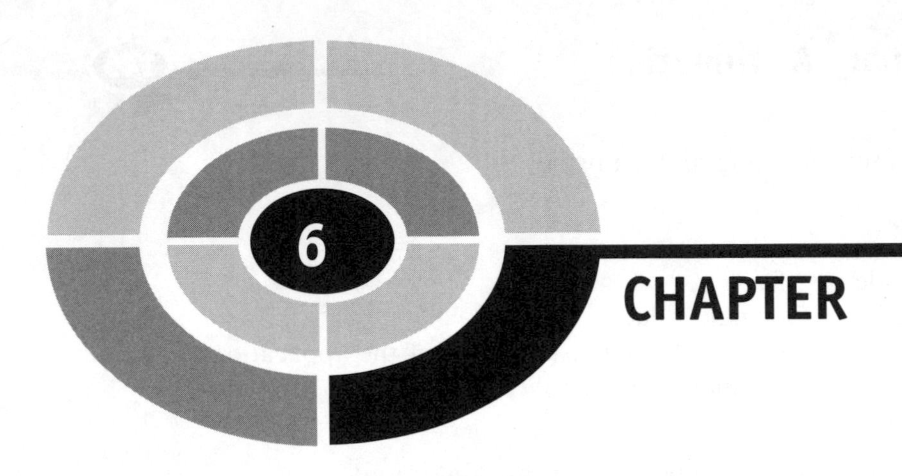

Practical Combinatorial Circuit Implementation

When you are designing your first application that is built from digital electronics, you will probably feel like you have just joined a never-ending role-playing game in which all the other players know more than you do. Later in the book, I will present some ideas on how to read a datasheet and what to look for in it, but for now, I would like to discuss a number of the options that you should be aware of and are thinking about when you first start designing your application.

Using the role-playing game analogy with digital electronics may seem to be facetious, but there are actually a lot of similarities that you should be aware of. First and foremost, each digital electronic chip that you can choose from has a number of characteristics that you will have to be aware of and

Table 6-1 Important characteristics of a digital electronic chip.

Characteristic	Comments
Function	Gate type, chip function
# Bits	The number of bits per gate input or number of bits used by the function
# Gates/functions	What does the chip do and how many are there
Technology	Electronic standard chip is implemented
Output type	Gate output type
Dependencies	Issues to be aware of
Manufacturer	Who makes the chip/where can it be purchased

choose from when you are specifying the parts used in your application. When choosing between the parts, it might be a good idea to come up with a card, similar to the cards used in role-playing games to explain the different characters, characteristics and strengths and weaknesses. A sample card for a digital electronic device might look something like Table 6-1.

The "function" of the chip is a brief description of the gates provided by the chip or the digital logic function provided by the chip (such as an adder or a magnitude comparator). At this point in the book, you might feel that it is sufficient to specify a chip for a needed function, but the following characteristics are critical for you to understand that you need to be able to select the right chip for the right application.

When I have presented simple gates, they have all (with the exception of the NOT gate) two inputs. Along with two inputs, there are a number of different inputs for a variety of different types of gates. For example, in standard TTL, you can get NAND gates with two, three, four and eight inputs. Four and eight bit adders as well as different chips with different bit counts are also available. When selecting a chip for an application, you should be cognizant of the bit options that are available to minimize the number of chips required.

The basic TTL chips have four two input gates and six one input gates, but if the number of bits changes, then the number of gates within the chip changes (or the plastic package type and the number of pins changes). As surprising as it seems, many complex functions can have more than one built into the package. Like the number of bits, the number of functions within the chip will help you plan out how many chips you will need in the application.

So far in the book, I have really just indicated that there are two types of technology used for standard logic devices. In actuality, there are dozens and in Table 6-2, I have listed the most popular ones with their input, output and operating characteristics. For the different varieties of "TTL", "C", "AC"

Table 6-2 Digital logic technologies with electrical characteristics.

Chip type	Power supply	Gate delay	Input threshold	"0" output	"1" output	Output sink
TTL	Vcc = 4.5 to 5.5 V	8 ns	N/A	0.3 V	3.3 V	12 mA
L TTL	Vcc = 4.5 to 5.5 V	15 ns	N/A	0.3 V	3.4 V	5 mA
LS TTL	Vcc = 4.5 to 5.5 V	10 ns	N/A	0.3 V	3.4 V	8 mA
S TTL	Vcc = 4.75 to 5.25 V	5 ns	N/A	0.5 V	3.4 V	40 mA
AS TTL	Vcc = 4.5 to 5.5 V	2 ns	N/A	0.3 V	Vcc – 2 V	20 mA
ALS TTL	Vcc = 4.5 to 5.5 V	4 ns	N/A	0.3 V	Vcc – 2 V	8 mA
F TTL	Vcc = 4.5 to 5.5 V	2 ns	N/A	0.3 V	3.5 V	20 mA
C CMOS	Vcc = 3 to 15 V	50 ns	0.7 Vcc	0.1 Vcc	0.9 Vcc	3.3 mA*
AC CMOS	Vcc = 2 to 6 V	8 ns	0.7 Vcc	0.1 V	Vcc – 0.1 V	50 mA
HC/HCT	Vcc = 2 to 6 V	9 ns	0.7 Vcc	0.1 V	Vcc – 0.1 V	25 mA
4000	Vdd = 3 to 15 V	30 ns	0.5 Vdd	0.1 V	Vdd – 0.1 V	0.8 mA*

and "HC/HCT" logic families, the part number starts with "74" and for the "4000" series of CMOS chips, it has a four digit part number, starting with "4". Table 6-2 lists the different aspects of the different types of logic chips that you will want to work with.

The "output sink" currents are specified for a power voltage of 5 volts. If you increase the power supply voltage of the indicated (with a "*") CMOS parts, you will also increase their output current source and sink capabilities considerably.

In Table 6-2, I marked TTL input threshold voltage as "not applicable" (N/A) because, as you know, TTL is current driven rather than voltage driven. You should assume that the current drawn from the TTL input is 1 mA for a "0" or "low" input. CMOS logic is voltage driven, so the input voltage threshold specification is an appropriate parameter.

The output current source capability is not specified because many early chips were just able to sink current. This was all that was required for TTL and it allowed external devices, such as LEDs, to be driven from the logic gate's output without any additional hardware and it simplified the design of the first MOSFET-based logic chips. The asterisk ("*") indicates that the sink current specification is for 5 volts power; changing the power supply voltage will change the maximum current sink capability as well.

There are three basic output types: totem pole, open collector and the tri-state driver (which is presented later in this chapter). In cases where multiple outputs are combined, different output types should *never* be combined due to possible bus contention.

Virtually all of the electrical dependencies that you should be aware of are listed in Table 6-2, but you may have a number of operating dependencies (such as making sure CMOS inputs are tied high or low) or physical design issues that you should be aware of. "Physical design" is the process of designing a printed circuit board with internal connections built into it that have the chips soldered onto it. The primary chip dependencies that you should be aware of when designing a printed circuit board are the location and type of the chip's pins as well as any heat removal (i.e. heat sinks) requirements that the chip may have.

Finally, you should know who makes the part and where you can purchase it. This point is often overlooked, but you will find many manufacturers that advertise parts that are seemingly designed just for your application. The first problem that you encounter is that your company has a policy of only buying parts that are available from multiple sources or you may discover that the manufacturer is not considered to be reliable and production quantity parts are difficult to come by. For your first designs, it is a good rule to only use parts that are easily obtainable and, ideally, built by multiple sources.

This may make the design operation a bit more difficult and the final product larger than it could have been, but chances are the product will go through manufacturing very smoothly and with few difficult "hiccups".

Race Conditions and Timing Analysis

As you begin to create digital electronic circuits that are more and more complex and run faster and faster, you are going to discover that they are going to stop working or they are going to start working unpredictably. In trying to find the problem, you will probably look at different parts of the circuit, ranging from the power supply to the wiring and maybe rebuilding it several times to see if it is being affected by other electrical devices running near the application. At some point you will give up and build as well as redesign the circuit, only to discover that the problem is still there.

So what's the problem? Chances are you have encountered a "race condition", which is normally defined as "A condition in digital electronics where two or more signals do not always arrive in the same order." Personally, I use a slightly different definition for race condition which states that "A race condition occurs in any digital electronic circuit where the output to input response time changes according to the inputs passed to it." My definition is a bit more specific and should give you some ideas on where to look for the problem.

Simply put, a race condition is a case where an expected event does not occur.

To illustrate the issue, assume that the application consists of a circuit that is designed to respond to an internal value at a specific time. If the digital electronics used to produce this internal value does not always complete within the specified time, what happens in the circuit that uses this value for input? Chances are the circuit will respond incorrectly, resulting in the problem that you are trying to debug.

An example circuit that has the capability of producing a race condition is shown in Fig. 6-1.

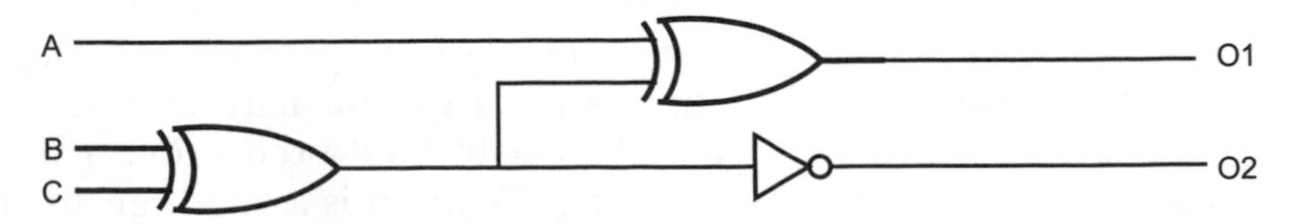

Fig. 6-1. Example race condition circuit.

Figure 6-2 is the waveform output of this circuit to a three bit incrementing signal and, in it, I have indicated the output bits ("O1" and "O2") and indicated where the operation of the circuit is "correct" (O2 is valid after O1) as well as a possible race condition (when O1 is valid after O2). I have also indicated times, using a shaded block, when both of the XOR inputs are changing and there could be a "glitch" caused by both inputs changing state simultaneously, at which time the output of the XOR gate is unknown.

The glitch produced by the XOR gate is an excellent example of a race condition. As I presented earlier in the book, the XOR gate is typically made up of five NAND gates in the configuration shown in Fig. 6-3. If one input changes, then the output will change state to either "1" or "0" without any glitches, but what happens when the two inputs change state simultaneously?

Quickly thinking about it, you might think that the output doesn't change state, but consider what happens at the NAND gate level of the XOR gate.

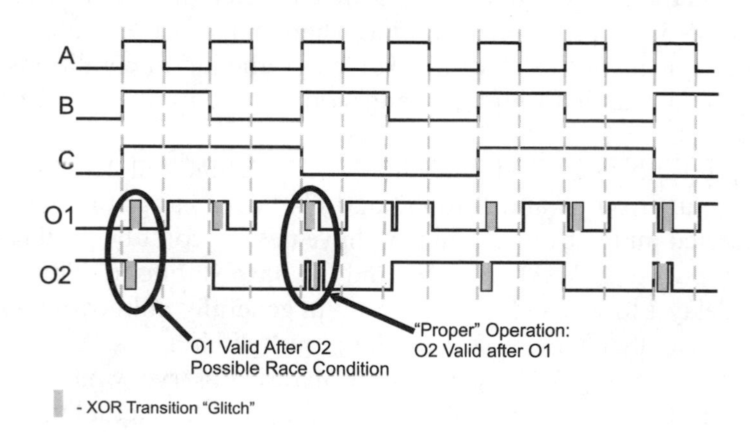

Fig. 6-2. Example race circuit wave forms.

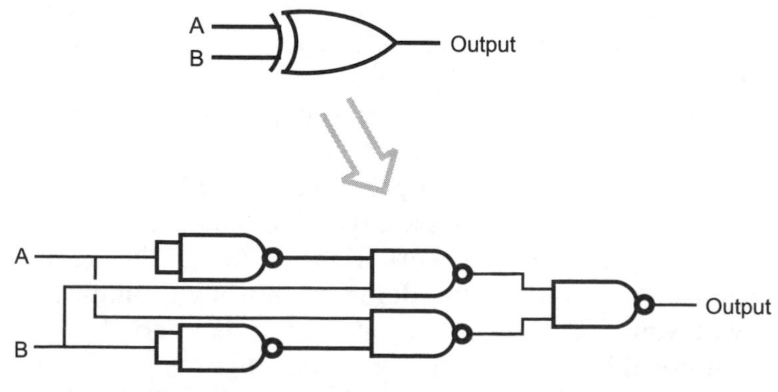

Fig. 6-3. XOR gate TTL implementation.

Table 6.3 XOR gate input change study.

Gate delay	Input "A"	Input "B"	G1	G2	G3	G4	Output
−1	1	0	0	1	1	0	1
0	0	1	0	1	1	0	1
1	0	1	1	0	1	1	0
2	0	1	1	0	0	1	1

Table 6-3 lists the NAND gate outputs for the different gates as I've marked them in Fig. 6-3. To help illustrate what's happening, I use "gate delays" as the time increments of this study. In Table 6-3, the initial conditions are one gate delay before the two inputs change value. The inputs change value at gate delay "0".

According to this study, at gate delay 1, the output will be a "0" because the direct inputs from A and B to G3 and G4 have changed at gate delay 0, *but* the inverted inputs from G1 and G2 have not. It won't be until gate delay 2 that the inputs to NAND gates G3 and G4 have stabilized. Thus, the time from gate delay 1 to gate delay 2 will result in generally unknown logic levels, which are normally characterized by the term "glitch".

Going back to Fig. 6-2, you can probably observe what I mean by the race condition, but I'm sure it seems very subtle. Actually, this is the point that I want to make: race conditions are very subtle and are very difficult to observe. For this section, I spent quite a bit of time with a 74C85 (quad XOR gate), a PIC16F627A (Microchip PIC Microcontroller to produce the "A", "B" and "C" inputs to the circuit) and an oscilloscope trying to capture the events shown in Fig. 6-2. I gave up after about 5 hours of trying to capture the event on the oscilloscope in a way that it would be easily seen.

Race conditions are dependent on part mix, applied voltage and ambient conditions. You may find some sample circuits which never have the problem while others will never seem to work right. Finding the actual event is extremely difficult and only after doing a thorough timing analysis of thc circuit will you find the opportunity for a race condition to occur. The prevention for this problem is quite simple – figure out what your worst case gate delay is through the circuit and only sample data after this time (even add a 10% margin to make sure there is no chance of marginal components causing problems).

Avoiding the opportunity is why chip designers work at making sure multiple outputs are active at the same time for changing inputs. Looking at the "A"/"B"/"C" waveform of Fig. 6-2, you might have thought that it is impossible to achieve the signals at precisely the same time, but it is very likely that if a single gate is producing incrementing outputs, the "edge" of each output bit will be precisely aligned with each other and will cause the glitch on the output of the XOR gate.

The process of determining what is the worst case gate delay is the same process I used for finding the "glitch" in the XOR gate and is known as "timing analysis". It is unusual for somebody to work through this analysis by hand as I have done, except for very simple circuits. When timing analysis is done on a commercial product, it is normally done using a logic simulator, which can find the longest delays and report on any problems.

Quick and Dirty Logic Gates

One of the most frustrating aspects of designing digital electronic circuits is that when you are almost finished, you often discover that you are a gate or two short and you are left with the question of whether or not you should add another chip to the circuit. The major problem with adding another chip to the circuit is the requirement for additional space to place the chip in the circuit. Along with the need for additional space, adding another chip will add to the costs of the application and the difficulty in assembling it. In Chapter 2, I discussed that by using the Boolean arithmetic laws and rules, you could produce various functions using different gates than the ones that are "best suited" for the requirements. In the cases where there are no leftover gates available, a gate can be "cobbled" together with a few resistors, diodes and maybe a transistor. These simple gates are often referred to as being "MML" or "Mickey Mouse logic" technology because they can generally be used in most situations and with different logic families when a quick and dirty solution is required.

To be used successfully, they must be matched to the inputs and outputs of the different logic families that you are using and should not result in long switching times, which will affect the operation of the application, or large current draws, which could damage other components. As a rule of thumb, do not use one of the simple gates presented here between differing

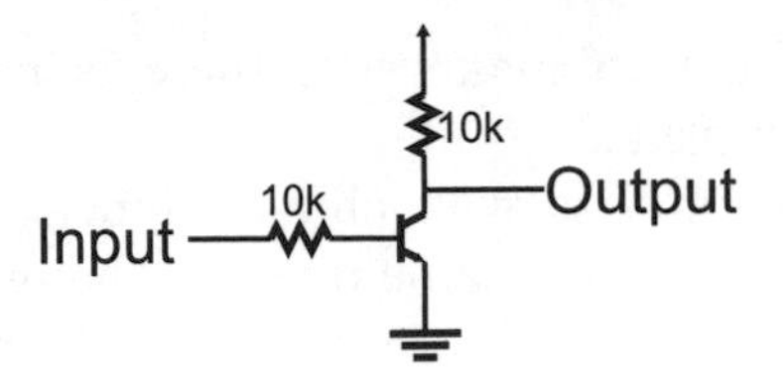

Fig. 6-4. RTL inverter.

technology gates; you will find that operation of different technologies can often be incompatible when you are adding resistors, diodes and transistors like the ones used in the sample gates presented here. Another rule of thumb is to make sure that each MML gate only drives one input – you can get into trouble with input fan-outs and multiple gate current sinking requirements very quickly. Along with trying to satisfy these requirements, there are cases where you will find that the MML gate will require at least as many pins as adding another chip and will be more difficult to wire. Generally speaking, adding MML gates to your application should be considered a "last" resort, not something you design in right from the start.

The most basic MML gate is the "inverter" and should not be a surprise. Figure 6-4 shows the circuit for the MML inverter, built out of two 10 k resistors and an NPN transistor. This inverter is actually a basic "RTL" (resistor–transistor logic) technology device and outputs a high voltage, when it is not being driven by any current. When current is passed to the gate, the transistor turns on and the output is pulled to ground (with good current sinking capability).

This circuit (as well as the other MML gates I discuss in this section) cannot handle high voltage or current inputs and outputs as well as commercially available logic gates and need to be "buffered". The need for buffering the MML's gate inputs and output is an important point to note when considering using an MML gate in an application. As a rule, MML gates must be placed in the middle of a logic "string" rather than at the input or output ends to ensure that if you are expecting certain characteristics (such as the ability to drive a LED), standard TTL or CMOS technology gates will provide you with it.

The inverter circuit can be simply modified by adding another transistor and resistor, as shown in Fig. 6-5, to create an RTL NOR gate. The RTL NAND gate is shown in Fig. 6-6. The NOR gate is considered the basis of RTL technology.

Implementing an AND or OR gate in MML is a bit more complex and requires a good understanding of the input/output parameters

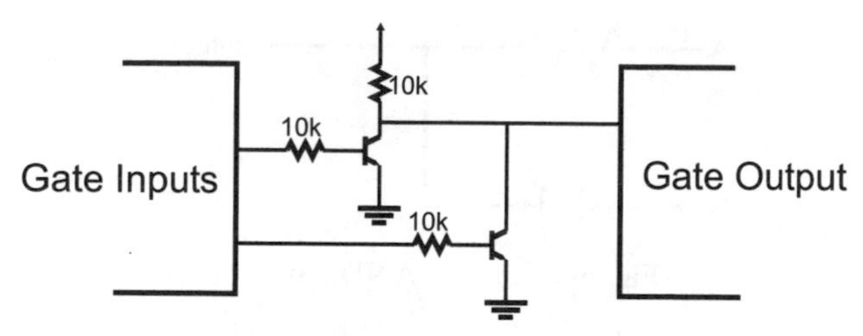

Fig. 6-5. RTL NOR gate.

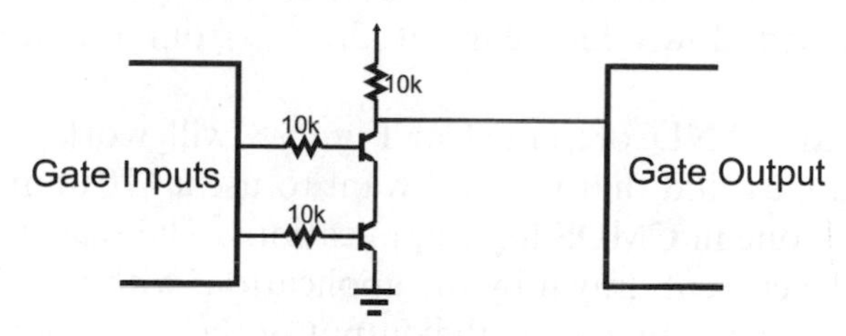

Fig. 6-6. RTL NAND gate.

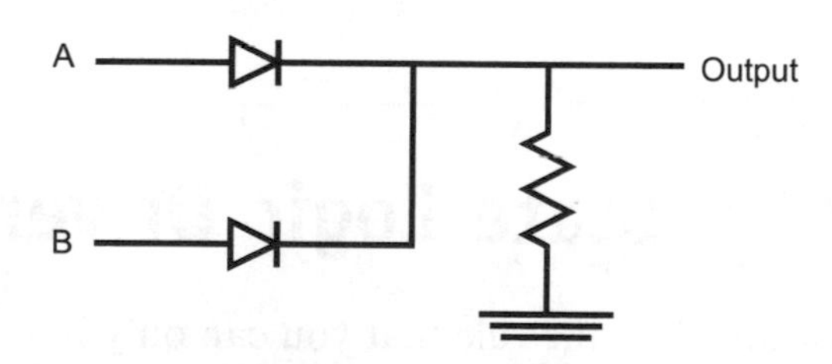

Fig. 6-7. MML OR gate.

of the logic families. In Fig. 6-7, I have shown a sample design for an OR using two diodes and a resistor. The use of a 470 ohm resistor is probably surprising, but it was chosen to allow the gate to be used with both CMOS and TTL logic. In this case, if neither input has a high voltage, then the output will pull the input to ground. If the input is a CMOS gate, then the input will behave as if it were tied to ground. The 470 ohm resistor will allow the TTL input current to pass through ground and it will behave as if the input was at a low logic level. When the resistor is connected to a CMOS input, it will be effectively tying the input to ground, even though no current is flowing through it. In either case, when one of the inputs is driven high, the input pin will be held high and the gate connected to the output of the OR gate will behave as if a high logic level was applied to it.

An MML AND gate (Fig. 6-8) is the simplest in terms of the number of components. The diode and resistor work together to provide a high voltage

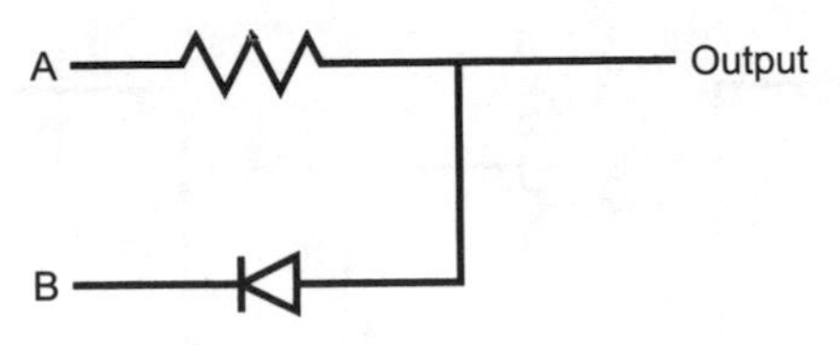

Fig. 6-8. MML AND gate.

when both inputs are high, but when one of them is pulled low, the voltage level will be pulled down and current drawn from the input gate it is connected to.

While the MML AND presented in Fig. 6-8 will work in virtually any application, you may find that you will want to use a 470 ohm resistor in the circuit and a 10 k one in CMOS logic applications. The reason for doing this is to minimize the current drawn by the application; with a 470 ohm resistor, roughly 10 mA will be drawn when the output of the gate is low. This current draw decreases to 100 μA when a 10 k resistor is used instead for the resistors in these two gates.

Dotted AND and Tri-State Logic Drivers

You may feel constrained by the rule that you can only have one driver on a single line (or net). In Chapter 3, I introduced you to the concept of the "dotted AND" bus in which there was a common pull up on the net along with a number of transistor switches, each one of which could "pull" the net to a low voltage/logic level (and draw the current from any TTL gates inputs connected to the dotted AND). The dotted AND works reasonably well and has the advantage that it can control output voltages greater than the power applied to the logic chips. Some more subtle advantages are that more than one output can be active (tying the net to ground) and the operation of the bus will not be affected and TTL open collector and CMOS logic open drain outputs can be placed on the bus along with mechanical switches and other devices which can pull the bus to ground.

The dotted AND bus's main disadvantage is its inability to source significant amounts of current. Smaller value pull up resistors can be used, but this increases the amount of current passed to ground when one of the open collector transistors is on. The dotted AND can be considered to be quite inefficient if it is low for a long period of time, because it is passing current directly to ground. The inability to source large amounts of current

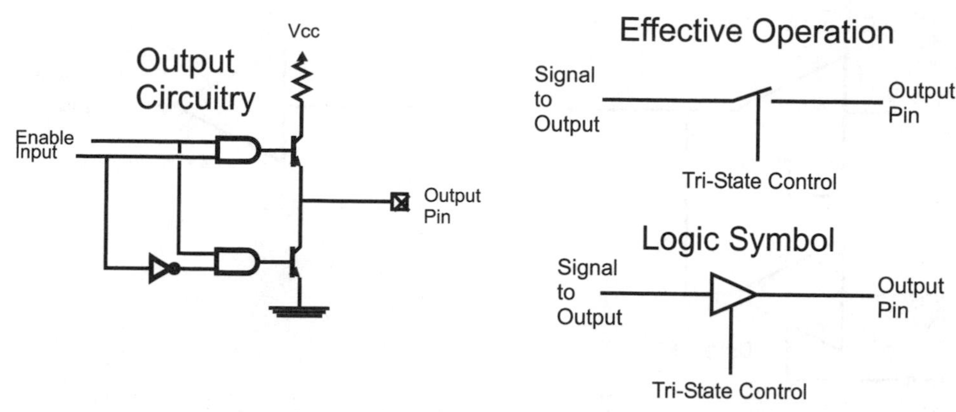

Fig. 6-9. Tri-state driver circuit diagram and operation.

is a drawback when high-speed signals are involved is the major disadvantage of the dotted AND bus. Changing an input from a high to a low, especially when there are some relatively large capacitances on the net, the switching time can become unreasonably slow.

A common error made by new circuit designers when they are adding a dotted AND bus to their designs is forgetting to add the pull up resistor. If the resistor has been forgotten, then the bus will never have a "high" voltage (although it will have a "low" voltage that can be detected). You will find that TTL inputs connected to a dotted AND but not having a pull up will work correctly, but CMOS logic inputs will not.

Another solution to the problem of wanting to have multiple drivers on the same net is to use "tri-state" drivers (Fig. 6-9). These drivers can "turn off" the transistors as effectively as if a switch were opened (the diagram marked "Effective Operation" in Fig. 6-9).

The left-hand side of circuit diagram of Fig. 6-9 shows how the tri-state driver works. If the tri-state control bit is inactive, the outputs of the two AND gates will always be low and the NPN output transistors can never be turned on. This "inactive" state is also known as the "high impedance state". When the tri-state control bit is active, then a high to either the top or bottom NPN transistor will allow the output to behave as an ordinary TTL output.

This ability to "turn off" allows multiple drivers, such as I have shown in Fig. 6-10, to be wired together. In this case, if data was to be placed on the net from Driver "B", the "Ctrl A" line would become inactive (the "high impedance state"), followed by the "Ctrl B" line becoming active. At this point, the bus would be driven with the data coming from Driver "B".

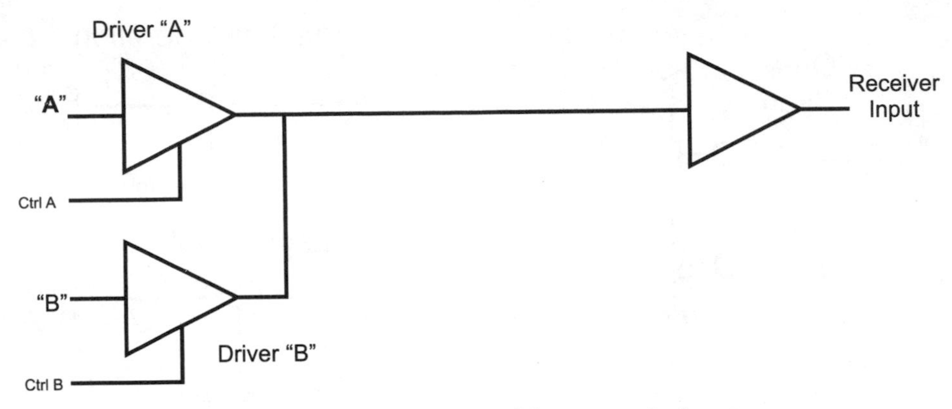

Fig. 6-10. Multiple tri-state drivers on a single net.

It is important that only one tri-state driver is active at one time because the voltage on the common net will be indeterminate, as will be the logic level. You may be thinking that the term "indeterminate" to be the case when two drivers are active and are attempting to drive the net at different levels. This is true, but it is often also the case if two different drivers are driving at the same level: CMOS logic and TTL drivers will attempt to drive the net to different voltage levels and even TTL will not give repeating answers when you are trying to understand what is happening. The technical term for the situation where two tri-state drivers are active at the same time is "bus contention" and it should be avoided at all costs – only one driver should be active on the net at any one time.

At the start of this section, I noted that there could be more than one output active on a net at the same time. Note that when I say "multiple active outputs", I mean more than one driver pulling the net low. I do not recommend this to be part of the design, however; multiple active outputs are impossible to differentiate and you will have problems figuring out which bits are active and what signal is being sent (with multiple outputs active, state changes from one output will most likely be masked by the active operation of others).

Before leaving this section, I do want to point out that tri-state drivers *can* be used on a dotted AND bus. This is probably surprising, considering the dire warnings I have put in regarding bus contention. The trick to adding a tri-state driver to a dotted AND bus is that it is normally disabled and only a low voltage can be put on the net by the tri-state driver. High values are output by simply disabling the tri-state driver and letting the net's pull up provide the high voltage.

Combining Functions on a Net

As a purely intellectual exercise, it can be interesting to see how many functions you can build into a single digital electronics net. From a practical point of view, cramming multiple functions on a single line will minimize the amount of effort that must be expended to build a prototype application. Many products carry out multiple functions on a single line; generally, this is done to allow the manufacture and sale of simpler products. Whatever the motivation, "stretching" a logic technology to allow multiple functions on a single net requires a strong knowledge of the technology's electrical parameters and the technology's normal operating conditions. The most important thing to remember is that the input/output devices attached to the net must be properly coordinated to make sure that data is read and written at the right times.

The most obvious ways of connecting two drivers together is to use dotted AND and tri-state drivers on a "bus", as I discussed in the previous section. These methods work well and should be considered as the primary method of implementing multiple devices on the net. The other methods discussed here work best for specific situations; but there is no reason why you can't modify your design to take advantage of these specific instances.

When interfacing the bi-directional digital I/O pin to a CMOS driver and a CMOS receiver (such as a memory with separate output and input pins), a resistor can be used to avoid bus contention at any of the pins, as is shown in Fig. 6-11.

Using this wiring, when the bi-directional I/O pin is driving an output, it will be driving the "Data In" pin register, regardless of the output of the "Data Out" pin. If the bi-directional and "Data Out" pins are driving different logic levels, the resistor will limit the current flowing between the bi-directional and the memory "Data Out" pin. The value received on the "Data In" pin will be the bi-directional device's output.

When the bi-directional digital I/O is receiving data from the memory, the I/O pin will be put in "input" (or "high impedance") mode and the

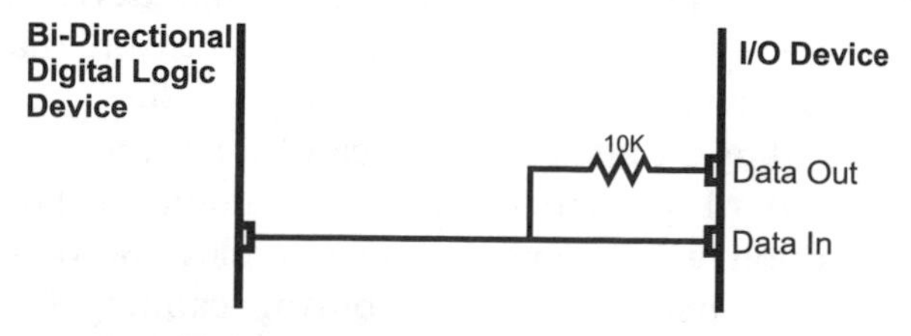

Fig. 6-11. Combining CMOS input and output pins to create a bidirectional bus.

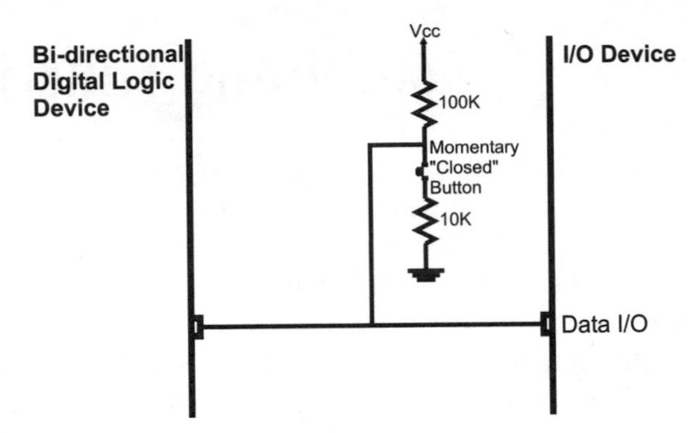

Fig. 6-12. Adding a button to a CMOS bidirectional net.

"Data Out" pin will drive its value to not only the bi-directional device's I/O pin, but the "Data In" pin, as I noted above. In this situation, the "Data In" pin should not be latching any data in; the simplest way to ensure this is to make the digital I/O pin part of the I/O control circuitry. This is an important point because it defines how this circuit works. A common use for this method of connection data in and data out pins is used in memory chips that have separate data input and output pins.

User buttons can be placed on the same net as logic signals as Fig. 6-12 shows.

When the button is open or closed, the bi-directional logic device can drive data to the input device, the 100 k and 10 k resistors will limit the current flow between Vcc and ground. If the bi-directional logic device is going to read the button "high" (switch open) or "low" (switch closed) it will be driven on the bus at low currents when the pin is in "Input Mode". If the button switch is open, then the 100 k resistor acts like a "pull up" and a "1" is returned. When the button switch is closed, there will be approximately a half volt across the 10 k resistor, which will be read as a "0".

The button with the two resistors tying the circuit to power and ground is like a low-current driver and the voltage produced is easily "overpowered" by active drivers. Like the first method, the external input device cannot receive data except when the bi-directional device is driving the circuit. A separate clock or enable should be used to ensure that input data is received when the bi-directional device is driving the line.

This method of adding a button to a net can be extrapolated to work with a switch matrix keyboard (presented later in the book), although the circuit and interface operation will become quite complex. Secondly, a resistor/capacitor network for "debouncing" the button cannot be used with this circuit as it will "slow down" the response of the bi-directional device driving the data input pin and will cause problems with the correct value being accepted.

For both of these methods of providing multiple features to a single net, you should only use CMOS logic as it is voltage controlled and not current controlled, like TTL. You *may* be able to use TTL drivers with these circuits, but they may be unreliable. To avoid problems with invalid currents being available to TTL receivers, just use the latter two circuits with CMOS digital logic.

Designing a circuit in which multiple functions are provided on a single net for an application is not always possible or even desirable. Like any design feature implemented in an application, before trying to combine multiple functions on a single net, you should understand the benefits as well as the costs. When it is possible, you can see some pretty spectacular results; my personal record was for a LCD driver in which I was able to combine five functions on a single net – LCD Data Write, LCD Data Read, Data In Strobe, Data Ready Poll and configuration switch poll.

Quiz

1. What parameter is not listed in the chip characteristic card?
   (a) Input fanout
   (b) Number of gates built into the chip
   (c) Electrical dependencies
   (d) Maximum operating speed
2. What is not a typical digital electronic output pin type?
   (a) Totem pole
   (b) Open collector
   (c) High-current
   (d) Tri-state driver
3. Other than the XOR gate, are any other of the six basic I/O gates capable of producing race conditions just by themselves?
   (a) Each one is capable of producing a race condition under certain circumstances
   (b) The NOR Gate in TTL
   (c) The AND Gate in CMOS Logic
   (d) No

4. What is not a factor in determining if a marginal circuit and component will produce a race condition?
   (a) Ambient temperature
   (b) Net length
   (c) Power voltage
   (d) The phases of the moon

5. Mickey Mouse logic should be used:
   (a) Never
   (b) When you are in a hurry to get the application finished
   (c) When you have board space, cost and available gate constraints that preclude adding a standard chip
   (d) When there is a need to pass a CMOS output to a TTL input

6. Each item is an advantage of a dotted AND bus except:
   (a) The dotted AND bus can have tri-state drivers on it as well as mechanical switches
   (b) The dotted AND bus can control voltages greater than the chip's Vdd/Vss
   (c) The dotted AND bus is cheaper than one manufactured with tri-state drivers
   (d) The dotted AND bus can consist of CMOS logic as well as TTL drivers

7. When tri-state drivers are inactive, another term that is used to describe the state is:
   (a) High resistance
   (b) High impedance
   (c) Low current output
   (d) Driver isolation

8. When should multiple tri-state drivers be active?
   (a) When more current is required on the net
   (b) When more speed is required on the net
   (c) When the receiver detects an ambiguous logic level
   (d) Never

9. When adding a push button to a net, can the 100 k resistor connected to positive power and the 10 k resistor connected to positive power be swapped?
   (a) Yes
   (b) No

(c) Only if TTL receivers and drivers are used.
(d) Yes, if you can ensure that the signals passing between the digital devices are still within specified operating margins.

10. When putting a receiver and driver on the same net, can the current limiting resistor be wired between the bi-directional logic device and the "Data In" pin, leaving a direct connection between the bi-directional logic device and "Data Out"?
 (a) Yes. There aren't any cases where it wouldn't work
 (b) Yes, if the resistor value is within 1 k and 10 k
 (c) Yes, for certain technologies of CMOS logic
 (d) No. This will cause bus contention

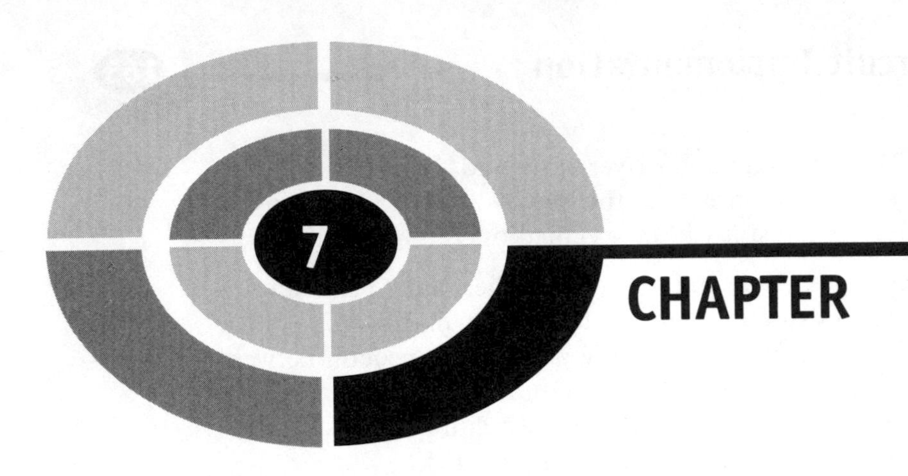

Feedback and Sequential Circuits

This chapter's title probably seems like a bit of a misnomer; you are probably wondering what feedback has to do with digital electronics? When I use the term "feedback", I am using it in the most literal sense, past state data are used to maintain current state data. These circuits built from the theory that I am going to provide in this chapter are commonly known as "memory devices". For digital electronic circuits to store information, that information will continually move through the circuit and is used to determine what the future value of the circuit is. Feedback is critical to provide digital electronics with the ability to "remember" previous states and data.

When you first hear the term "feedback", you probably think of an amplifier with its microphone input brought close to its speaker output (Fig. 7-1). You also probably involuntarily wince at the thought of the term "feedback" because it brings back the memory of the horrible sound the amplifier made when the microphone was too close. This type of feedback

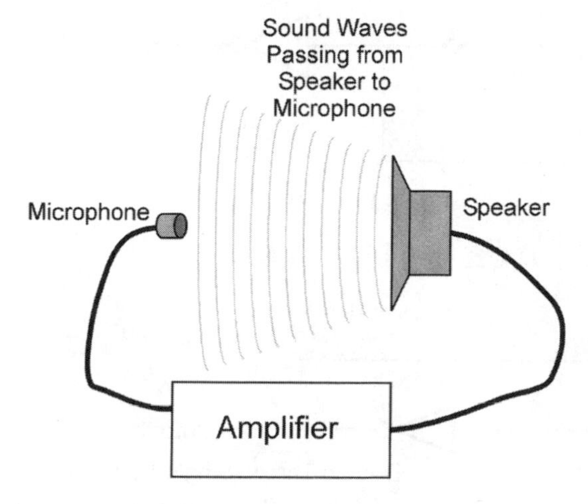

Fig. 7-1. Analog feedback.

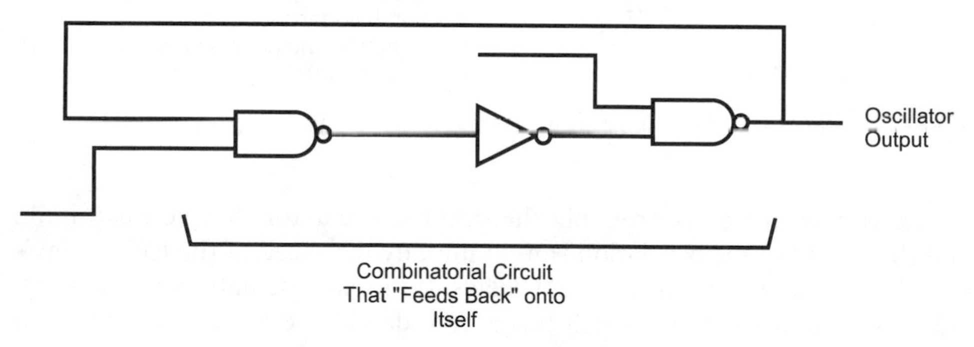

Fig. 7-2. Digital feedback circuit.

cannot save information; the uncontrolled amplification of the signal distorts and destroys information in a very short order.

When I introduced combinatorial circuits at the start of the book, I noted that an important part of combinatorial circuits was that data could only travel in one path; no outputs were passed back to earlier inputs in a logic chain, like the one shown in Fig. 7-2. The reason for specifying that outputs were not to be passed to inputs was to make sure that an inadvertent oscillator, known as a "ring oscillator" (Fig. 7-3), was not created.

You should know that the inverter in Fig. 7-3 is going to invert its input. The problem arises when the input is tied to the output, as it is in this case. When the input is passed to the inverter, it outputs the inverted value, which is then immediately passed back to the input and the gate inverts the value again, and again, and so on.

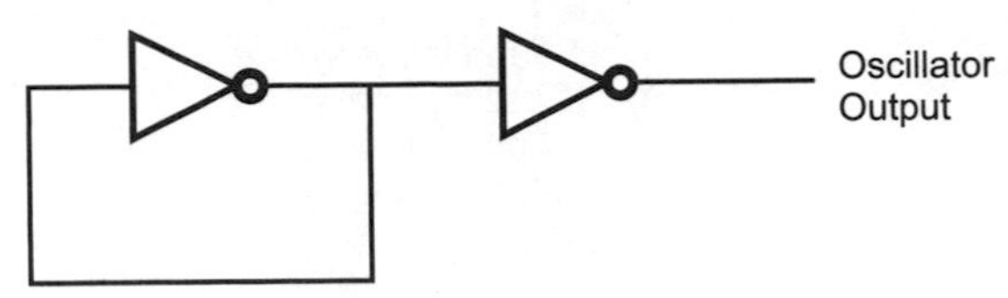

Fig. 7-3. Ring oscillator.

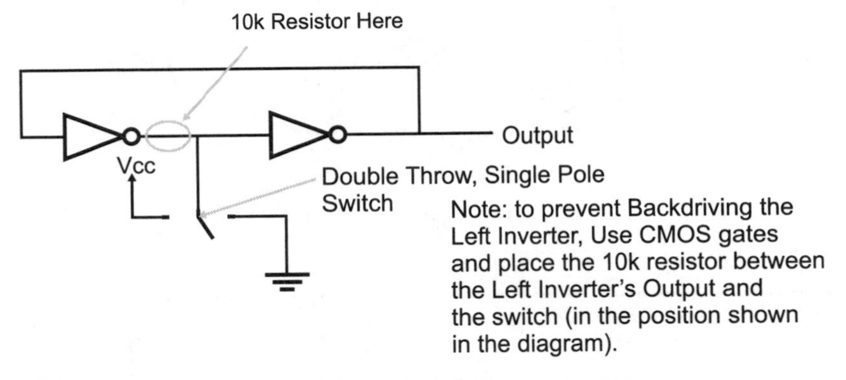

Fig. 7-4. Single switch debounce circuit.

The ring oscillator is probably the simplest oscillator that you can build and the period of the oscillation runs at literally the speed of the technology's gate delay times the number of gate delays. If the ring oscillator shown in Fig. 7-3 was built from TTL (which has a gate delay of 8 ns) you would see a "square wave" with a period of 62.5 MHz. One of the functions that ring oscillators perform is the measurement of a logic technology's gate delay.

Extrapolating from what has been discussed here, you could build a simple memory circuit using the two inverters and a double throw switch, wired as shown in Fig. 7-4. This circuit is used to "debounce" a switch input. As I will discuss in later chapters, when a mechanical switch is thrown, the physical contacts within the switch literally bounce against each other before a hard, stable contact is made. This bouncing can cause quite a bit of grief when you are trying to respond to a single switch movement.

The circuit in Fig. 7-4 will pass a signal continuously between the two inverters (the output of the two inverters is the same as the input, so there is no chance for a ring oscillator) until the switch comes in contact with a connection that forces the state to change. If the switch was originally at the ground position, the signal coming from the inverter to its left would be a "0". When the switch was moved to the "Vcc" position, the signal going to the inverter to the right would be changed and its output would change.

The beauty of this circuit is that when the switch is in between contacts, the output state of the circuit remains constant.

When the switch is not touching either contact, the two inverters are maintaining the previous bit value and the circuit behaves essentially as a memory device.

There is a downside to the button debounce circuit in Fig. 7-4 and that is when the switch is thrown, it connects the left inverter output to the opposite power supply that it is driving out. This is known as "backdriving" and it should always be avoided.

Backdriving a gate will shorten its life in the best case and could burn it out in very short order. As noted in Fig. 7-4, you should only use CMOS inverters (which are voltage, rather than current controlled) and place the 10 k resistor between the switch and the output of the left inverter. By using this circuit, there will be no chance that the left inverter's output is tied directly to power or ground (which will be the opposite value that it's at) and the 10 k resistor will limit the amount of current that is passed. I did not put the resistor into Fig. 7-4 as it is a basic circuit that I have seen in a number of references and I wanted to point out that it does backdrive a gate output and there are ways of avoiding this problem.

The other term used in this chapter's title, "sequential circuits", is used to identify the class of digital electronic circuits that have memory devices within them and use their data, along with combinatorial circuits, to produce applications. A digital clock (Fig. 7-5) is an excellent example of a sequential circuit. The data output from the memory circuits of the clock are passed to

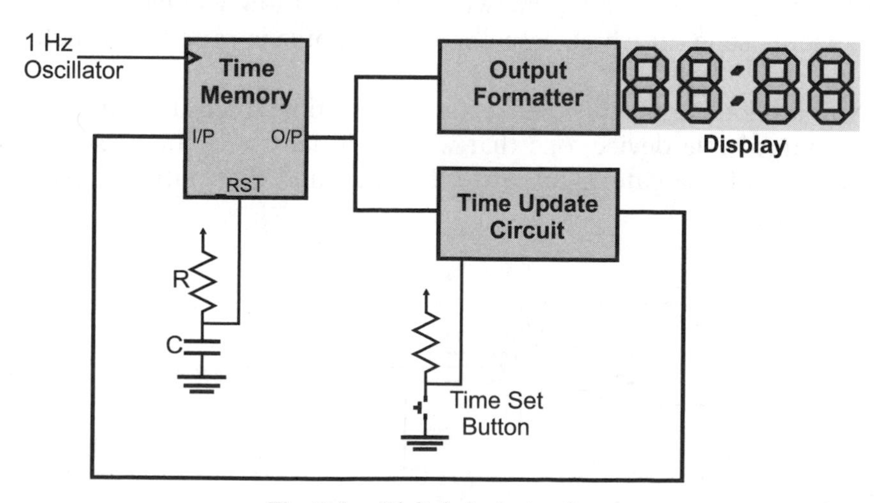

Fig. 7-5. Digital clock circuit.

combinatorial logic circuits and the outputs of the combinatorial circuits are passed back to the inputs of the "time memory" circuits.

Any time memory circuits, like the ones presented in this chapter, are used in a digital electronics application, the circuit is called a "sequential circuit".

Flip Flops (RS and JK)

The best analogy I can find for a simple, one bit "memory device" is the two coiled relay of Fig. 7-6. The relay coil does not have a return spring that only one coil pulls against; when the relay's wiper is placed in a position it stays there. This memory device is set to one of two states, depending on which relay coil was last energized, pulling the wiper contact into connection with it. Once electricity to the coil is stopped, the memory device will stay in this state until the other coil is energized and the wiper is pulled towards it. This device works very similarly to the most basic electronic memory device that you will work with, the "reset-set" (RS) "flip flop".

The term "flip flop" is indicative of the operation of the memory device: it is either "flipped" to one value or "flopped" to another. Where the relay device relies on friction to keep the saved value constant, the electronic memory unit takes advantage of feedback to store the value. Digital feedback can only be one of two values, so its use in circuits probably seems like it is much more limited than that of analog feedback. This is true, except when it is used as a method to store a result in a circuit like the "NOR flip flop", shown in Fig. 7-7. Normally, the two inputs are at low voltage levels, except to change its state, in which case one of the inputs is raised to a high logic level.

If you are looking at this circuit for the first time, then it probably seems like an improbable device, one that will potentially oscillate because if the output value of one gate is passed to the other and that output is passed to

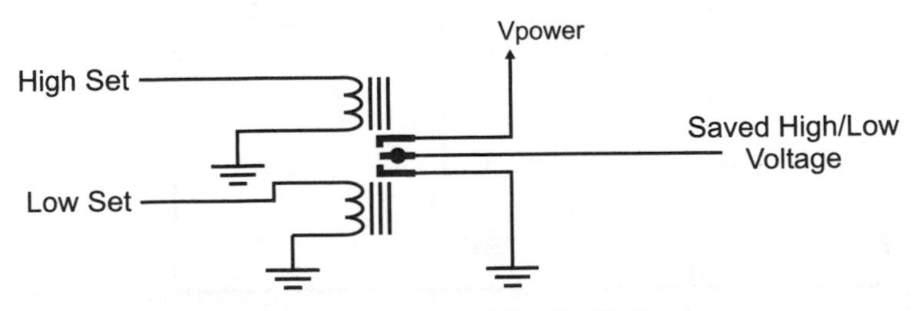

Fig. 7-6. Relay-based "flip flop" circuit.

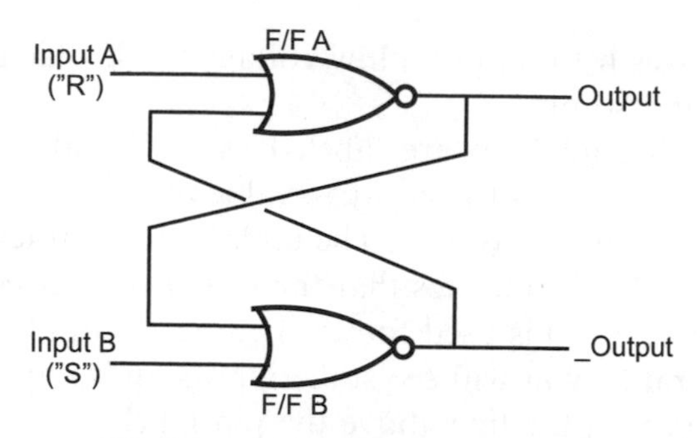

Fig. 7-7. NOR-based RS flip flop.

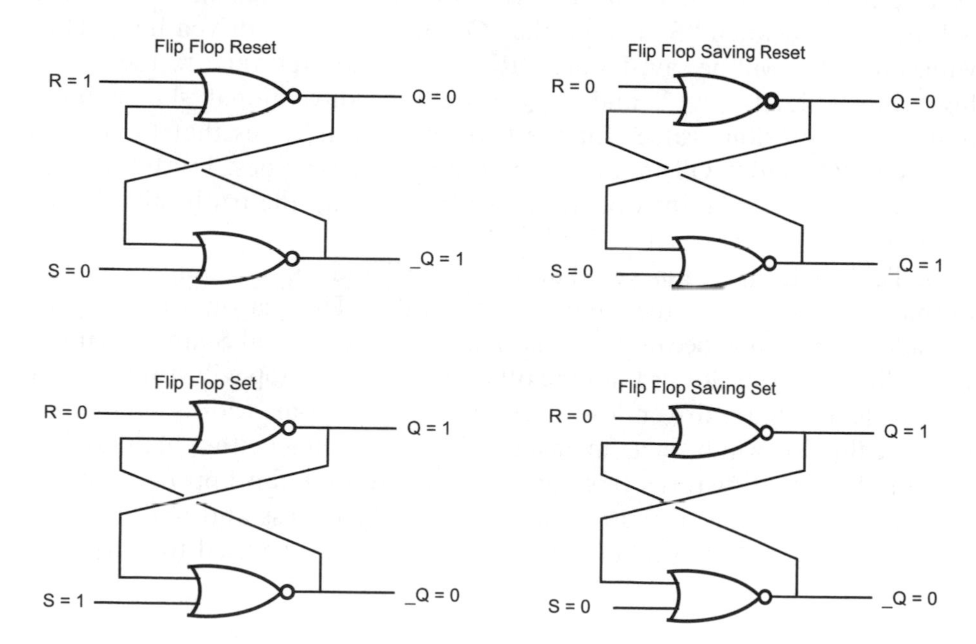

Fig. 7-8. RS flip flop state changes.

the original, it seems logical that a changing value will loop between the two gates. Fortunately, this is not the case; instead, once a value is placed in this circuit, it will stay there until it is changed or power to the circuit is taken away. Figure 7-8 shows how by raising one pin at a time, the output values of the two NOR gates are changed.

When the "R" and "S" inputs to their respective NOR gates are low, there is only one signal left that will affect the output of the NOR gates and that is output of the other NOR gate. When "Q" is low, then a low voltage will be passed to the other NOR gate. The other NOR gate outputs a high voltage because its other input is low. This high signal is passed to the original

NOR gate and causes it to output a low voltage level, which is passed to the other NOR gate and so on....

The outputs of the flip flop are labeled as "Q" and "_Q". "Q" is the positive output while "_Q" is the negative value of "Q" – exactly the same as if it were passed through an inverter. The underscore character ("_") in front of the output label ("Q") indicates that the signal is inverted (the same as if an exclamation mark ("!") is used for an inverter's output). When you look at some chip diagrams, you will see some inputs and outputs that have the underscore before or on the line above the pin label.

The "R" and "S" input pins of the flip flop are known as the "reset" and "set" pins, respectively. When the "R" input is driven high the "Q" output will be low and when "S" is high the "Q" output will be driven high. These values for "Q" will be saved when "R" and "S" are returned to the normal low voltage levels. "Q_0" and "$\_Q_0$" are the conventional shorthand to indicate the previous values for the two bits and indicates that the current values of "Q" and "_Q" are the same as the previous values. Truth tables are often used to describe the operation of flip flops and the truth table for the NOR RS flip flop is given in Table 7-1.

In Table 7-1, I have marked that if both "R" and "S" were high, while the outputs are both low, the inputs were invalid. The reason why they are considered invalid is because of what happens when R and S are driven low. If one line is driven slower than the other, then the flip flop will store its state. If both R and S are driven low at exactly the same time (not a trivial feat), then the flip flop will be in a "metastable" state, Q being neither high nor low, but anything that disturbs this balance will cause the flip flop to change to that state. The metastable state, while seemingly useless and undesirable is actually very effective as a "charge amplifier" – it can be used to detect very

Table 7-1 NOR RS flip flop truth table.

R	S	Q	_Q	Comments
0	0	Q_0	$\_Q_0$	Store current value
1	0	0	1	Reset flip flop
0	1	1	0	Set flip flop
1	1	0	0	Invalid input condition

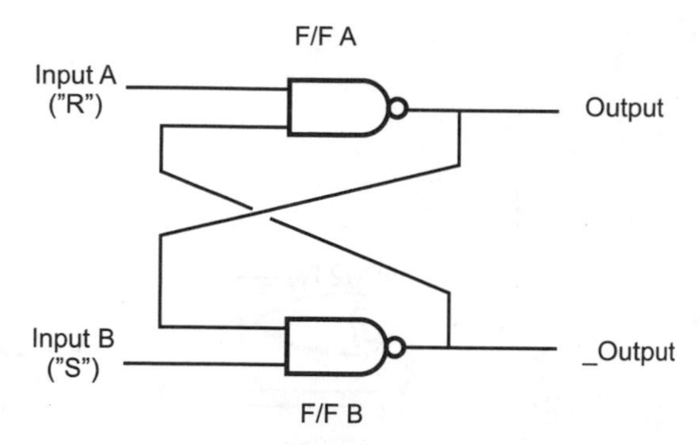

Fig. 7-9. NAND-based RS flip flop.

Table 7-2 NAND RS flip flop truth table.

R	S	Q	_Q	Comments
0	0	1	1	Metastable input state
0	1	0	1	Reset flip flop
1	0	1	0	Set flip flop
1	1	Q_0	$\_Q_0$	Save current value

small charges in capacitors. This is an important mode of operation that is taken advantage of for DRAM and SDRAM memories.

Along with building a flip flop out of NOR gates, you can also build one out of NAND gates (Fig. 7-9). This circuit works similarly to the NOR gate, except that its metastable state occurs when both inputs are low, and the inputs are active at low voltage levels, as I have shown in Table 7-2, which is the NAND RS flip flop's truth table.

You can build your own NOR RS flip flop, which has its state set by two switches as I show in Fig. 7-10 and is wired according to Fig. 7-11. I suggest that you test out the circuit in as many different ways as possible – especially investigating the metastable and post-metastable states. Unless you were to wire the R and S inputs to one switch, you will find it impossible to achieve the metastable state. The parts that are needed to build the RS flip flop are listed in Table 7-3.

Before going on, there is one additional point about flip flops that may not be immediately obvious but will be something that you will have to consider

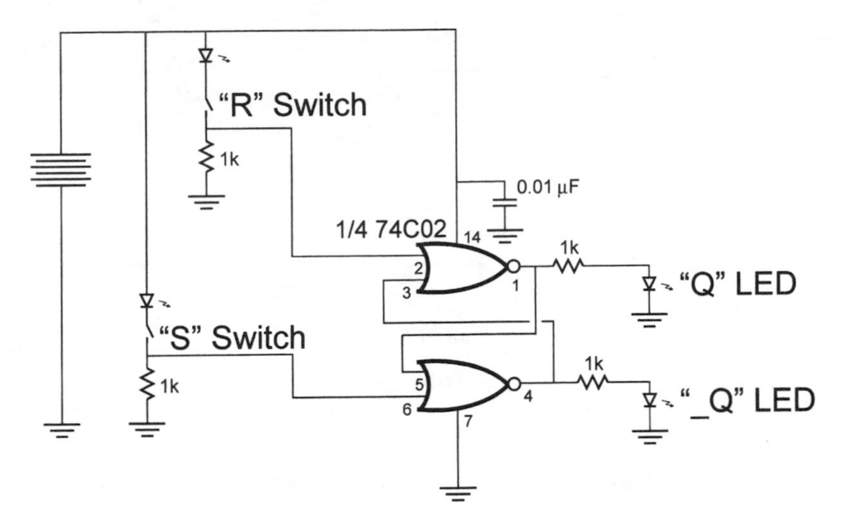

Fig. 7-10. NOR RS flip flop test circuit.

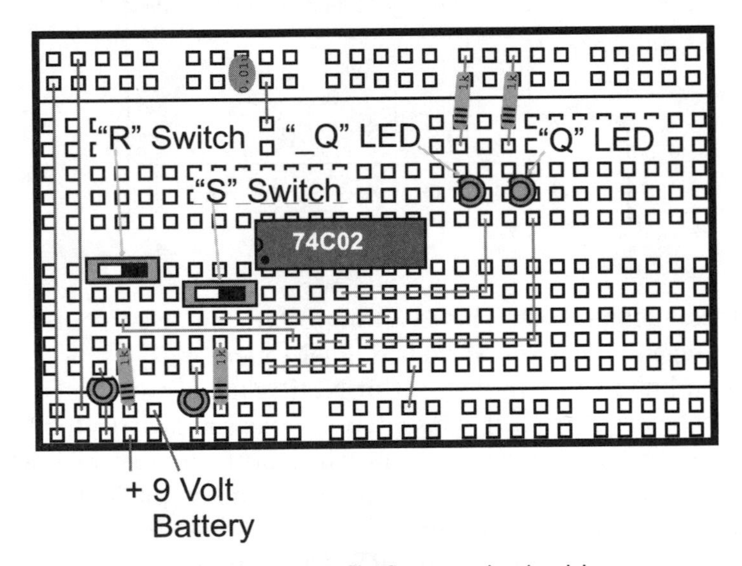

Fig. 7-11. NOR RS flip flop test circuit wiring.

in your career as a designer of digital electronic devices; when power is removed, the flip flops will lose the bit information contained within them. The term used to describe this phenomena is "volatility"; flip flops are considered "volatile" devices. Flash memory (like the flash used in your PC) does not lose its information when power is shut off and is known as "non-volatile" memory.

Table 7-3 Parts needed to build NOR gate based RS flip flop.

Part	Description
74C02	CMOS quad, 2 input NOR gate
4 × LEDs	Any color
4 × 1 k resistors	1/4 watt
2 × breadboard mountable switches	Digikey EG1903-ND
0.01 μF capacitor	Any type
Breadboard	
9 volt battery with clip	

Edge Triggered Flip Flops

The RS flip flop is useful for many ad hoc types of sequential circuits in which the flip flop state is changed asynchronously (or whenever the appropriate inputs are active). For most advanced sequential circuits (like a microprocessor), the RS flip flop is a challenge to work with and is very rarely used. Instead, most circuits use an "edge triggered" flip flop which only stores a bit when it is required. You will probably discover the edge triggered flip flop (which may also be known as a "clocked latch") to be very useful in your own applications and easier to design with than a simple RS flip flop.

The most basic type of edge triggered flip flop is the "JK" (Fig. 7-12), which provides a similar function to the RS flip flop except that it changes state when the "clock" input is "rising" (changing from "0" to "1"), as shown in the waveform diagram of Fig. 7-13.

There are a few points about Fig. 7-13 that should be discussed. I have assumed that in the initial state for this example, the output value "Q" is "1". When the first rising edge of the clock ("Clk") is encountered, both J and K are 1, so Q "toggles" or changes state. Next, when the rising edge of the clock is encountered, J is 1 and K = 0, so Q becomes 1 and the opposite is true for the next rising edge. In the final rising edge, both J and K are 0 and the value of Q remains the same. There is no metastable state for the JK flip flop. The operation of the JK flip flop is outlined in Table 7-4.

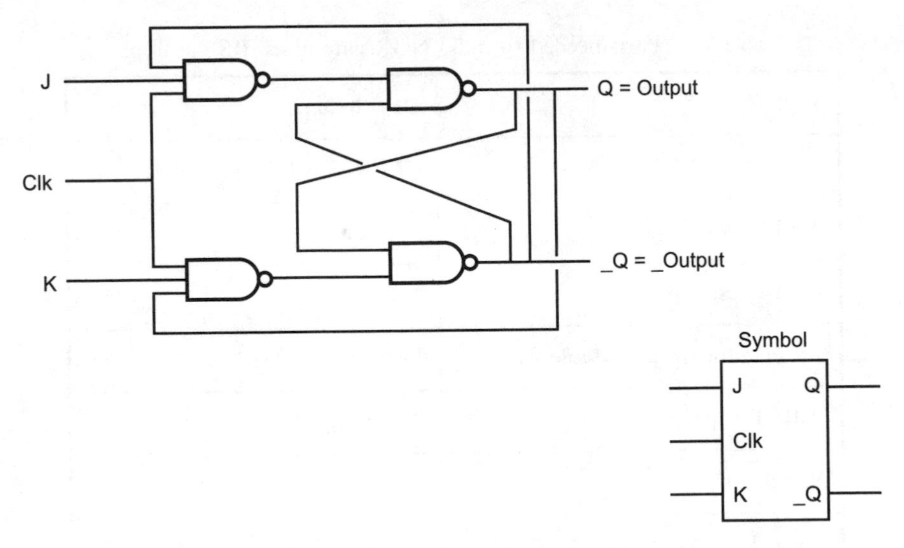

Fig. 7-12. Edge triggered JK flip flop.

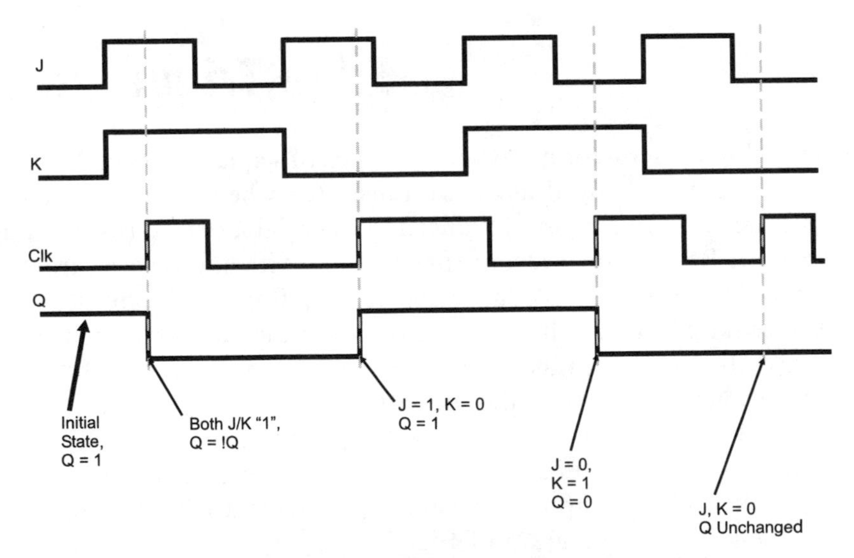

Fig. 7-13. Edge triggered JK flip flop operating waveform.

Just as a small circle on an input or an output of a logic gate indicates that the value is inverted, the clock pin on some chip diagrams is indicated by a small triangle. This convention helps minimize the clutter present in a logic diagram.

The JK flip flop is useful in general digital electronics applications, but it does not provide the necessary function for a computer register. Ideally,

Table 7-4 JK flip flop truth table. "Q" and "_Q" change when a rising clock edge is received.

J	K	Q	_Q	Comments
0	0	Q_0	$\_Q_0$	No change
0	1	0	1	Reset flip flop
1	0	1	0	Set flip flop
1	1	$\_Q_0$	Q_0	JK flip flop toggles

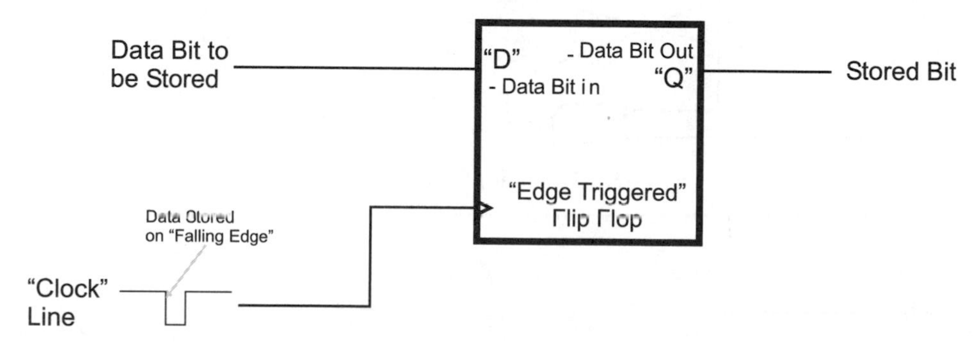

Fig. 7-14. "D" flip flop.

a clocked register's block diagram is quite simple (Fig. 7-14), consisting of a data line passed to the flip flop along with a "clock" line. While the data line stays constant, the contents of the flip flop doesn't change. When the clock line goes from high to low, the data is stored in the flip flop – this is known as a "falling edge clocked flip flop" or a "falling edge clocked register" and it is probably the most common type of flip flop that you will work with.

The edge triggered flip flop (Fig. 7-15) is based on the RS flip flop. Instead of always calling this circuit a "falling edge triggered flip flop" or "clocked register", this circuit is normally known as a "D flip flop". The organization of the flip flops used in this circuit may seem complex, but their operation is actually quite simple: the two "input" flip flops "condition" the clock and data lines and only pass a changing signal when the clock is falling, as I show in Fig. 7-16. To try and make it easier for you to understand, I have marked the outputs of the RS flip flops in Fig. 7-15 and showed the waveforms at these points.

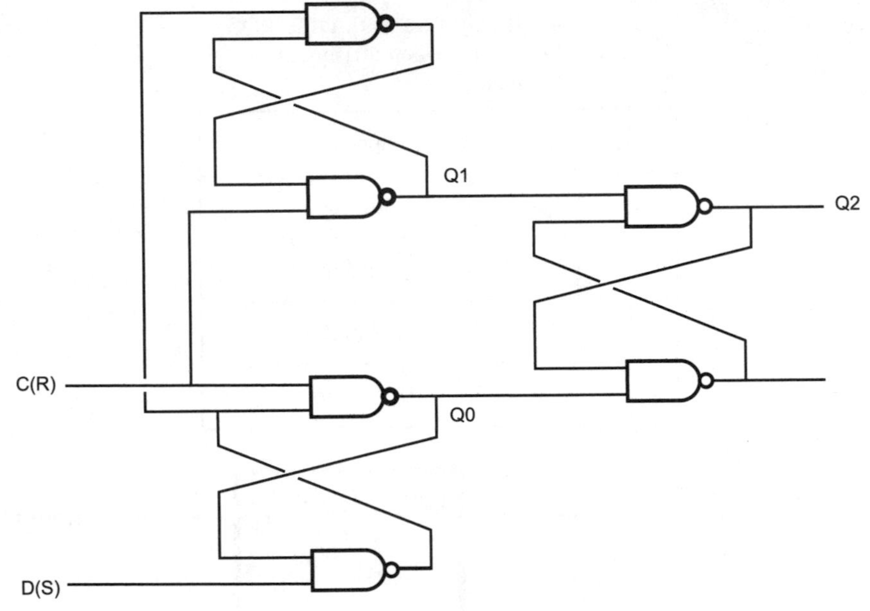

Fig. 7-15. D flip flop logic diagram.

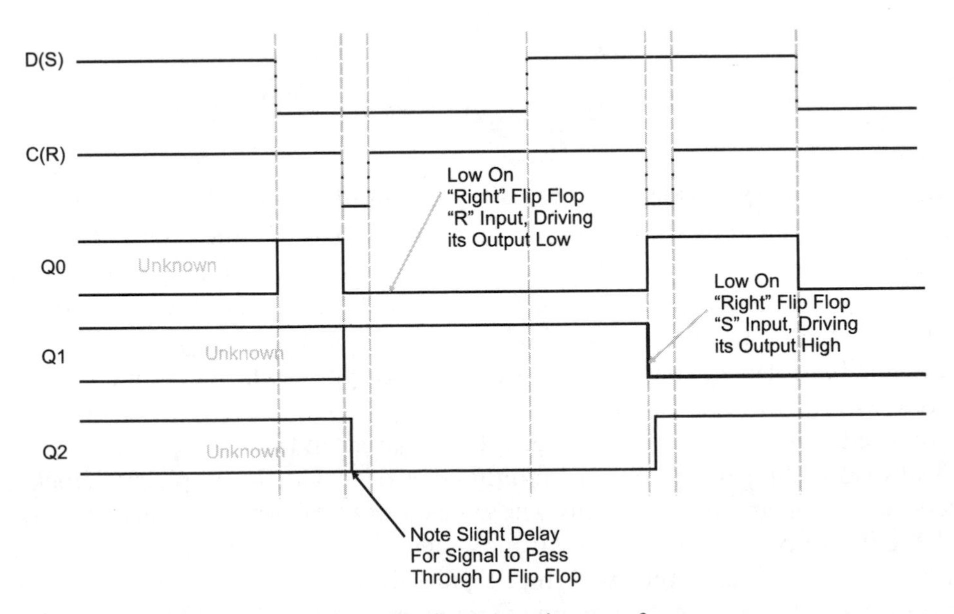

Fig. 7-16. D flip flop operating waveform.

Note that in Fig. 7-16, I have marked the flip flop states *before* the first clock pulse as being "unknown" (in Fig. 7-14, the initial state was assumed). This is actually a very important point and one that you will have to keep in mind when you are designing your own circuits. You cannot expect a flip flop to be at a specific state unless it is set there by some kind of "reset" circuit (which is discussed in the next section). The output of the edge triggered flip flop stays "unknown" until some value is written in it. If you look at the signals being passed to the right flip flop (Output Q0), you will see that the inputs are unknown until the "data" line becomes low, at which point the two inputs to the right flip flop become high and the "unknown" bit value is stored properly in the flip flop.

The first value written into the D flip flop is "zero", the "data" line's value for the write is changed before the "clock" line goes negative. When the "clock" line goes low, it forces out a "1" to be passed to the "right" flip flop, keeping it in its current state. The operation of the edge triggered flip flop should become very obvious if you were to build it (it would require two 74C00s).

I find the D flip flop to be the flip flop that I build into my circuits most often. It is simple to work with and can interface to microcontrollers and microprocessors very easily. It is, however, quite awkward to wire, especially when you want to work with the "full circuit", which is shown in Fig. 7-17. This circuit not only has data stored on the rising edge of the clock line but also two other lines "_Clr" and "_Pre" will force the flip flop's output to a "0" (low voltage) or a "1" (high voltage), respectively, when they are pulled low. This allows for a number of different options for using the D flip flop in your circuit that can allow you to pull off some amazing feats of digital logic.

If you want to experiment with this circuit using two input NANDs (74C00s), I must warn you that it will be quite difficult and complex for you to wire. If you were to use three gates to produce one three input NAND gate, 18 NAND gates would be required to implement the full D flip flop function, which would require four and a half 7400 chips. To demonstrate the operation of the circuit, you could build it out of two 7410 (three, three input NAND gates) or be lazy like I am and just use one 74LS74 (Fig. 7-18) to experiment with the different functions of the full D flip flop.

The 7474 chip consists of two D flip flops with both the "Q" and "_Q" outputs passed to the chip pins. All four inputs shown in Fig. 7-18 (Data and Clock as well as two pins that provide you with the ability to set or reset the state of the flip flop without the use of the data or clock pins) are provided for each of the two flip flops built into the chip. The 7474 is a very versatile chip and can be used for a wide range of applications.

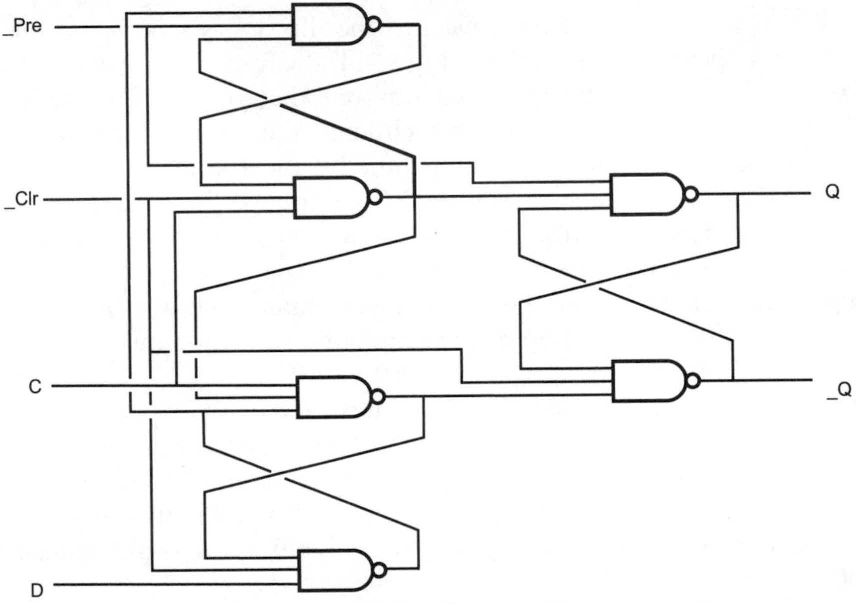

Fig. 7-17. Logic diagram of full D flip flop with negative active preset and negative active clear pins.

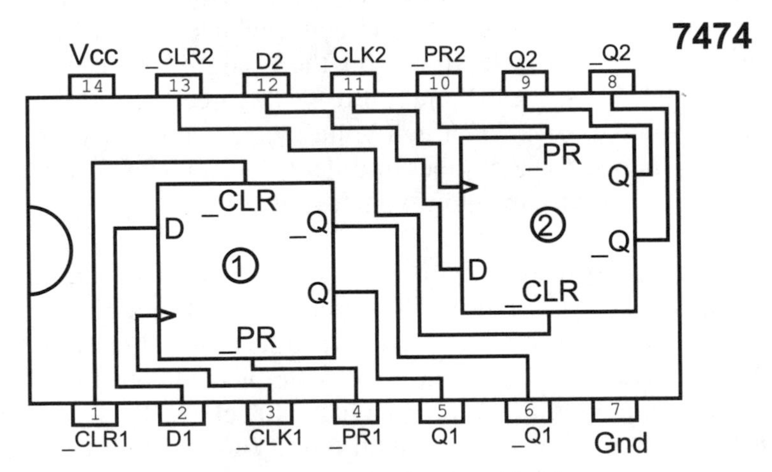

Fig. 7-18. 7474 dual D flip flop chip pinout.

Latches Versus Registers

Two terms that are often used interchangeably are "register" and "latch". In the previous section, I introduced you to the "register", which is another term for an edge triggered flip flop. When you look at parts lists and datasheets, you will see parts that are identified as "registers" and others as "latches" and these parts will have the same pinouts with no obvious differentiation in operation between the devices. Furthermore, I have found many chip manufacturers that have labeled their parts as "latches" when in fact they were "registers" and vice versa.

Quite simply put, "registers" are flip flops that store data when the rising (low to high or 0 to 1) or falling (high to low or 1 to 0) edge (whichever is used by the device) is received on the "clock" (or, my abbreviation, "Clk") pin. Registers are aptly named because they are normally used as simple data storage devices for microprocessor memory. Latches are often used in microprocessor applications to save an address on a multi-purpose bus.

The best analogy for the "latch" that I can think of is a latch on a barn door: when the latch is not engaged, animals and whatever can wander in. Once the latch is closed, what is in the barn stays in. The "latch" flip flop works similarly to this; with one state of the clock line, the input data is passed to the output directly and can be changed at any time (i.e. there isn't any storage) but once the clock line changes state, the last value of the data is stored in the latch until the clock changes value.

In the previous section, I introduced you to the edge triggered D flip flop "registers". The D flip flop "latch" is actually quite a bit simpler (Fig. 7-19), but what is interesting about it is that it doesn't work anything like its edge triggered cousin. In Fig. 7-20, I have drawn a data input along with a clock and the "Q" (output pin) values for an edge-triggered D flip flop register and a D flip flop latch.

You will probably be surprised to see that waveforms for the two memory devices are completely different. The edge triggered D flip flop register stores data in a very consistent and logical way – every time the clock pin

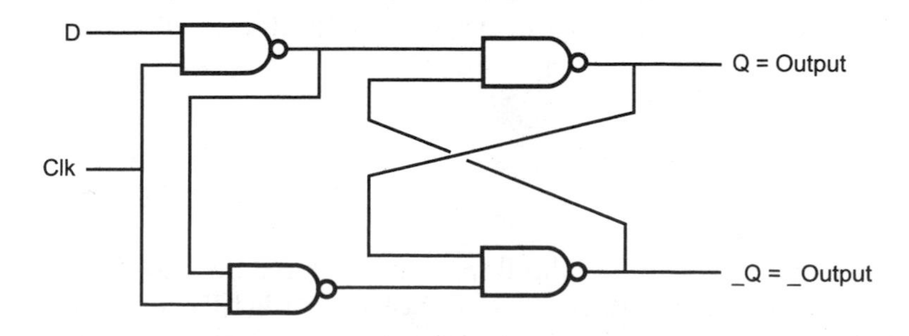

Fig. 7-19. D flip flop latch.

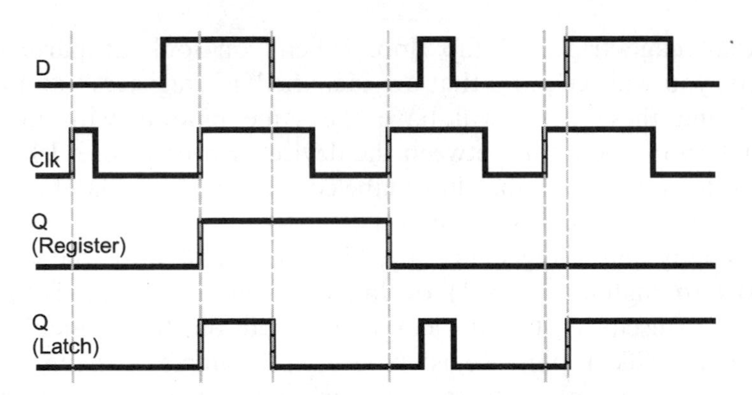

Fig. 7-20. D flip flop latch operating wave form.

rises, the value of "D" is stored in the flip flop and nothing changes until the next rising edge of the clock pin.

The latch, on the other hand, seems to operate more like an AND gate than a memory storage device. The storage function tends to be obfuscated in the example of Fig. 7-20 because in many cases, I show the D pin changing state before the clock line returns low. This is an important point because many people consider the two devices to be interchangeable and this is simply not the case. Latches and registers have different applications and it is critical for you to understand what they are. You cannot put a latch chip in place of a register simply because they are pin compatible; you must make sure that the incoming data does not change state until the clock goes low.

Interestingly enough, latches do not need as much time to save data as a register; there are 9 fewer gate levels for a signal to pass through and even though I show the data save operations being instantaneous in Fig. 7-20, they are not. The latch can take as little as one-third of the time to save data as a register and only requires two gate delays before passing the data along (after which the data can be stored). This makes the latch an important chip for working with microprocessors with a "multiplexed" address bus.

Reset

If you cycle the power to any flip flop, you will have noticed that the initial "state" (or value) can be either "0" (LED off) or "1" (LED on), with no way of predicting which value it will be. This is normal because when power is applied to the flip flop it will start executing in the metastable state, and for any kind of imbalance in the circuit (e.g. residual charge or induced voltage) on the inputs of either NAND/NOR gate, the flip flop will respond and this will be its initial state. Often, this random initial state is not desired – instead, the circuitry should power up into a specific known state for it to work properly. This is why throughout this chapter I have taken pains to note that the initial state of a flip flop is not known. You may find that an application with one flip flop usually powers up the same way; if you were to do a statistical analysis of the power up values, you might even find that a single power up state approaches 100%, but you cannot guarantee this for all occurrences of the chip, or even that all other similar chips in the same application circuit will power up the same way.

Specifying the state when the circuit is powered up is known as "initialization" (just as it is for programming) and is required for more than just sequential logic circuits. Initialization normally takes place when the application is "reset", or waiting to start executing. To avoid confusion later, I should point out there are two types of "reset" described in this book when I talk about digital circuits. Earlier, when I was talking about simple combinatorial circuits, I also called a "low" or "0" voltage level "reset" (and "high" or "1" as "set"). Now, when the term "reset" is used, I am describing the state when the circuit is first powered up or stopped to restart it from the beginning. When you read the term "reset" later in the book (as well as in other books), remember that if a single bit or pin is being described, the term "reset" means that it is "0" or at a low level. If a sequential circuit (like a microcontroller) is "held reset" or "powering up from reset", I mean that it is being allowed to execute from a known state.

The "_CLR" pin on the full D flip flop (like the 7474) is known as a "negative active control" and is active when the input is at a "0" logic level. To make this pin active during power up, yet allow the chip to function normally, a resistor/capacitor network on the TTL input pin delays the rise of the pin (as shown in Fig. 7-21) so that the pin is active low while power is good. When the signal on "_CLR" goes high, and the clear function is no longer active, the chip can operate normally, with it being in an initial known state.

The time for the RC (resistor/capacitor) network to reach the threshold voltage can be approximated using the equation:

$$\text{Delay time} = 2.2 \times R \times C$$

When you work with microprocessors and microcontrollers, you will want to implement a more sophisticated reset circuit. Many microprocessor manufacturers recommend an analog comparator based reset circuit like the one shown in Fig. 7-22. This circuit controls an open collector (or open drain)

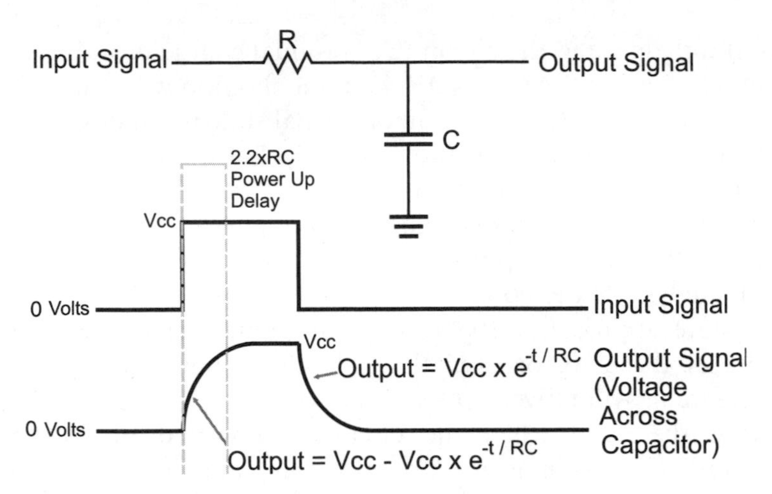

Fig. 7-21. RC network operation.

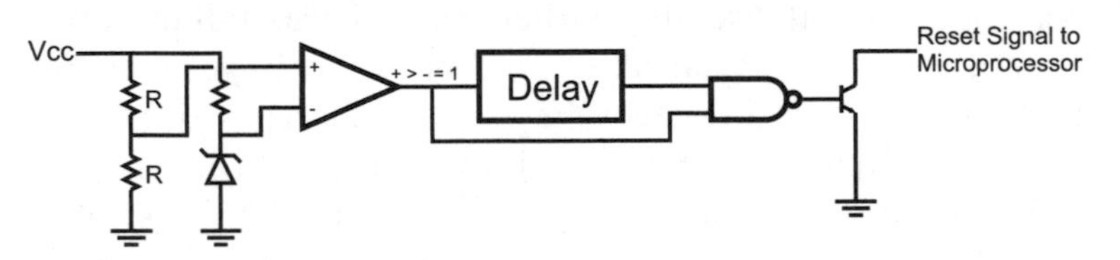

Fig. 7-22. Block diagram of commercial reset control chip.

transistor output pin that will pull down a negative active reset pin when power dips below some threshold value. This circuit is often available as a "processor reset control" chip and is put into the same black plastic package as a small transistor (known as a TO-92).

Processor reset control chips are available for a very wide variety of different "cut off" voltages, ranging from 2.2 volts and upwards. Figure 7-23 shows the operation of the internal parts of the processor reset control chip when the input voltage drops below the set value; the comparator stops outputting a "1" and a delay line is activated. This delay line is used to filter out any subsequent "glitches" in the power line and makes sure that the power line is stable before allowing the processor to return from reset and continue executing. When the comparator outputs a low value or the delay line is continuing to output a low value, the output of the NAND gate they are connected to is high and it turns on the open collector output transistor, pulling the circuit to ground.

The Panasonic MN1381 line of chips is a very popular processor reset control and can be used to control a sequential circuits reset using

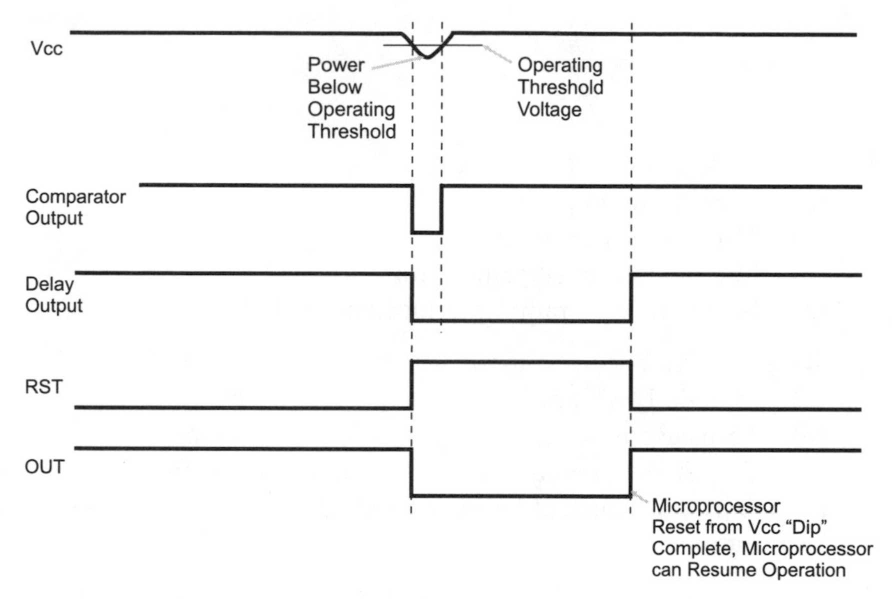

Fig. 7-23. Reset circuit operation.

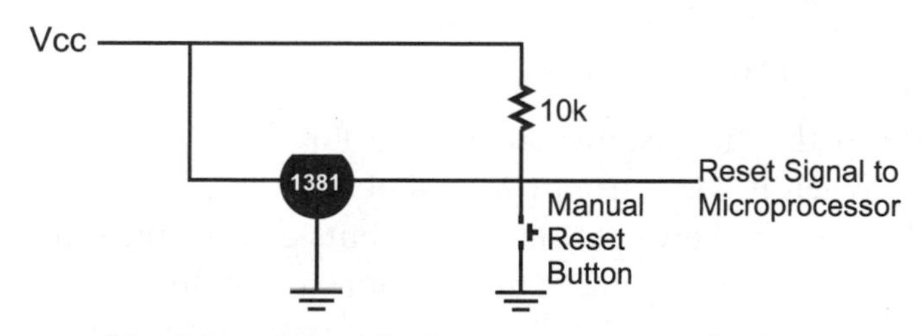

Fig. 7-24. Practical microprocessor reset circuit.

a circuit similar to the one shown in Fig. 7-24. This circuit will take advantage of the RC network delaying the rise of the control signal, provide you with the ability to reset or stop the operation of the microprocessor and halt the operation of the robot if the battery falls below a safe minimum.

If you power on and off a circuit quickly, you may find that it does not power up properly. This is due to the capacitor in the reset circuit not discharging fully – it may take as much as 10 seconds for it to discharge completely. This was actually an issue with the original IBM PC; if you had a situation where the PC "hung", you would have to power down and wait at least 15 seconds to make sure that the reset circuit would allow the computer to power up properly.

Quiz

1. Feedback in digital electronics:
   (a) Is built into every gate
   (b) Must always be avoided
   (c) Can be used to store bit data
   (d) Is only used in radio interface circuitry
2. Ring oscillators can be used:
   (a) In digital watches
   (b) To measure the gate delay of a logic technology
   (c) To test the operation of a combinatorial circuit
   (d) Only when current limiting resistors are in place to protect gate outputs
3. What do the letters "R" and "S" stand for in the RS flip flop?
   (a) "Recessive" and "Static"
   (b) "Reset" and "Set"
   (c) "Rothchild" and "Stanislav"
   (d) "Receive" and "Send"
4. What is the "metastable state" of a flip flop?
   (a) When it has started to oscillate
   (b) The time between when the inputs change the output is correct
   (c) The state in which the outputs of a flip flop are half way between "0" and "1" and can be easily "pushed" into a specific state
   (d) The state in which "Q_0" is unknown
5. "Toggling" a bit means:
   (a) Setting (making the output a 1) of a bit
   (b) Leaving the bit in its current state
   (c) Inverting the bit's state
   (d) Resetting (making the output a 0) of a bit
6. A "Register" can be used in:
   (a) Nowhere, it is a thought experiment used to show feedback in a digital application
   (b) Just computer processors
   (c) Just sequential digital electronics application
   (d) Just about any digital electronics application

7. The "_Pre" pin of a D flip flop will:
   (a) Set the bit
   (b) Reset the bit
   (c) Nothing
   (d) Toggle the state of the bit
8. Which formula specifies the RC network response to a sudden voltage input?
   (a) $V = 2.2 \times R \times C$
   (b) $V = Vcc - Vcc \times e^{-t/RC}$
   (c) $V = Vcc \times e^{-t/RC}$
   (d) $V = i \times R$
9. Why are latches like barn doors?
   (a) They provide a secure environment for what's inside them
   (b) They are both relatively heavy
   (c) They allow free passage until the latch is engaged
   (d) They are the fastest method for passing things in and out
10. Which application is a latch best suited for?
    (a) Main memory in a computer system
    (b) Bicycle lock combinations
    (c) Stopping and saving data mid-stream
    (d) Temporary storage of data in a microprocessor

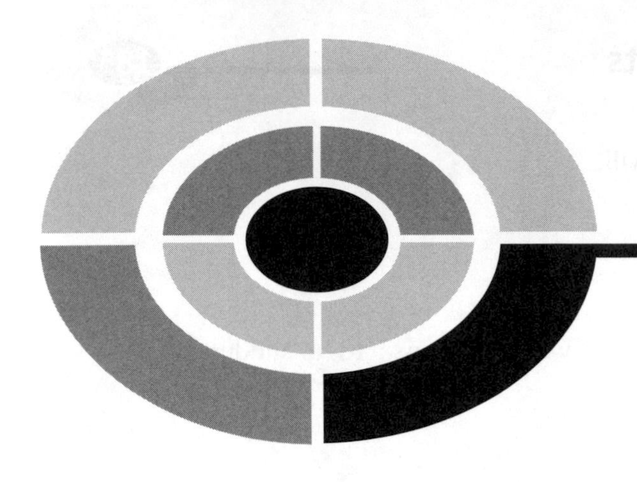

Test for Part One

Do not refer to the text when taking this test. You may draw diagrams or use a calculator if necessary. A good score is at least 38 correct answers. Answers are in the back of the book. It's best to have a friend check your score the first time so you won't memorize the answers if you want to take the test again.

1. The assertion "John is going to go out with the boys tonight or date Mary" is an example of:
   (a) Negative logic
   (b) The AND operation
   (c) The inclusive OR operation
   (d) The exclusive OR operation

2. Which symbol does not represent AND, OR or NOT?
   (a) "*"
   (b) %
   (c) "+"
   (d) "!"

3. How would you create a three input NOR gate from two input NOR gates?
   (a) Invert each input by passing them to a two input NOR gate and then combining it like the three input AND gate
   (b) Pass two inputs to a NOR gate and pass this input to a second NOR gate along with the remaining input and invert the final result
   (c) Pass two inputs to a NOR gate, use a second NOR gate to invert this NOR gate's output and pass this result, along with the third input to a third NOR gate
   (d) Pass two inputs to a NOR gate and pass this input to a second NOR gate along with the remaining input

4. "Product of sums" combinatorial logic circuits are not as common as "sum of products" because:
   (a) They rely on "negative logic", which makes their operation more difficult to understand by simply looking at the circuit
   (b) They are not as fast as product of sums combinatorial circuits
   (c) Automated design tools are typically not programmed to work with product of sums circuits
   (d) Product of sums combinatorial logic circuits cannot produce the same functions as sum of product combinatorial logic circuits

5. Which one of the following statements is false?
   (a) Combinatorial logic circuits are drawn with inputs entering the gates from the left and exiting from the right
   (b) Outputs from some of the gates in the combinatorial are passed back so that they are part of their own inputs
   (c) Combinatorial logic circuits can be designed to have true or false outputs for given inputs
   (d) The function of the individual gates in a combinatorial logic circuit does not change, even if the gates are used to provide a function which is radically different

6. Idealized waveform diagrams do not show:
   (a) Potential "glitches" caused by gates changing state
   (b) Delays in gates, responding to changes in inputs
   (c) What happens with wiring problems such as when multiple outputs are connected to the same input
   (d) All of the above

7. If an application has a critical speed requirement, you should design your circuit:
   (a) To be as simple as possible, as this will minimize the delay a signal has passing through the circuit
   (b) With as few gate delays as possible, while keeping an eye on the number of gates required as well as whether or not it can be efficiently implemented in the technology that you are using
   (c) Using the fastest technology available
   (d) Using computerized design systems

8. The NOR equivalent to an AND gate is:
   (a) Built from two NOR gates and requires two gate delays for a signal to pass through
   (b) Built from three NOR gates and requires three gate delays for a signal to pass through
   (c) Built from three NOR gates and requires two gate delays for a signal to pass through
   (d) Built from one NOR gate as well as a NOT gate and requires two gate delays for a signal to pass through

9. When circling "1" outputs in a Karnaugh map:
   (a) A maximum of two bits can only be circled at one time
   (b) No bits can be circled more than once
   (c) Each circle should be around a power of two number of bits
   (d) Single bits on one side of a the map cannot be circled with bits on the other side

10. The four bit Karnaugh map

	AB			
	00	01	11	10
CD – 00	0	0	1	0
01	1	1	1	1
11	0	1	1	0
10	1	0	1	1

has the optimized sum of product equation:
   (a) Output = (!B · !D) + (!A · !B) + (C · !D) + (B · !C · D)
   (b) Output = (A · B) + (!C · D) + (B · D) + (!B · C · !D)

(c) Output = $(!C \cdot !D) + (A \cdot B) + (B \cdot D) + (!B \cdot C \cdot !D)$
(d) Output = $(A \cdot B) + (!C \cdot D) + (B \cdot !D) + (!B \cdot C \cdot !D)$

11. Benjamin Franklin postulated:
    (a) Electricity flows from positive to negative
    (b) Lightning is dangerous
    (c) Keys had to be charged before they would open doors
    (d) Thomas Edison unfairly copied his work

12. The term "net" is used for:
    (a) Wires in a circuit and lines on a circuit diagram
    (b) Indicating the active signal lines of an ethernet cable
    (c) A search tool through which information is passed through and relevant "hits" sticks to
    (d) Nylon webbing used to protect a circuit against falling metal components.

13. Knowing Ohm's law and the resistance of a load and the voltage of a battery powering it, you can determine:
    (a) The current passing through it
    (b) The amount of water coming through an analogous pipe/tap/hose
    (c) Its equivalent parallel resistance
    (d) The Thevenin equivalent circuit

14. Using the SI numbering methodology and symbols, 10,000,000 volts would be written out as:
    (a) 10 million V
    (b) 10 MV
    (c) 10 μV
    (d) 10,000,000 V

15. The voltage drop across a resistor in a series circuit:
    (a) Cannot be calculated
    (b) Is proportional to the power dissipated in the circuit
    (c) Is always zero
    (d) Is proportional to the resistor's value relative to the total resistance in the circuit multiplied by the applied voltage

16. If a 10 ohm and 20 ohm resistor are in series, the equivalent resistance:
    (a) Cannot be calculated without knowing the voltage applied
    (b) 30 ohms
    (c) 7.5 ohms
    (d) 6.7 ohms

17. A 0.01 μF capacitor is most often used in:
    (a) Radio applications; it has no use in digital electronics
    (b) Decoupling digital electronic chips
    (c) Filtering power supply "noise"
    (d) Ballast in fluorescent lighting

18. The value at a given time for the capacitor voltage in a resistor–capacitor low-pass filter circuit responding to a rising step input is:
    (a) Infinite
    (b) Defined by the formula $V(t) = V - V \times e^{-t/\tau}$
    (c) Defined by the formula $V(t) = V \times e^{-t/\tau}$
    (d) Zero; the capacitor has no voltage drop across it

19. The NOR gate was chosen as the basic CMOS logic gate because:
    (a) It can be built most efficiently using MOSFET transistors
    (b) It provides the fastest logic functions in CMOS logic
    (c) It helps the circuit designer differentiate the functions provided by TTL and CMOS logic circuitry
    (d) The NOR gate minimizes the power lost in the chip

20. Which statement is not a reason cited for using resistor pull ups and resistor/NOT gates for pull downs?
    (a) The resistor can be connected to negative voltage without damaging the circuit
    (b) Test equipment can easily change the state of logic pin inputs
    (c) TTL and CMOS logic operate optimally with these circuits
    (d) The resistor pull ups and resistor/NOT gate pull downs will work for both TTL and CMOS logic

21. If a silicon diode was passing 2 A of current, it would be dissipating:
    (a) 14 watts of power
    (b) 0.2 watts of power
    (c) 0 watts of power
    (d) 1.4 watts of power

22. In a 5 volt powered circuit, you have two LEDs in series and want to pass approximately 5 mA through them. What is the best current limiting resistor value should you use?
    (a) 47 ohms
    (b) 100 k ohms
    (c) 5 ohms
    (d) 1 k ohms

23. A bipolar transistor is best suited for:
    (a) Radios and high-fidelity sound systems
    (b) Small, high-density chips
    (c) Memory circuits
    (d) Low-power, high-density chips

24. The basic CMOS logic gate is:
    (a) The NOT gate
    (b) The AND gate
    (c) The NOR gate
    (d) The NAND gate

25. TTL is:
    (a) Sound controlled
    (b) Resistor controlled
    (c) Current controlled
    (d) Voltage controlled

26. When a TTL input is low:
    (a) Current is being drawn from it
    (b) A low voltage is being applied to it
    (c) A "0" is being passed to it
    (d) Electrons are being drawn from the emitter of the input gate's NPN transistor

27. TTL/CMOS logic outputs:
    (a) Can be used to drive neon lamps
    (b) Can source/sink roughly 20 mA
    (c) Cannot be used with different technology inputs
    (d) Are limited to driving inputs less than 20 m away

28. "Fanout" is the term applied to:
    (a) The number of outputs that can be driven by one input
    (b) The number of fans required to cool a set number of chips
    (c) The speed a signal travels through multiple paths of a logic chain
    (d) The number of inputs that can be driven by one output

29. CMOS logic has the following characteristics:
    (a) They are low speed, low power
    (b) Require just about no power, regardless of the speed they operate at

(c) The current required is a function of the speed of operation
(d) Require less power than TTL because MOSFETs cannot be packed as tightly as bipolar transistors

30. LEDs are used in beginner digital electronic circuits:
(a) To indicate analog voltage levels
(b) To indicate a part is overheating
(c) To indicate input and output binary values
(d) To communicate with other circuits

31. The difference between 74Cxx and 74xx chips is:
(a) The 74Cxx is built from CMOS logic while the 74xx is TTL
(b) Signals in the 74Cxx propagate at the speed of light (as indicated by the "C" in the part number)
(c) The 74xx can work from 5 to 9 volts while the 74Cxx can only work with 5 volts
(d) The 74Cxx is built with a "compacted" chip

32. Gray codes were invented:
(a) To make your life miserable
(b) For simplifying Boolean logic statements
(c) For simplifying the task of determining the position of a device
(d) As a method of counting that was faster than binary

33. Adding 6 to 5 and getting the result 11 is the same as:
(a) Adding 7 to 4 and getting the result 11
(b) Adding 3 to 4 and getting the result 7 because in both cases, a prime number is produced
(c) Adding 5 to 6 using the commutative law and getting the result 11
(d) Adding 6 to 5, writing down "1" and then "10 × 1" because a carry digit is produced

34. The term "ripple" as applied to addition and subtraction is:
(a) The carry and borrow bits
(b) The result of the two single digit operation passed to the next significant digit
(c) The affect the operation has on its surrounding digits
(d) The oscillations caused by the need to carry and borrow data

35. Using the negated addition for subtraction, the borrow (negated carry) bit for the operation 5 – 6 is:
(a) Not required
(b) 1

(c) 0
(d) Indeterminate

36. A small circle on a gate's input indicates:
(a) That the signal can only be used for output.
(b) That the signal is inverted before being passed to the gate
(c) Only open collector drivers can be used with this input
(d) The I/O can be used for monitoring the passage of the signal output in the gate

37. Magnitude comparators are based on:
(a) Three initial input state values
(b) Two four bit inputs
(c) Two subtracters
(d) One subtracter and one adder

38. Cascading chips is usually required because:
(a) Faster speed is required than a single chip can provide
(b) More bits must be processed than a single chip can handle
(c) The only chips that can provide all the necessary function require too much power
(d) It minimizes the cost of a circuit

39. Dividing a binary number by 8 can be accomplished by:
(a) Clearing the least significant three bits
(b) Shifting left three bits
(c) Shifting right three bits
(d) Setting the least significant three bits

40. Mickey Mouse logic solutions should be placed in the circuit:
(a) In the middle of a logic string
(b) On the inputs of a logic string
(c) On the outputs of a logic string
(d) Where high-current I/O is required

41. The resistor used in the Mickey Mouse logic AND gate shown in Fig. Test 1-1 should be:
(a) 10 k for TTL applications
(b) 10 k for CMOS applications
(c) The complementary one specified by the diode's manufacturer
(d) Power rated for the load current of the application

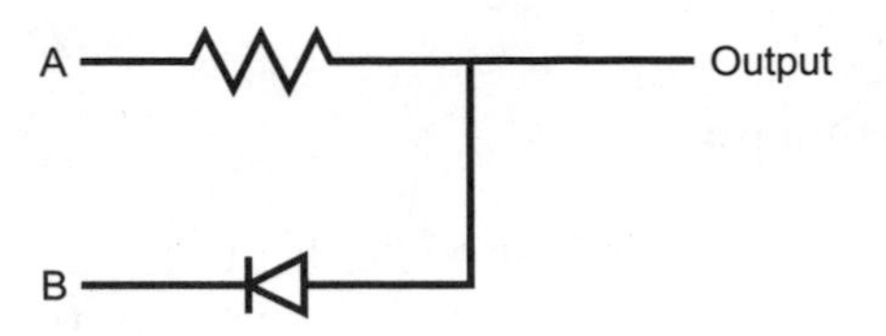

Fig. Test 1-1.

42. Each item below is a disadvantage of a dotted AND bus except:
    (a) High power consumption when the output is low
    (b) The dotted AND bus has a slower response than tri-state buffers
    (c) The dotted AND bus is cheaper than one manufactured with tri-state drivers
    (d) It is very difficult to find open collector output chips

43. When providing multiple functions to a net, what logic technology/technologies should be used?
    (a) Just CMOS
    (b) Just TTL
    (c) CMOS receivers and TTL drivers
    (d) TTL receivers and CMOS drivers

44. Sequential circuits contain:
    (a) Memory devices
    (b) Power supplies
    (c) Input and output devices
    (d) CMOS logic

45. Backdriving gates can:
    (a) Simplify your application design
    (b) Speed up gate operation
    (c) Change the input of a downstream device
    (d) Burn out the gate's output transistors

46. What is the difference between "Q_0" and "_Q_0"?
    (a) There is no difference
    (b) "Q_0" is correct earlier than "_Q_0"
    (c) "Q_0" is current state of the flip flop and "_Q_0" is the previous
    (d) "_Q_0" is the inverted value of "Q_0"

47. The "_Clr" pin of a D flip flop will:
    (a) Set the bit
    (b) Reset the bit
    (c) Nothing

(d) Toggle the state of the bit

48. Which full D flip flop input pin is typically connected to the RC delay circuitry?
    (a) D
    (b) Clk
    (c) _Clr
    (d) _Pre

49. Which application is a register best suited for?
    (a) Main memory in a computer system
    (b) Permanently storing access passwords
    (c) LED output states
    (d) Temporary storage of data in a microprocessor

50. "Volatile memory" means:
    (a) The contents of the memory device will not be lost when power is taken away
    (b) The contents of the memory device will be lost when power is taken away
    (c) The memory device is made up of a liquid which will evaporate if the chip package is broken
    (d) Data is stored as patterns of a condensed gas

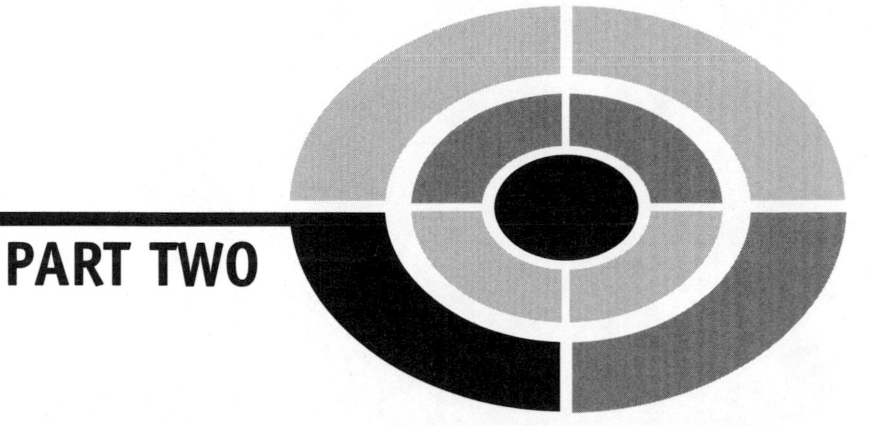

PART TWO

Digital Electronics Applications

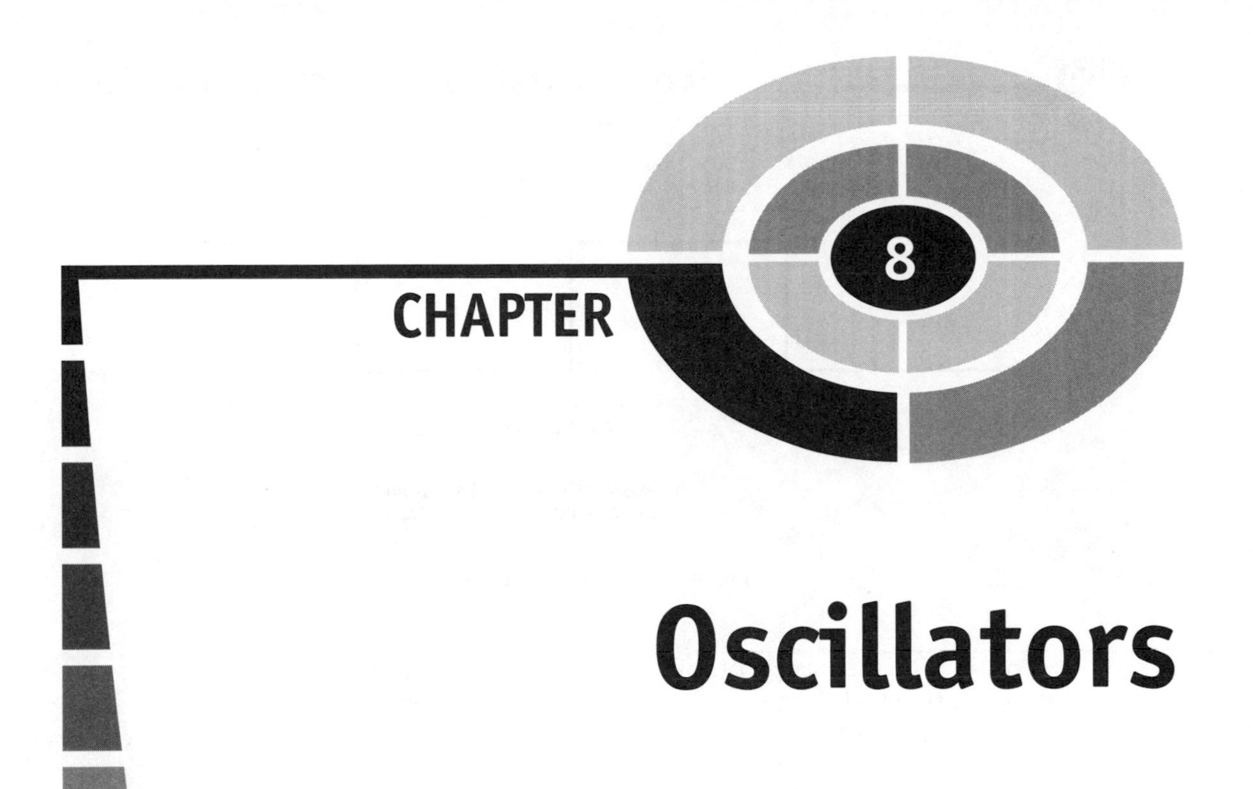

Oscillators

In the introduction to the previous chapter of this book, I presented you with a simple block diagram of a "complete" digital electronic device: the digital clock. This device has all three of the necessary components for a functional, "stand alone" device: combinatorial logic for converting binary data as required; a memory function which "remembers" the last state it was in; and a clock or "oscillator" which synchronizes the functions together. The science of oscillator design is extremely rich and, as I will show in this chapter, there are a lot of options that you can choose from to make sure your application operates most efficiently.

It could also be argued that there is a fourth component to producing a complete, stand alone digital device – the power supply. Power supply design is a facet of electronics which is just as rich and sophisticated as digital electronics or any other major study in electronics. While I introduce you to some of the basic types of power supplies that are available to you later in the book, this does little more than just scratch the surface of this complex topic.

The application's "clock" is a set of repeating pulses (ideally with the same "on" and "off" time) which is input into the sequential circuits of an application to carry out the operations within them. Figure 8-1 shows an ideal digital electronic clock waveform with the important features marked

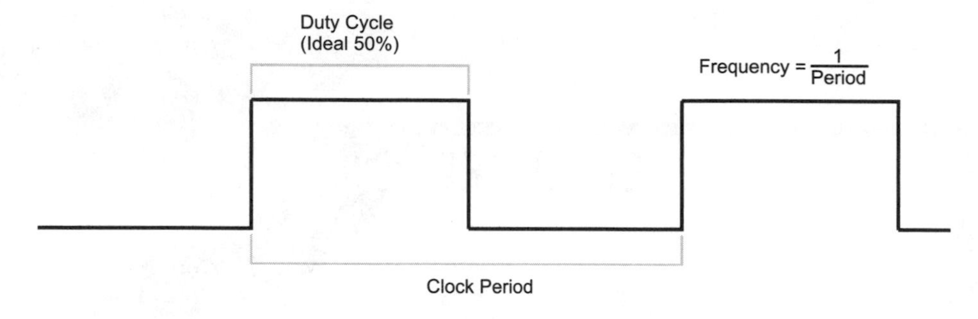

Fig. 8-1. Features of a clocking signal.

on it. The frequency output of a clock is measured in "hertz" (cycles per second), which is the reciprocal of the time the pulse is on added to the time the pulse is off:

$$\texttt{Frequency} = 1/(\texttt{Time "On"} + \texttt{Time "Off"})$$

A rather obvious example of this would be a 1 pulse per second signal that is used to drive the sequential circuits of a timekeeping clock. The term "clock" is probably confusing, as the digital "clock" that I have defined is only very rarely used to tell the time. The clock signal that I am discussing in this section is used to drive the digital counters of the timepiece.

The "clock" in a digital circuit is driven from an "oscillator" that uses some form of feedback to toggle the clock line in a consistent manner. In this book, you will find that I use the terms "clock" and "oscillator" interchangeably, with the clock or oscillator signal being responsible for the operation of the digital circuit.

In Fig. 8-1, I noted the important features of a clock signal to be its constant period as well as its 50% duty cycle. If the clock is not constant and you were to look at the operation of the clock and the digital circuit on an oscilloscope, you would see the signals blur as they "jittered" back and forth. The shortened or lengthened features of a clock waveform are called "jitter" and are shown in Fig. 8-2. Jitter is a problem from a couple of perspectives. First, it makes it very difficult to observe the operation of a circuit using an oscilloscope, making it hard to debug an application. Secondly, the combinatorial logic parts of sequential circuits are often designed to have completed their bit data processing by the time the clock has completed; a shortened clock cycle, like the one shown in Fig. 8-2 could result in incorrect data being stored in a circuit.

Fig. 8-2. Clock waveform with "jitter."

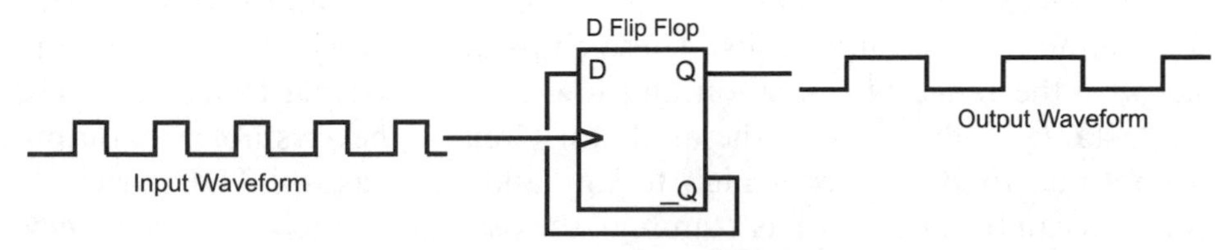

Fig. 8-3. Clock divide by 2 conditioning circuit.

Incorrect logic operation caused by jitter is extremely difficult to capture on an oscilloscope or a logic analyzer. If you have errors and have a clock with jitter, you would be well advised to assume that the jitter is the problem and look towards another clocking scheme. Jitter is often caused by voltage transients caused by different parts of a digital electronic circuit changing state; often it can be reduced or eliminated by providing better noise filtering between the oscillator circuit and the digital logic devices.

In Fig. 8-1, I indicated that the ideal "duty cycle" of a clock signal is 50%. I will discuss duty cycles in more detail later in the book when I present PWMs, but for now, you should understand that the duty cycle is the percentage of the period in which the waveform is high. Ideally, the clock should have a 50% duty cycle (be high for half the period) to minimize harmonics, simplify using an oscilloscope or logic analyzer to observe the operation of the circuit and, because some digital electronic devices (most notably microprocessors) poll both the high and low of the clock signal, to help speed up its operation.

In some oscillator circuit designs, the output does not come close to having a 50% duty cycle and in these cases, some kind of signal "conditioning" is required. The most basic way of ensuring the clock signal has a 50% duty cycle is to use an edge triggered D flip flop as I show in Fig. 8-3. The D flip flop (called a "toggle flip flop" when wired this way) will only change its output when a triggering edge on the input clock has been received. The only drawback to this circuit is that it halves the frequency of the clock, so, in some cases, to use this circuit you will have to double the clock output frequency.

There are many different designs of oscillators that you can choose from. In this chapter I will introduce you to many of the most common ones along with their characteristics and the formulas required to work with them.

Transistor Astable Oscillators

When I was growing up, all educational and hobbyist circuits were built up from individual transistors – it wasn't until the mid to late 1970s that "building block" chips such as the 555, LM339, LM386, and LM741 started to be commonly used in circuits. These chips are all very configurable, but none offer the range of operation and low cost of discrete transistors. The term "astable" indicates that the oscillator circuit is never stable; its output will continue to switch from high to low and back again. The science of oscillators can be thought of as "taming" the oscillator in terms of frequency, duty cycle and jitter.

A very common and relatively simple oscillator circuit that I am going to examine is the basic "relaxation oscillator" circuit shown in Fig. 8-4. Included in Fig. 8-4 are the defining formulas for the time that the output is high and low as well as an important formula indicating that the value of R1 and R2 (the time defining resistors) must be the transistor h_{FE} multiplied by the value of the pull up resistors (Rpu). If R1 or R2 is less than this product, then you will find that the oscillator will not start reliably and not run at a constant frequency.

The operation of the relaxation oscillator is illustrated in Figs. 8.5 through 8.7. In Fig. 8-5, I show an initial condition where one transistor is on and the other is off. In this case, the capacitor by the on transistor is charging because its cathode is being pulled to ground by the "on" transistor. The other capacitor is unable to be charged because the transistor connected to its cathode is off, holding the voltage at the cathode at the same voltage as the anode.

In Fig. 8-6, the capacitor that was charging in Fig. 8-5 has finished and any current passing through the resistor is passed to the other transistor, turning it on. By turning on this transistor, the capacitor's cathode connected to its collector is now tied to ground and it is able to be charged. With this capacitor now charging, the current that was once available to the

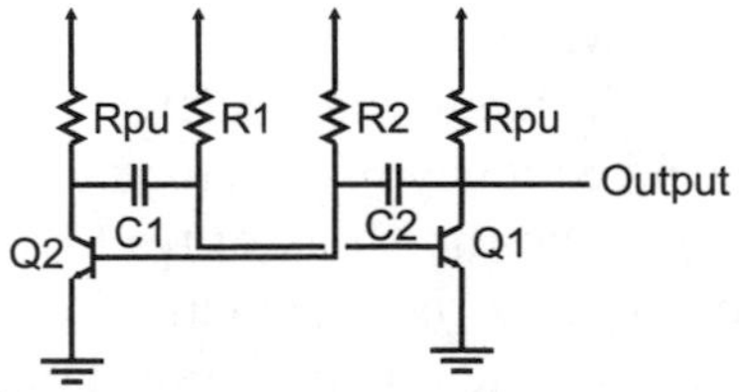

Fig. 8-4. Basic transistor relaxation oscillator.

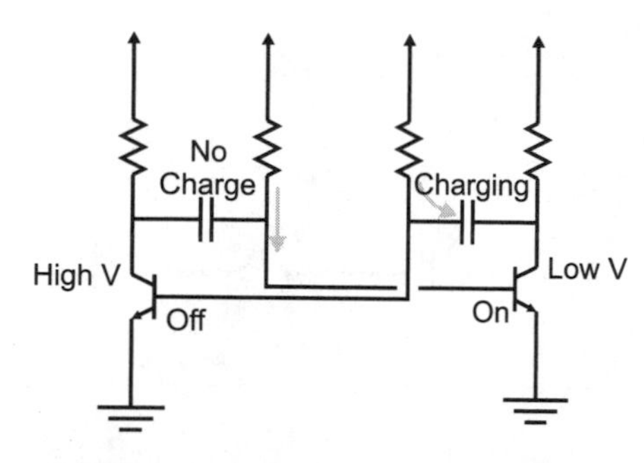

Fig. 8-5. Relaxation oscillator charging right-side capacitor.

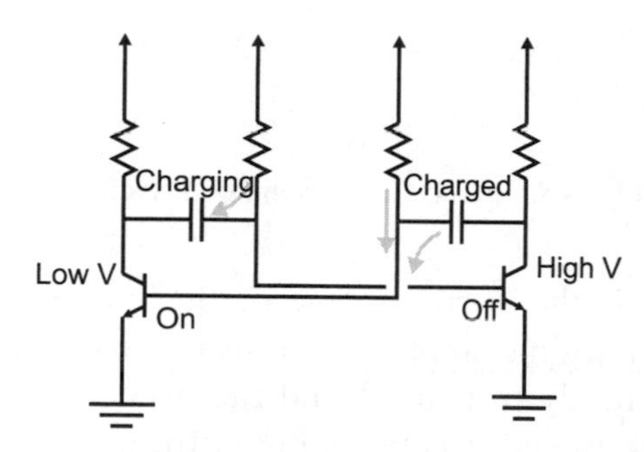

Fig. 8-6. Relaxation oscillator discharging right side.

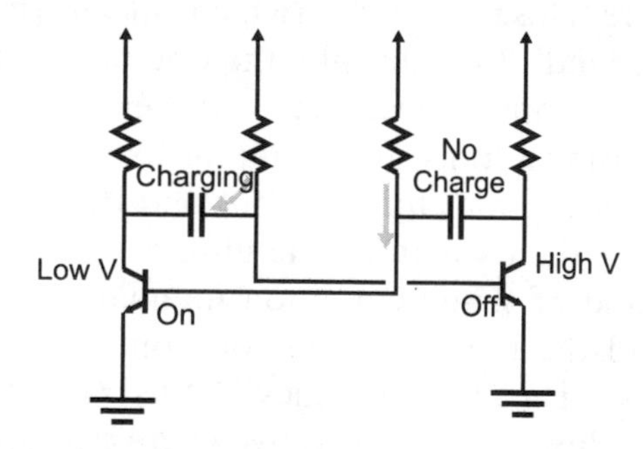

Fig. 8-7. Relaxation oscillator charging left side.

transistor's base to turn it on is no longer available and the transistor turns off. This raises the collector of this transistor to the applied voltage and, along with this, the cathode of the capacitor connected to it. This places the charge in the capacitor tied to the collector of the transistor just turned off at a voltage higher than the applied voltage, so its charge is now passed to the transistor that was just turned on. In the final case (Fig. 8-7), the operation of

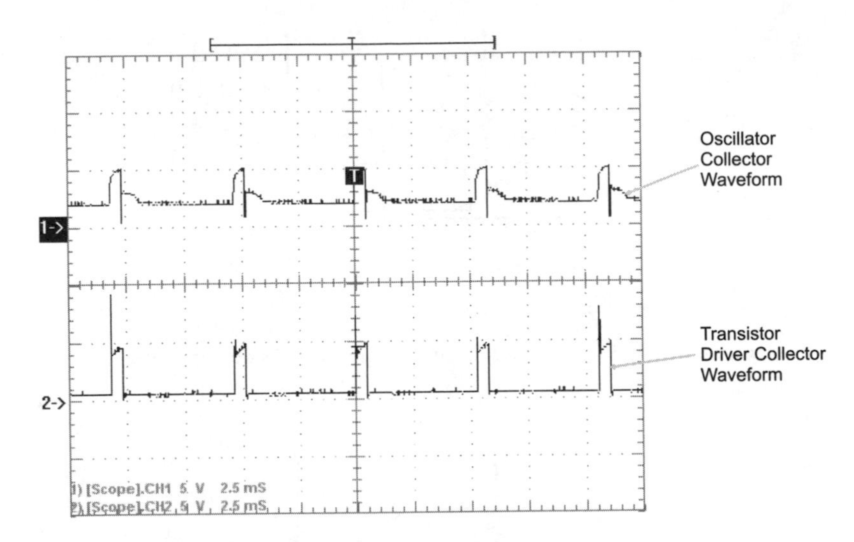

Fig. 8-8. Relaxation oscillator waveforms.

the oscillator circuit is the mirror image of the initial conditions shown in Fig. 8-5. When the charging capacitor is finished, its current is passed to the transistor that is currently turned off and the process repeats itself.

The output of this oscillator is probably nothing that you would expect – Fig. 8-8 is an oscilloscope display of a sample NPN transistor relaxation oscillator output as well as the collector voltage. The duty cycle of the waveform is nowhere close to 50% (which means it will have to be conditioned by some kind of circuit, like the one in Fig. 8-3. However, this circuit is quite good in terms of accuracy, with very little jitter.

The drawbacks to using a transistor oscillator like the one presented in this section include the unusual waveform output and the use of discrete analog components for timing the oscillator. The unusual waveform output makes the need for some kind of signal condition mandatory when working with digital electronics and the use of analog components makes the frequency output quite imprecise. The characteristics of the transistor-based oscillator make it best suited for low-cost applications where clocking accurate to 20% is acceptable.

Ring Oscillators

In the previous chapter, I introduced the concept of "ring oscillators" as being a digital electronic device in which an inverted output signal is fed back to the input of a combinatorial circuit and showed that it could be created inadvertently (Fig. 8-9) or purposely using a single logic inverter (Fig. 8-10). One of the useful characteristics of the ring oscillator is that it will always produce a 50% duty cycle and its output is literally the maximum speed of the technology.

I should say that the ring oscillator's maximum output is the maximum speed of the technology. In Fig. 8-10, I have drawn two ring oscillators, the first outputting the signal from a single inverter – the period of the output of this circuit will be 1 gate delay. In the lower diagram of Fig. 8-10, I show that you are not limited to just running at the technology's maximum speed; by adding an even number of additional inverters to the ring oscillator, the output signal's period can be lengthened.

The ring oscillator's actual frequency output can be "tuned" by varying the number of inverters in the ring oscillator along with the technology used in the inverters. Knowing this, along with a couple of operating rules, provides you with an inexpensive, high-speed oscillator that is quite reliable and robust.

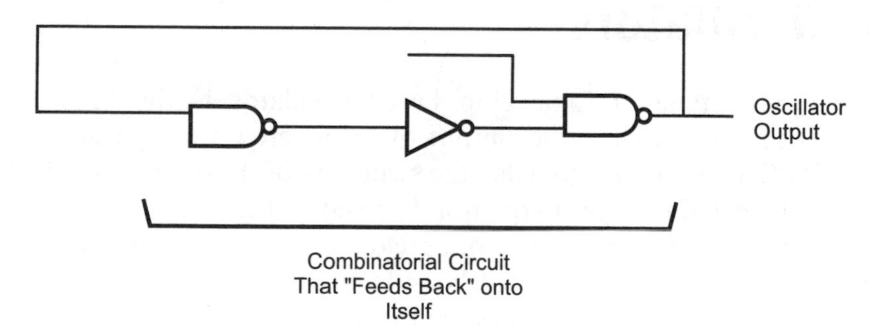

Fig. 8-9. Inadvertent ring oscillator.

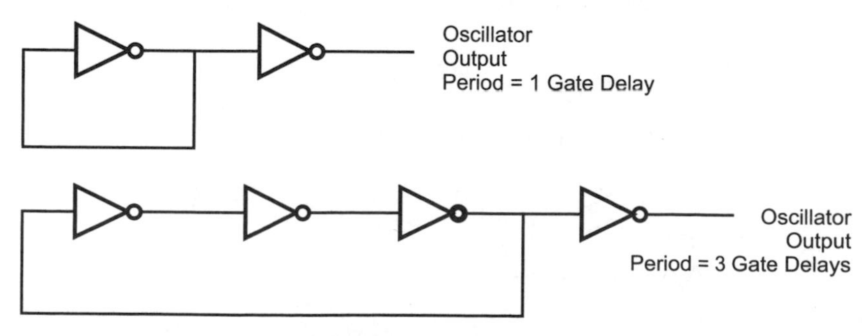

Fig. 8-10. Ring oscillators using gate delays for frequency determination.

The first of the operating rules should not be surprising because I have alluded to it in the text above: the number of inverters in a ring oscillator should always be odd. If an even number of inverters is used in a ring oscillator, there will be no signal which cannot be resolved (which is the cause of the "astable" operation of an oscillator) and the circuit will not oscillate. The second operating rule is that no other functions should be used by the leftover gates in the chip and the chip's power pins should have both small (0.1 μF or less) and large (1.0 μF or greater) decoupling and filter capacitors on its power supply. The oscillating gates within the chip are experiencing significant transients which could affect the operation of other devices in the application.

While I have only used ring oscillators a couple of times over the years, they are fascinating circuits to build and watch executing. The ring oscillator is what I consider to be a "hip pocket" circuit: something to be pulled out only when nothing else seems to work or have the characteristics that you require.

Relaxation Oscillators

The most basic type of logic chip based oscillator is the "relaxation" oscillator which feeds back the output of an inverter through a "resistor/capacitor" ("RC") network to delay the switching of the oscillator. The basic circuit and its defining output equation is given in Fig. 8-11.

In this circuit, the R1 and C network are driven by the first inverter and the characteristic "RC" response is fed back to the first inverter's input. When the voltage on the capacitor reaches the threshold voltage of the left inverter input, the inverter changes state and drives a new output voltage. This voltage is again passed through the R1, C network and delayed until the threshold voltage is reached again.

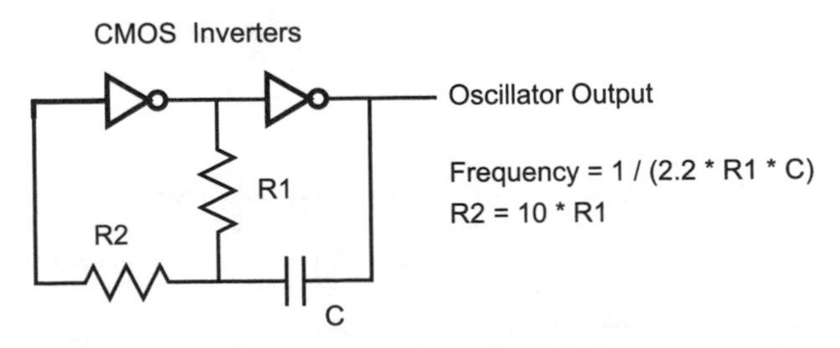

Fig. 8-11. CMOS logic technology relaxation oscillator.

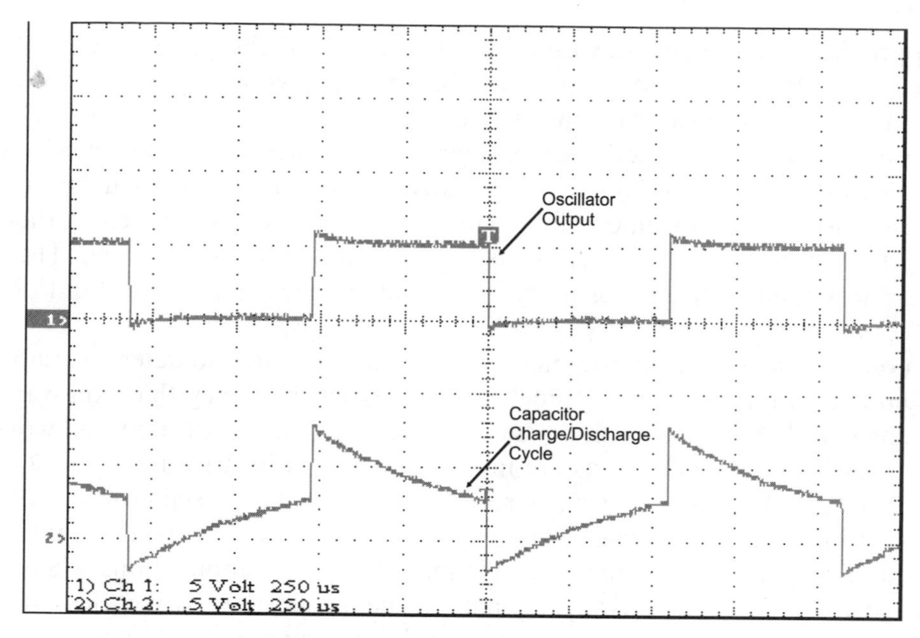

Fig. 8-12. CMOS relaxation oscillator waveform.

In Fig. 8-12, I have shown the voltage waveforms at the R1, R2 and C junction of this circuit as well as the output voltage signal. Note that the R1, R2 and C junction voltage exceeds the Vcc/Gnd (+5 volts and 0.0 volt) limits. This is due to the capacitor being connected to the output driver.

Having the capacitor wired to the output driver "moves" the charge (and the capacitor voltage) by 5 volts each time the state changes. Observing the circuit's operation from the capacitor, the output value changes the charge within the capacitor until it is back at the threshold voltage for the CMOS inverter, which is 2.5 volts (one-half applied power). You can see in Fig. 8-12 that the transitions take place every time the voltage across the capacitor is at 2.5 volts relative to Gnd.

CMOS inverters are used in this circuit because they are voltage controlled rather than current controlled and this makes the oscillator's operation easier to understand. A TTL inverter cannot be used in this circuit because of the current drain operation required by the input when a "0" is input will affect the operation of the oscillator. A Schmidt trigger input device (i.e. the 74HC14) could be used, but it is not necessary because the reference voltage of the capacitor is changing with every transition.

You may want to test out this circuit with a 74C04 or 74HC04 with a 4.7 k resistor, a 47 k resistor and 0.1 μF capacitor to create an oscillator that produces a clock signal of approximately 1 kHz. I say that the output is "approximately" 1 kHz because of the tolerances of the parts used in the circuit. For the circuit used to produce the signal shown in Fig. 8-12, I used a 0.1 μF tantalum capacitor for the "C" in the relaxation oscillator circuit. This is probably not a "correct" use of a tantalum capacitor, as they can have tolerances approaching 30% of their rated value – I only used it because I have a lot of them around. Along with the tolerance of the capacitor, there are also the tolerances of the resistors in the circuit to consider as well. These tolerances result in the opportunity for the actual clock signal to be "out" by 40% or more.

Your immediate response may be to add a potentiometer (variable resistor) into the circuit and "tune" it to the exact frequency that you want. Personally, I would discourage this practice as it involves a lot of work (especially if production parts are involved), which will drive up the cost of the product. If you are using a simple RC relaxation oscillator in your application, then additional costs are something that you would want to avoid. The relaxation oscillator is adequate for many applications where a low-cost oscillator of an approximate value is required. Like the NPN transistor astable oscillator, I recommend that the circuit should not be used in any applications where any kind of precision is required.

Another aspect of this circuit that you must be aware of is the potential for large current transients within the chip that are produced to change and discharge the capacitor. These transients are similar to the transients discussed in the ring oscillator. For most circuits, this is not a problem, but if you have other sensitive circuits built into an application, you will want to keep the relaxation oscillator (as well as any other oscillators in the circuit) as electrically removed as possible from the other chips by using both large and small decoupling and filtering capacitors. Also like in the ring oscillator, as a rule of thumb, no other gates should be used in a chip if it is being used as an oscillator.

Crystals and Ceramic Resonators

For the best clock accuracy, a *quartz crystal* should be used in an oscillator circuit like the one shown in Fig. 8-13. A quartz crystal is a *piezo-electric* device that provides a constant delay between one side of the piece of quartz within the part to the other. The term piezo-electric refers to the property of quartz (and some other compounds) to mechanically deform when a current is applied to it or produce a voltage potential when it is mechanically deformed.

In an oscillator, the quartz crystal will have a voltage applied to one end of it and this will cause the quartz crystal to deform. The rate at which this deformation takes place is known and will cause a voltage potential to be produced at the other end of the quartz crystal after a known delay. This voltage is used as a feedback value to an inverter built into the oscillator circuit. The NPN bipolar transistor-based inverter can be seen in Fig. 8-13.

The circuit in Fig. 8-13 is somewhat "fiddly" to build and to get working reliably. There are some formulas that can be used to specify the different resistor, capacitor and inductor values, but, personally, I would never use this circuit in my own applications. This is why I did not put in any component values on the diagram; instead, I would use the inverter-based oscillator shown in Fig. 8-14. In this circuit, instead of understanding a circuit well enough to specify the correct different analog values, you can simply put a crystal across the input and output of a CMOS inverter.

The capacitors and resistors are necessary to ensure that the oscillator runs reliably and there are not any large over- or under-voltage spikes (caused by the operation of the piezo-electric producing its own voltage output). For most MHz range oscillator circuits, 15–33 pF capacitors are

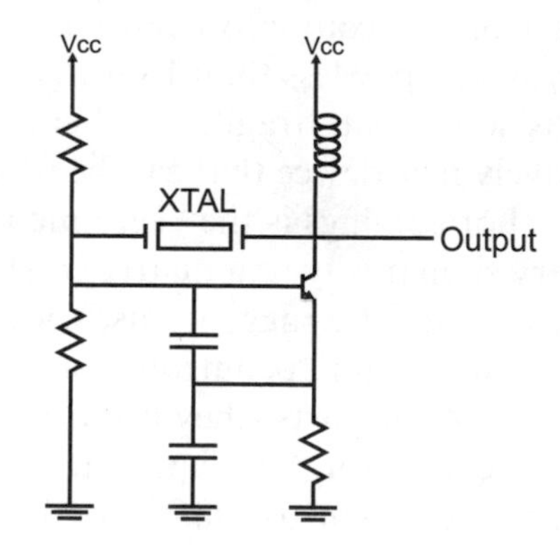

Fig. 8-13. Crystal oscillator circuit.

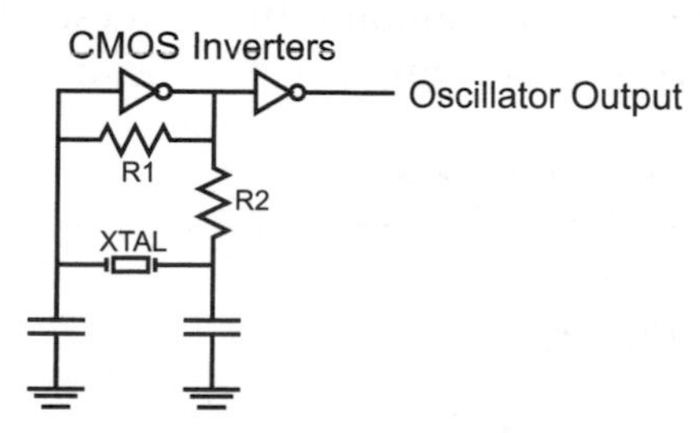

Fig. 8-14. CMOS logic gate crystal oscillator circuit.

adequate, 1–10 MΩ for R1 and 100 to 100 kΩ for R2 is appropriate. You may find that depending on the frequency of the crystal that you choose and how it is wired, you may have to vary these parts.

The second inverter in the circuit is not "strictly" required, but I like to have it in place to ensure the crystal is not "loaded down" by other devices and the operation of the oscillator doesn't change. Any large loads on the output side of the oscillator's primary inverter will affect the amount of current/voltage available to the crystal to pass the signal to the other side (and the oscillator's frequency will drop or the oscillator won't work at all).

Changing the capacitance on the inverter output side of the primary inverter results in small (1–2%) changes to the output and to help ensure the absolutely correct frequency output is produced a variable capacitor is used in place of the fixed capacitors. I do not feel this is practical and the nominal 0.01% or less error rate of the crystal should be accepted. I realize that there are applications (like digital clocks) where these changes are critical, but for the most part you should not have to "tune" the oscillator for the application.

Crystals work quite well, although there are two drawbacks that you should be aware of. Crystals are relatively expensive parts (especially compared to the RC network relaxation oscillator). You can pay up to $10 for a crystal (although you can pay less than 1$ for common frequencies). In addition, the oscillator is somewhat "fragile" and can be easily damaged by rough handling. A relatively new device that can be used in place of a crystal and does not have these shortcomings is the "ceramic resonator". A ceramic resonator is used in a very similar way to a quartz crystal (Fig. 8-15), but it is usually much less expensive and very rugged. I use ceramic resonators almost exclusively for clocking all my microcontroller applications. Despite the somewhat poorer accuracy of the parts (they are usually accurate to 0.5%), they really are the part of choice for most applications.

Many designers eschew the use of oscillators built from discrete parts as I have shown in this section. These circuits are rarely used because of the difficulty in specifying the correct parts for an application, the cost of the crystals and the potentially large amount of "real estate" that they can take up. Instead, oscillators are usually implemented using some kind of "canned"

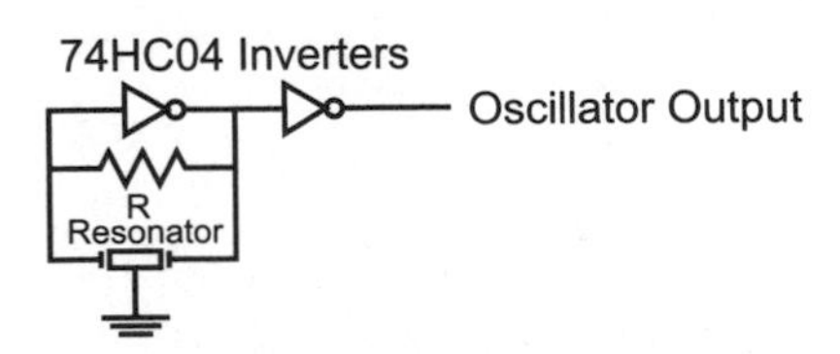

Fig. 8-15. Ceramic resonator oscillator.

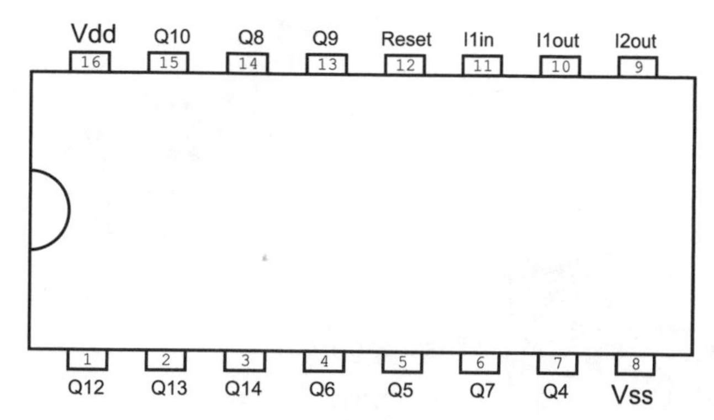

Fig. 8-16. CMOS 4060 chip pinout.

solution. These parts are designed to take up the same "footprint" as an eight or 14 pin "DIP" package and normally have four pins – power, ground, oscillator output and oscillator enable.

One of the most common chips used in digital logic applications is the CMOS 4060 (shown in Fig. 8-16). This chip can be used with the different oscillator types listed in this section and the "divide by" outputs are very handy in many circuits (often eliminating the need for separate counters). The crystal, ceramic resonator and relaxation oscillator circuits that I have shown in this chapter can be used with this part. The "Q4" through "Q13" outputs are divided by counters (i.e. "Q4" is the clock divided by 2 to the 4 or 16 times).

When using the 4060, note that Pin 11 is the input to the inverter used in the oscillator circuits shown in this chapter, Pin 10 is output of the first inverter and input of the second inverter while Pin 9 is the output of the second inverter. If Pin 12 ("Reset") is pulled high, the oscillator is stopped and the counters in the chip are reset.

555 Timer Chip

The 555 timer chip is probably the most versatile non-programmable part I have ever seen. Over the past 40 years, many people have created at least hundreds probably thousands of applications that have used this chip in ways I'm sure the original designer never would have thought possible; the original function of the chip was to provide a regular train of pulses. In this section, I will show how the chip is used in a circuit, along with some of the tricks that can be performed with it.

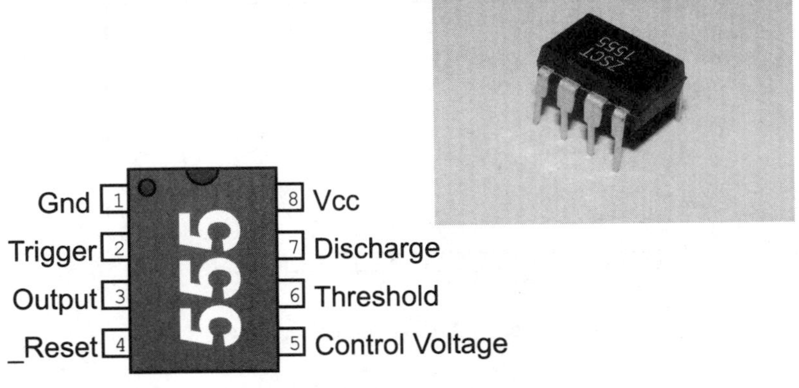

Fig. 8-17. 555 pinout.

In the previous sections, I have shown you the "pinout" of a number of different components – each one of them having a unique form factor. The 555 is usually built into an eight pin "dual in-line package" that is commonly used for chips. In Fig. 8-17, I have put in an "overhead" view of the 555, along with a photograph of an actual 555 chip.

Looking at the labels for each of the pins, most of them do not make a lot of sense. What should jump out at you is the "Gnd" (ground) at Pin 1 and the "Vcc" (positive power) at Pin 8. These two pins are used to provide power for the part; they match the power pins I've presented elsewhere for digital devices elsewhere in the book.

To try and get a better understanding of a chip, one of the first things I do is look for its block diagram and try to understand it. In Fig. 8-18, I have drawn out the block diagram for the 555 timer.

There should be two parts to the block diagram that you should recognize immediately. The first is the transistor at the bottom middle of the diagram. This transistor is wired in an open collector configuration and is acting as a switch that will pass current to ground. The next piece that you should recognize is the voltage divider running along the left side of the block diagram that I have separated out into Fig. 8-19. If you were to work out the voltages at "Vcontrol" and "Vtrig", you would discover that they are at 2/3 Vcc and 1/3 Vcc, respectively. This is actually an important clue as to how the chip works.

One aspect of the 555's voltage divider circuit that you may find confusing is its connection to an outside pin called "Control Voltage". As I have shown in Fig. 8-20, this connection allows the circuit designer to change the voltage

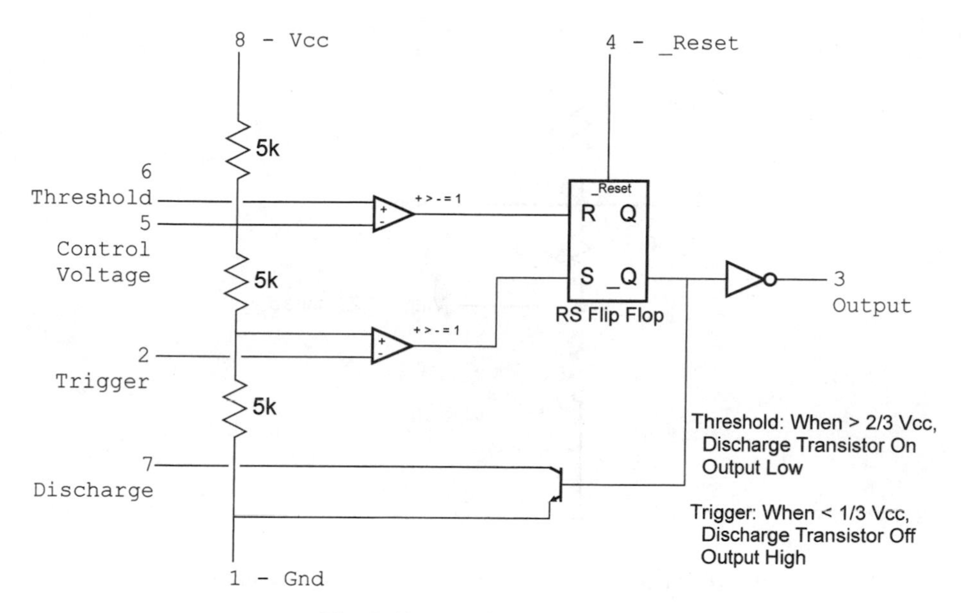

Fig. 8-18. 555 block diagram.

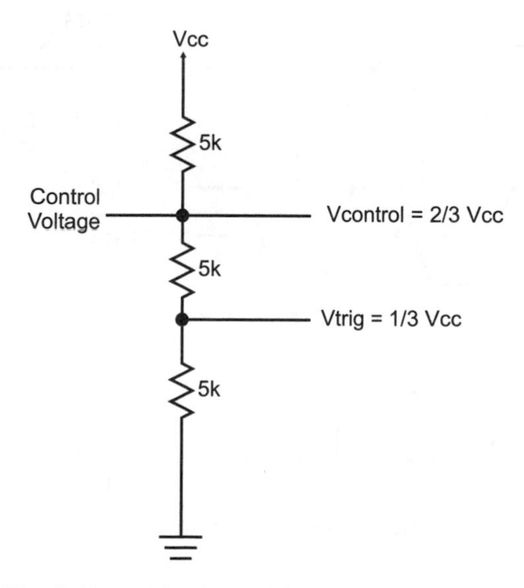

Fig. 8-19. 555 voltage reference/voltage divider.

levels of the voltage divider circuit. Rather than "Vcontrol" being 2/3 Vcc, it can now be any value (less than Vcc) that the designer would like. Changing "Vcontrol" also changes "Vtrig" to 1/2 "Vcontrol", as I have shown in Fig. 8-20.

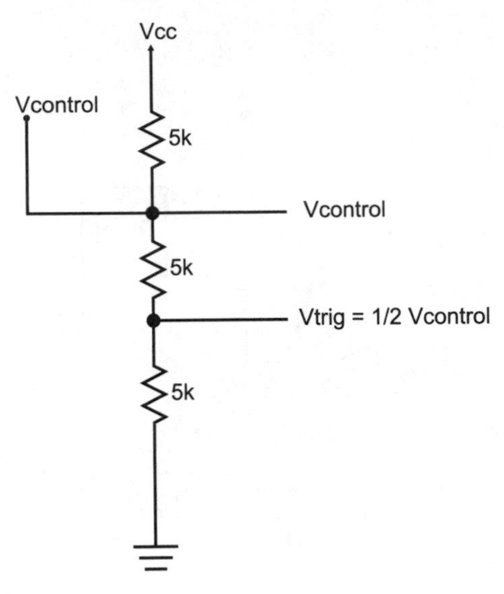

Fig. 8-20. Modified 555 operating voltage threshold.

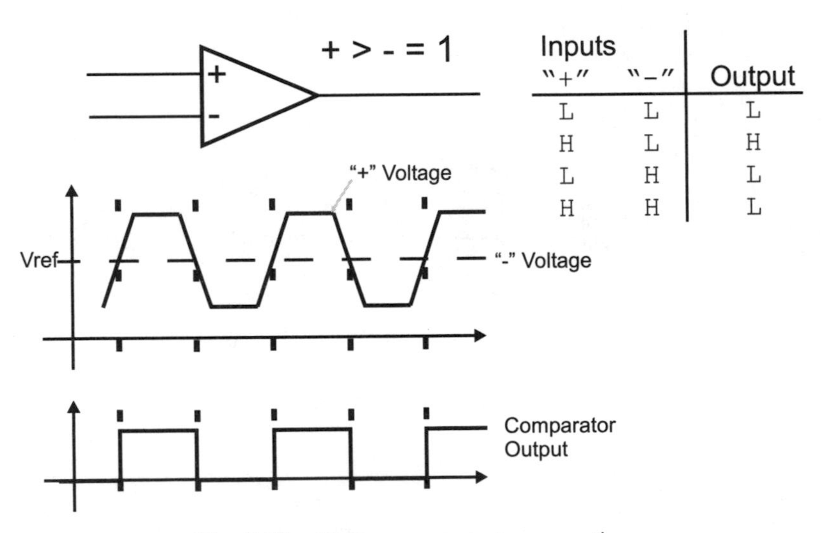

Inputs "+"	Inputs "−"	Output
L	L	L
H	L	H
L	H	L
H	H	L

Fig. 8-21. Voltage comparator operation.

The voltages at the "Vcontrol" and "Vtrig" are passed to two triangular boxes with a "+" and "−" along with a funny looking equation. These boxes are representations for voltage "comparators" and, as I have shown in Fig. 8-21, the comparators output a high voltage level when the voltage at the "+" input is greater than the voltage at the "−" input. The 555 uses the two comparators to continuously compare two external voltage levels to "Vcontrol" and "Vtrig" and pass the results to a box labeled "RS flip flop".

The 555's RS flip flop saves an indication of which comparator last passed a high voltage to it. If the comparator connected to the "threshold" pin of the 555 and "Vcontrol" of the voltage divider output a high voltage, then the flip flop will output a high voltage at "_Q", which turns on the transistor at the bottom of the block diagram. If the other comparator passes a high voltage to the "RS flip/flop", then the voltage at "_Q" is driven low and the transistor is turned off. This is a fairly complete explanation of how the 555 works and I'm sure that you are at least as confused as you were when I first showed you the block diagram of the chip. The individual parts are quite easy to understand, but I'm sure you're mystified how they work together.

When I described the operation of the 555 chip, I neglected to take into account the components that would be wired to it. The timing delay that is integral to the operation of the 555 is produced by resistors and capacitors wired in the "RC networks" that I have described earlier in the book. What I didn't go into detail on in the previous sections of the book is that as you change the value of the resistor or capacitor in the circuit, you will change the delay produced by the two components (Fig. 8-22).

In Fig. 8-23, I have drawn a 555 oscillator circuit; when this circuit starts running, the 555 will be an "astable" oscillator with the output toggling,

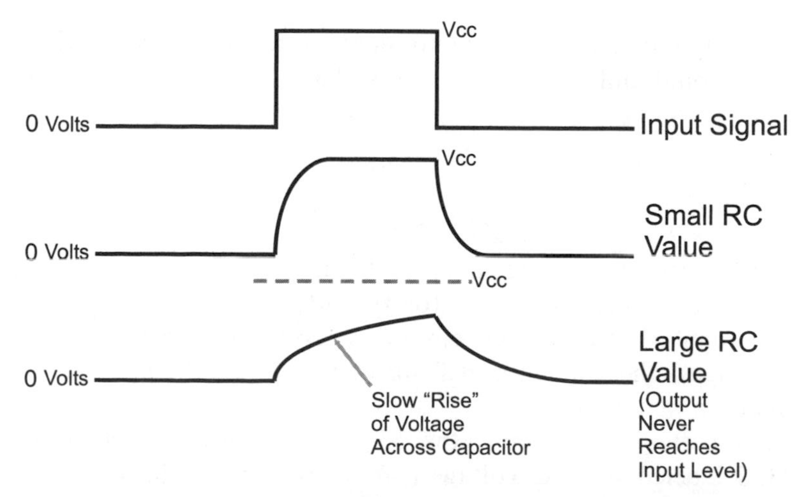

Fig. 8-22. RC network operation for ranging values of RC.

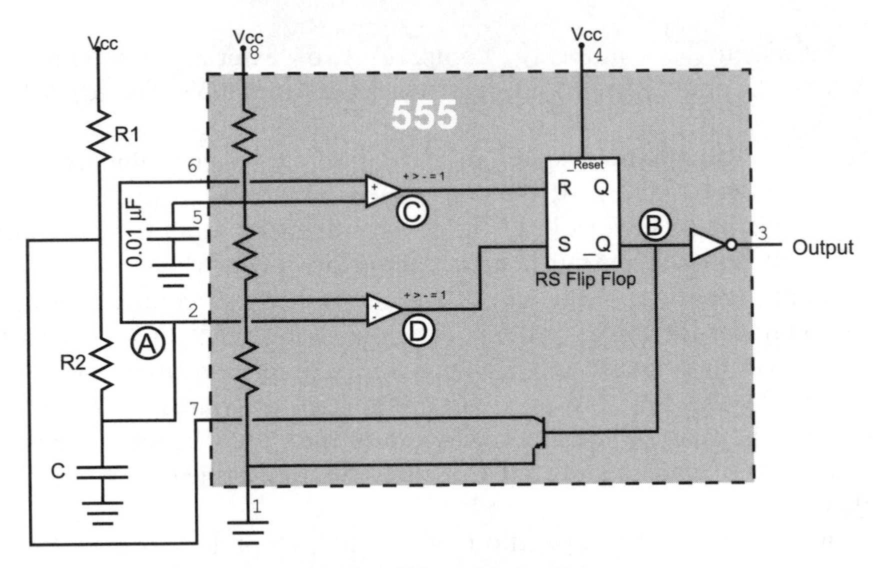

Fig. 8-23. 555 astable oscillator.

according to the values R1, R2 and C. The 555's output times are defined by the formulas:

$$T_{high} = 0.693 \times C \times (R1 + R2)$$

$$T_{low} = 0.693 \times C \times R2$$

$$Period = 0.693 \times C \times (R1 + 2R2)$$

$$Frequency = 1.443/(C \times (R1 + 2R2))$$

When choosing the resistor and component values for working with the 555 timer, you should only use components that are within the ranges listed below:

$$10k \leq R \leq 14M$$

$$100\,pF \leq C \leq 1{,}000\mu F$$

The 0.01 µF capacitor wired to the "control voltage" pin of the 555 (in Fig. 8-23) is used as a "filter" for the internal voltages. This capacitor works very similarly to the logic circuit's decoupling capacitor; if the input voltage changes, the capacitor will absorb or release charge to keep the voltage as even as possible.

To get a better idea of how the 555 timer works as an oscillator, in Fig. 8-23, I labeled the RC voltage ("A"), the RS flip flop output ("B" – which is in the inverted 555 output), the "threshold" comparator voltage

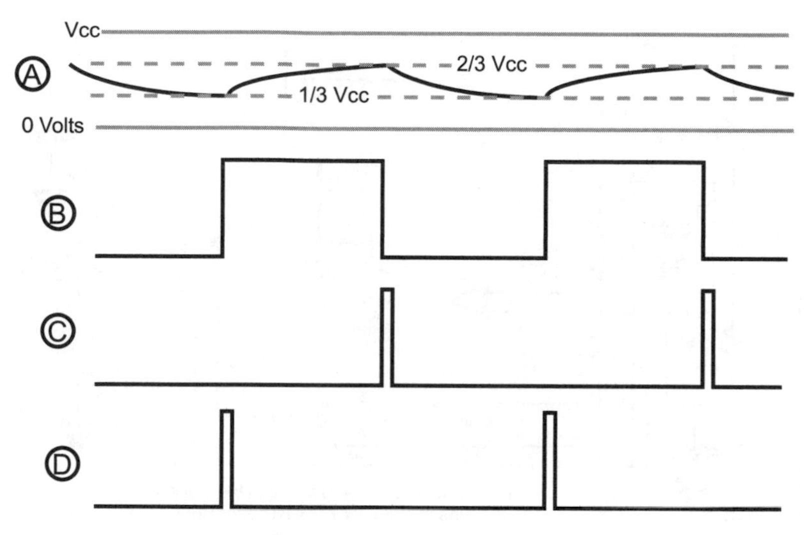

Fig. 8-24. 555 astable operation.

("C") and the "trigger" comparator voltage ("D"). Figure 8-24 shows the waveforms for each of these parts marked in Fig. 8-23, so you can see the changing RC waveform, the output from the two comparators and the action of the RS flip flop.

Before going on, I want to share with you some of the more clever and useful circuits that have been created using the 555 timer. The few I will show in this chapter are just a small fraction of the number that is possible or has already been developed. If you are in a used book store, you should look for a copy of Don Lancaster's "555 Timer Cookbook"; it will really open your eyes to the incredible variety of applications this chip can help implement.

When stretching the envelope and using the 555 timer in a way that it wasn't originally designed for, you generally look at the different input pins and see how they can be given a completely different function. If you wanted to make a circuit that drove out a tone for a set amount of time, you could use two cascaded 555 timers (or a single "556", which consists of two 555 timer chips in a single package) or the circuit shown in Fig. 8-25.

When this circuit is first turned on, the voltage at Pin 4 (the RS flip flop reset pin) is low, stopping the circuit from oscillating. When the momentary on button is pressed, current is passed to the 10 μF capacitor, charging it and driving up the voltage on Pin 4. This happens quite quickly and when Pin 4 reaches the logic threshold to stop holding the RS flip flop reset, then the 555 will start oscillating, driving out a signal that oscillates at 464 times per second.

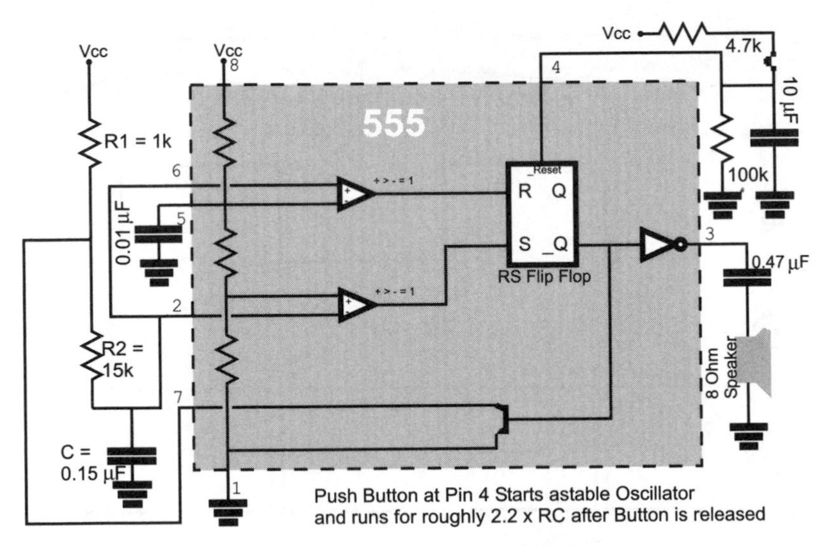

Fig. 8-25. 555 tone output.

This tone will stay on as long as the button is pressed and then will continue until the 10 µF capacitor discharges through the 100 k resistor. By varying the values for these two parts, you can vary the length of time the 555 continues to oscillate. Remember the rule of thumb that is used to approximate the time for an RC network, like this one, to discharge:

$$T_{discharge} = 2.2 \times R \times C$$

I take advantage of the ability of the human ear to distinguish between different sounds and categorize them with a circuit design for a continuity tester that you will probably find is a lot more useful than the simple instrument built into your digital multi-meter.

This continuity tester circuit shown in Fig. 8-26, is useful in a variety of different situations. Instead of just driving out a simple tone when an electrical path between the two probes has been found, it provides you with different tones and sounds, based on the resistance between the probes as well as an indication of whether or not a diode is between the probes. Because the circuit is self-powered, you can use it with circuits that are already working, without worrying about having a valid ground connection.

When there is no connection between the leads, the 5.1k resistor is part of the RC network that provides the delay for the 555 timer wired as an astable oscillator. With no connections, you will find that the oscillator outputs a tone that is at about 440 cycles per second ("hertz" or "Hz"). If there is a direct connection (or short circuit) between the two probes the 5.1 resistor is

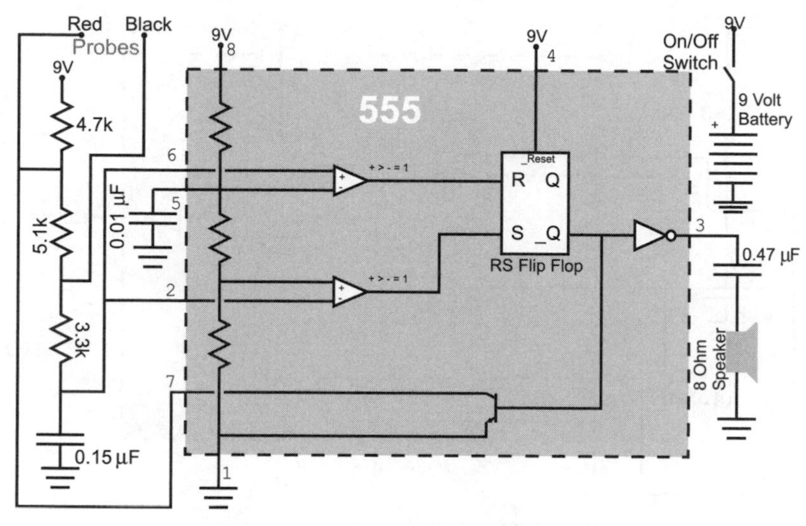

Fig. 8-26. 555 based continuity tester.

not used in the delay and the frequency output is around 880 hertz, there is a full octave difference between the two signals.

What I like about this circuit is that you get a different frequency based on the resistance across the probes – the probe resistance is in parallel with the 5.1 k resistor, changing its frequency and the frequency output from it. This can be useful in finding "almost" short circuits such as a "just touching" connection rather than a hard soldered connection or a "high impedance short", when you expect no connection at all. In addition, try out a diode in a forward biased and reversed bias connection; you will be able to hear a noticeable difference here as well.

Delay Circuits

Another basic function of the 555 is use as a "monostable". In the previous section, I alluded to this function and noted that the 555 could do more than just be part of the "astable" oscillator which will run for ever – the monostable, on the other hand, will only execute once and requires triggering. The monostable is very useful for a variety of different applications and works similarly to digital logic chips that can provide a similar delay. Figure 8-27 shows the 555 wired as a monostable generator and Fig. 8-28 shows the response waveforms to the input pulse.

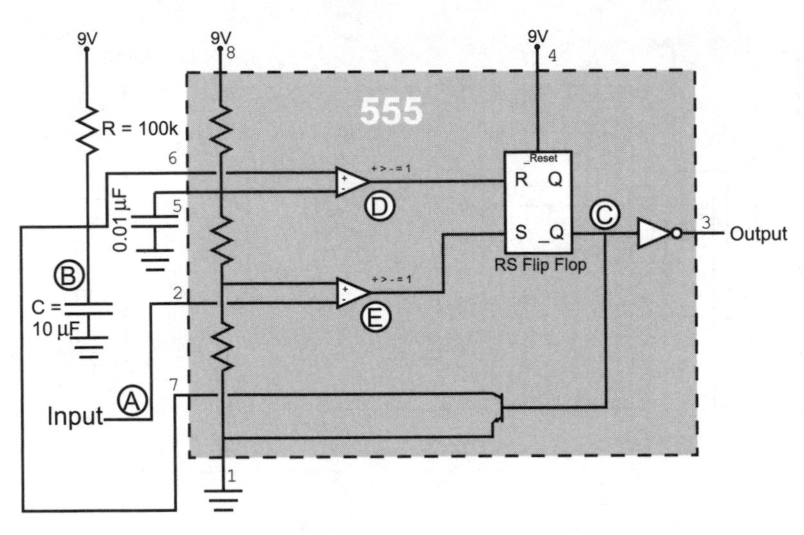

Fig. 8-27. 555 monostable circuit.

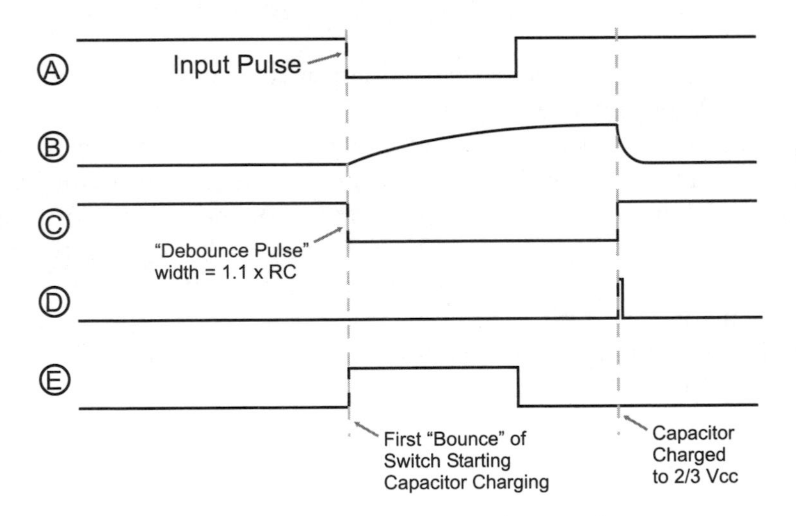

Fig. 8-28. 555 monostable operation waveforms.

When the input goes low, the pulse output from the 555 timer is determined using the formula:

$$T_{pulse} = 1.1 \times R \times C$$
$$= 1.1 \times 100\,k \times 10\,\mu F = 1.1\text{ seconds}$$

The 555's RS flip flop is initially "reset", and the transistor that passes capacitor charge to ground is turned on. When the input ("A") goes low and the "trigger" input receives a low voltage input, its comparator signal ("E") goes high, changing the state of the RS flip flop ("C") and changing the state of the output pin. When the RS flip flop state changes again, the transistor is turned off and the capacitor charges through the resistor. The capacitor charges according to the formula:

$$\texttt{Output} = \texttt{Vcc} - \texttt{Vcc} \times e^{-t/RC}$$

until its voltage reaches 2/3 Vcc. When it reaches 2/3 Vcc, the "threshold" comparator ("D") goes high and the RS flip flop changes state again, changing the value of the output pin ("C"). At this point, the 555 is back to its original state.

This circuit works quite well except for one point – the input pulse must *always* be shorter than the calculated output pulse. If the input pulse is longer than the calculated output pulse duration, you will find that the output will stay active, but it will pulse periodically. When the capacitor charges to 2/3 Vcc and the input is low down, both of the comparators will be driving a high voltage to the RS flip flop. This is an invalid condition for the RS flip flop and the output from the flip flop is "indeterminate", resulting in the transistor tying the capacitor to ground periodically. To avoid this behavior, you should always make sure that the length of time for the pulse output from the 555 is longer than the expected input.

Instead of using the 555 timer as a monostable delay generator, there are a number of logic chips that perform the same function using a resistor and capacitor. The 74123 incorporates two monostable delays that are programmed using a resistor and capacitor. The 74123 and other logic family chips have the advantage that the voltage level transitions do not cause as much disruption to the surrounding circuitry.

In some applications, there is a delay that is either more precise, shorter or longer than can be practically created using the RC-controlled monostables that I have presented so far. In the next chapter, I will be introducing you to "counters" which will either count continuously or stop when a specific value has been reached. The counter, driven by one of the oscillators presented in this chapter, is used to produce either a very long or very precise delay.

For shorter delays, there are two methods that you can consider. The first method is to use a "canned" delay line. These components usually consist of an inverting buffer, driving a long copper line. At different points along the line, inverting "taps" are put in place to drive out the signal. Figure 8-29 shows how these components are used.

When you see an actual "delay line" component, you will probably refer to it as a "chip". I hesitate to do so because I consider a chip to be simply a silicon chip bonded to a "lead frame" and "encapsulated" in some manner. In a delay line component (or "module"), the wire delay is wound in a coil

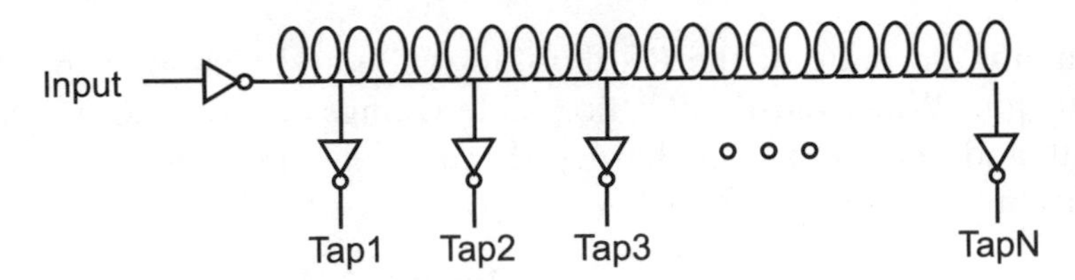

Fig. 8-29. Delay circuit.

with the tap inverter inputs soldered to it at different intervals. The actual device requires very high precision mechanical assembly (more than a standard plastic encapsulated chip) and any errors in assembly or encapsulation will result in a useless part.

The wire in the part passes the digital signal to the various taps within the part, with a delay of roughly 1 nanosecond per foot (30 cm) of wire. When you design high-speed applications, this rule of thumb is very important when you are designing high-speed digital electronic circuits. Chances are you will ask that the traces on PCBs are "routed" with 0.1 inch (2.54 mm) precision to ensure that parallel signals all "show up" at the same place at the same time in the high-speed application circuit.

The advantage of the delay line module is that timing delay can be very precise. Custom-made delay line modules are available (the manufacturer solders the taps at specified points in the coil rather than at standard positions), which can be critical in some applications. This high level of assembly/encapsulation precision has a price that you will have to pay. If you can buy a 74LS04 for less than a quarter in single units, you should not be surprised to discover that a delay line module will cost you over $10.00. The delay line provides you with the best control over different delays required in a circuit, but at quite a significant cost. Delay line modules should only be considered if no other options are available to you when you are designing a circuit.

Another method of delaying signals is to take advantage of the natural delays of digital electronic gates and simply "chain" a number of them together to get a needed delay. In Fig. 8-30, I have shown a 20, 40 and 60 nsec delay built out of a 74LS04 TTL chip.

The advantage of this method is that it is quite low cost and reasonable precision can be built into the circuit. When you are designing delays for your applications, you should consult with the technology operational characteristics chart that I provided earlier in the book (see Table 6-2).

Working with different technologies, you should be able to get quite accurate delays quite inexpensively. The disadvantage of this method is that

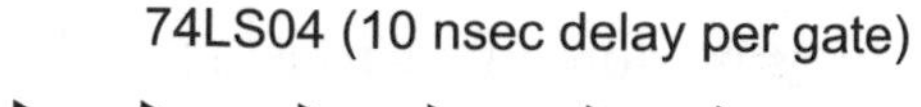

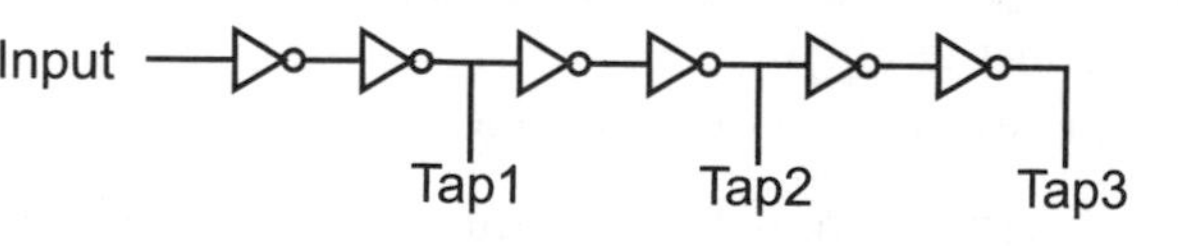

Fig. 8-30. Using logic gates for specific delays.

it can take up a lot of space on a board (at which point the canned delay line may have to be considered).

Quiz

1. An ideal digital electronic clock waveform has:
   (a) A constant period with a 50% duty cycle
   (b) A selectable speed range
   (c) A period that is less than the gate delay of the logic technology being used with it
   (d) A varying period that takes advantage of the operation of the combinatorial circuitry in the sequential circuit
2. Each of the following are important characteristics of astable oscillators except for
   (a) Period
   (b) Jitter
   (c) Duty cycle
   (d) Power required
3. If the R1 or R2 resistor values are less than the product of h_{FE} and Rpu, the NPN transistor relaxation oscillator will:
   (a) Not be reliable and may not start up
   (b) Not work correctly
   (c) Get very hot because the transistors are continuously in saturation
   (d) Produce a perfect, 50% duty cycle output
4. The practice of putting ring oscillators in leftover gates:
   (a) Helps minimize the cost of a digital electronics application
   (b) Helps synchronize other gates in the chip

(c) Should only be done if there is an odd number of inverting gates left in the chip
(d) Should be strenuously discouraged

5. A relaxation oscillator has an R1 value of 10 k, C of 0.1 μF and R2 equal to 1 k. What frequency will it oscillate at?
   (a) It won't oscillate
   (b) 4.54 kHz
   (c) 4.54 MHz
   (d) 4.54 Hz

6. What type of application is the relaxation oscillator best suited for?
   (a) High power
   (b) High cost, not requiring accurate operation
   (c) Low cost, high accuracy
   (d) Low cost, not requiring accurate operation

7. Canned oscillators are used because:
   (a) They are more accurate than discrete component solutions
   (b) They are cheaper than discrete oscillator solutions
   (c) They are simple and accurate
   (d) They are more reliable than discrete component solutions

8. The only type of oscillator that the 4060 cannot implement is:
   (a) Ring oscillator
   (b) Relaxation oscillator
   (c) Crystal/ceramic resonator oscillator
   (d) NPN transistor relaxation oscillator

9. Increasing the value of a resistor or capacitor in a 555 astable oscillator will:
   (a) Lower its operating frequency
   (b) Raise its operating frequency
   (c) Increase the output voltage
   (d) Lower the output voltage

10. Connecting six 74AS inverters (gate delay 2 ns) end to end will produce a delay circuit that is:
    (a) 24 ns
    (b) 12 ns
    (c) 72 ns
    (d) 48 ns

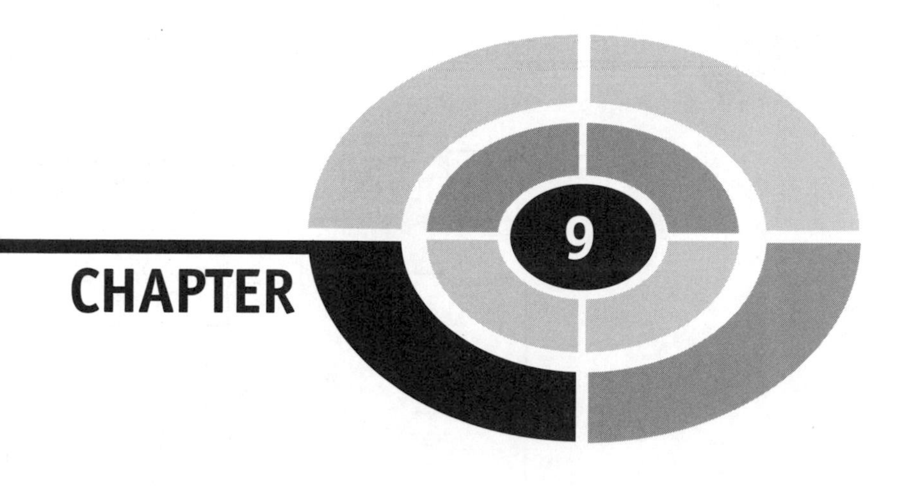

Complex Sequential Circuits

As the saying goes: "You now know enough to be dangerous." You should be fairly comfortable with working with logic functions and equations, have an understanding of electronics and how to interface logic chips (of different families together), understand the basics of memory and have gone through a number of clocking schemes. I'm sure that you now feel you are ready to start bringing these pieces together into some interesting applications. I'm sure that you have some ideas of things you would like to have your hand at dcsigning. Before being set free to wreak havoc on an unsuspecting world, I want to spend some time presenting you with some chips and tools that will make your plans for world domination much easier. In this chapter, I want to go through some of the subsystems that are available in chips that will make your design work easier.

In Chapter 7, I introduced you to the digital clock block diagram shown in Fig. 9-1. There shouldn't be any part of this diagram that is a surprise to you; the "time memory" consists of a number of flip flop registers that are reset

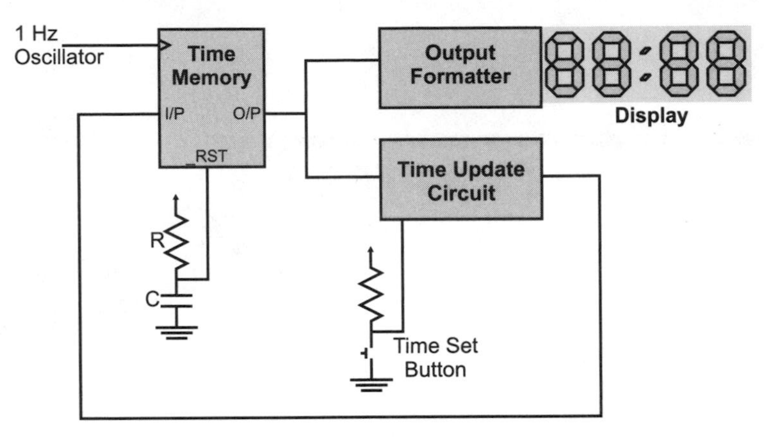

Fig. 9-1. Digital clock block diagram.

upon power up. The "time update circuit" consists of an adder along with logic and a user push button to determine what the *next* time will be. The "output formatter" consists of logic to decode the time memory and display it on a set of LEDs. Finally, the 1 Hz clock can be produced a number of different ways that are covered in the previous chapter.

While I must admit that it would be cool to see a digital clock designed using two input logic gates, I want to point out that there are a number of commonly available chips that provide major subsystems needed for such an endeavor. For the rest of this book I will be focusing on these chips and how they are interconnected to form "real world" applications.

Virtually all of these chips are sequential circuits in their own right; consider a "counter" chip that increments its internal memory devices each time a rising clock edge is received. The counter chip consists of several flip flop bits and combinatorial logic that processes input data, provides the value increment and outputs the data in a specific format. These functions are very similar to that provided by the digital clock in Fig. 9-1.

To show what I mean, consider the sequential circuit block diagram in Fig. 9-2. Superficially, the diagram has a very strong resemblance to Fig. 9-1 because many of the same basic functions and capabilities are required in both instances.

The "state memory" is the current operating state of the chip. The term "state" simply means at what operating point the chip is at. For a counter or other arithmetic function chip, the term "state" probably seems somewhat grandiose, but it is an accurate way of describing the current value in a counter. For a microprocessor, the term "state memory" is almost an understatement, as it includes not only the program counter (which points to the next instruction to execute) but also data and status register information.

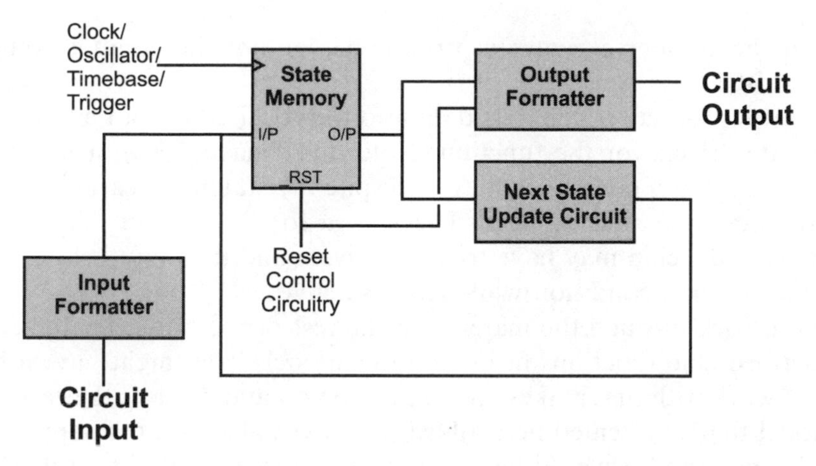

Fig. 9-2. General case sequential circuit.

The state memory can be reset (as shown in the previous chapter) and, more importantly, it is usually the only point in which the clock/oscillator input is received. The philosophy behind most sequential circuits is that the combinatorial logic processing input, output and the next state information, regardless of the circumstances, will be available in time for the next active clock cycle. The term normally used to describe this methodology of design is "synchronous" because a central clock is keeping track of the operations within the chip.

This philosophy is currently under challenge from scientists interested in investigating "asynchronous" digital logic design. This effort involves designing sequential circuits that are not "paced" by a central clock, but the length of time of each operation. For example, moving data from one register to another should take much less time than an instruction which stores data in the main memory. Asynchronous digital logic design holds the promise of faster computers that use much less power because the only active circuitry are the required gates and flip flops of the current time – nothing else needs to be active, nor do other circuits need to be clocked.

The "next state update circuit" and "input formatter" blocks process the current bit data and any relevant input for storage in the state memory. In Fig. 9-1, I combined both of these functions into the "time update circuit" because the only input required for this clock is whether or not the "time set button" is pressed – if it is, then the time update circuit will increment the hours, minutes and seconds stored in the digital clock's "time memory".

The "input formatter" circuitry can be processing different inputs controlling what the next state is going to be – this is why I link it to the "next state update circuit". For a counter, this information could be the direction the counter executes in or whether or not the counter counts in binary or BCD.

The output formatter converts data into the required output and provides appropriate drivers for the function. Note that I have drawn a link to this box from the "reset control circuitry" despite my statement earlier that only the state memory could be reset. The reason for drawing in this link is to indicate that the chip may have tri-state drivers and these are held in a high impedance ("off") condition while reset is active.

As you work through the material in the rest of the book, try to see how the described chip functions fit in with this model. You might have a better model to work with that makes more sense to you and if this is the case use it. The model that's presented here allows *me* to visualize what is happening in an application, and it would be arrogant of me to assume that it works for everyone else.

Counters

One of the most useful functions that you will use when you develop digital electronic circuits is the counter. The counter is actually a smaller piece of many complex chips, as it provides a basic way of maintaining the current operating state along with a method of progressing to the next one. The basic counter circuit consists of a set of flip flops that drive into and are driven from an adder. A counter circuit is shown in Fig. 9-3.

The use of "edge triggered" flip flops is a very important aspect of the circuit shown in Fig. 9-3 and one that you should keep in mind. When the "counter clock" changes state, the output value of the adder (which is the D flip flop value plus 1) is presented to the inputs of the D flip flop register bits as the next value to be saved.

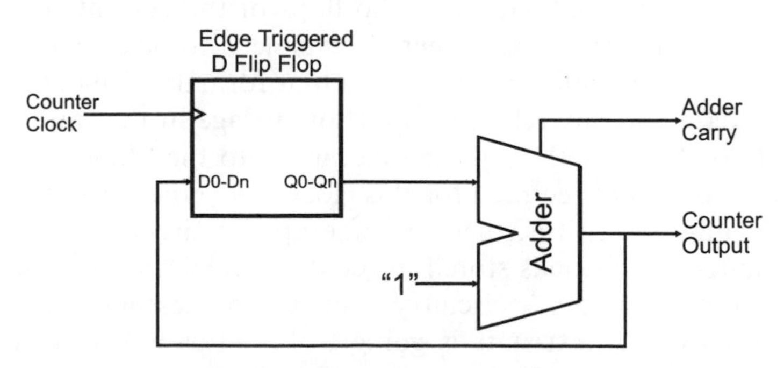

Fig. 9-3. Basic counter circuit.

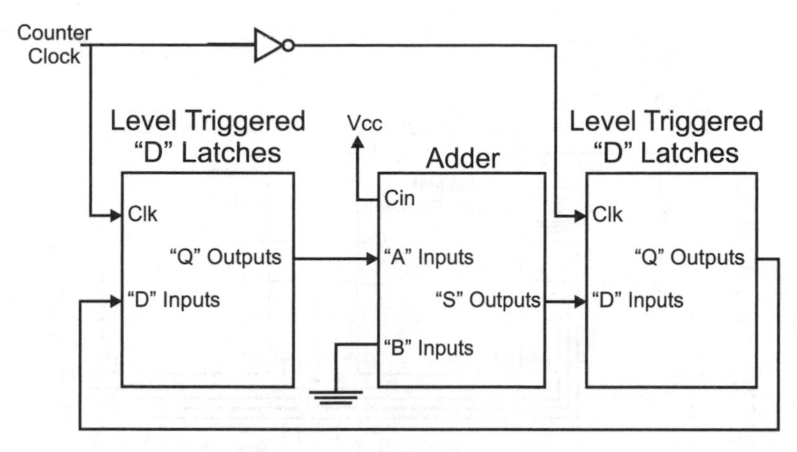

Fig. 9-4. Counter circuit built from level triggered flip flops.

If an edge triggered flip flop register wasn't used in the circuit, then you would have to use latches and design the counter something like the one shown in Fig. 9-4. In this circuit, I have put in two latches, each one "out of phase" with each other. This is to say that when the clock is high, one latch is storing the data while the other is passing through the value presented at its inputs. When the clock changes value, the latches change from passing data to storing and vice versa. This method of implementing a counter is unnecessarily complex and potentially very slow – the extra set of flip flops will slow down the performance of the counter and limit its maximum speed.

The counter circuit of Fig. 9-3 can be built using a 74C174 hex D flip flop and a 74C283 four bit adder circuit. The circuit shown in Fig. 9-5 will demonstrate how the counter works. When the term "floating" is used with respect to pins, it means that the pins are left unconnected.

When you try out this circuit, the first thing that you will probably notice is that when you press the button, the LEDs will not "increment" by 1, but by 2, 3 or even 4. The reason for this is known as "switch bounce". Earlier in the book, I showed a two inverter circuit for eliminating switch bounce, and later I will discuss a number of other strategies for minimizing the problem. For now, if you wire a 0.1 μF tantalum capacitor as shown in Fig. 9-5, you should minimize this problem (although you will probably not eliminate it).

The counter circuit should work well for you. As with the previous projects, a single chip can be used where multiple required. The counter chip that I usually work with is the 74LS193 (Fig. 9-6) which combines a four bit D flip flop register and adder along with the ability to decrement the result. Later in the book, I will show how this chip can be used with others to "cascade" from a 4 bit counter to an 8 and 16 bit counter.

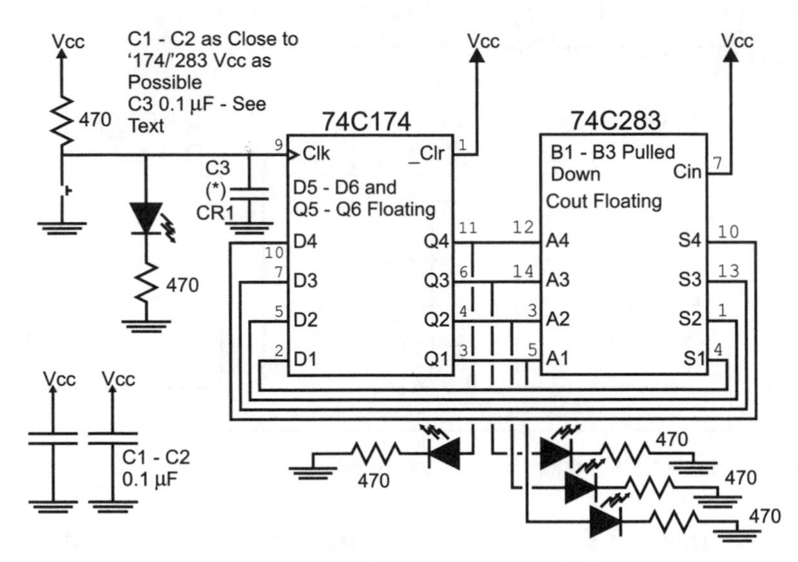

Fig. 9-5. Discrete component counter.

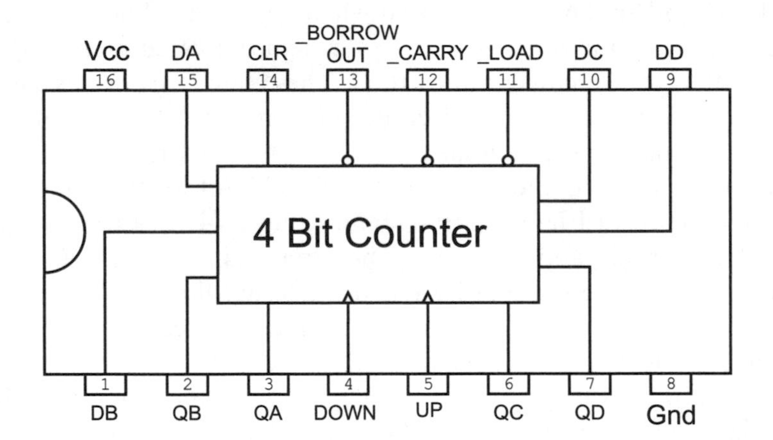

Fig. 9-6. 74193 counter chip pinout.

The "_Carry" out bit of the 74193 can be passed from one counter to the clock input of another to provide the ability to count more bits, as I show in Fig. 9-7. The carry bit can be thought of as an overflow to the more significant counter, indicating that it should increment its value. If by looking at this circuit you recognize it as being similar to the "ripple" adder presented earlier

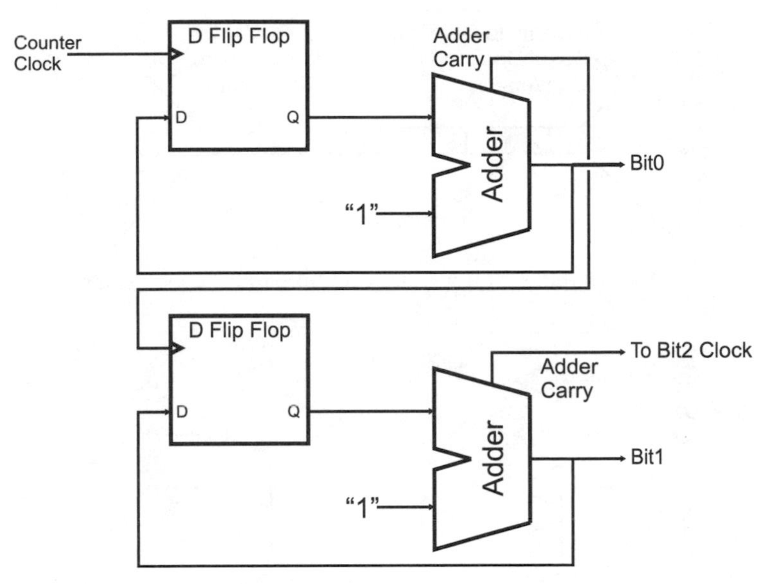

Fig. 9-7. Ripple counter block diagram.

in the book, go to the head of the class. This circuit is known as a "ripple counter" and does not have the same high speed as a counter built from look-ahead carry adders.

Along with the 74193, you might want to consider the 74161 counter, which can only count up and changes to the count value must be clocked in (the 74193 allows changes to the count value asynchronously, which is to say without the clock). The 74160 and 74192 chips are identical to the 74161 and 74193, respectively, but only count up to 9 and are known as "decade" counters. The 74160 and 74192 are useful in circuits in which the digits 0 through 9 are required for counting.

Shift Registers

Most intersystem (or intercomputer) communications are done serially. This means that a byte of data is sent over a single wire, one bit at a time, with the timing coordinated between the sender and the receiver. So far in this book, if you were to transfer a number of bits at the same time, you would send them in "parallel", one connection for each bit. The basis for serial communications is the "shift register", which converts a number of "parallel" bits into a time-dependent single string of bits and converts these strings of bits back

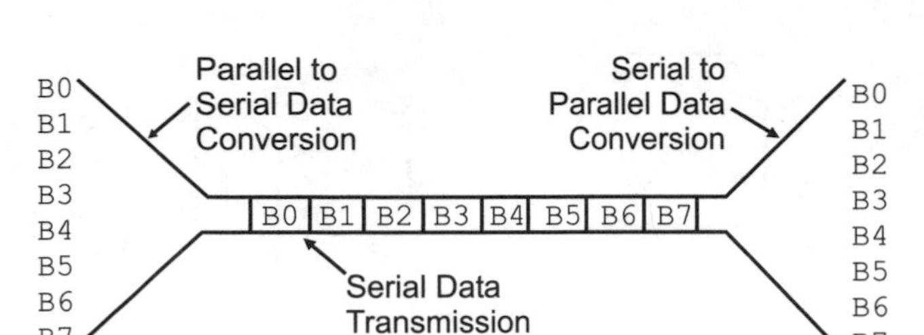

Fig. 9-8. Parallel to serial and back to parallel conversions.

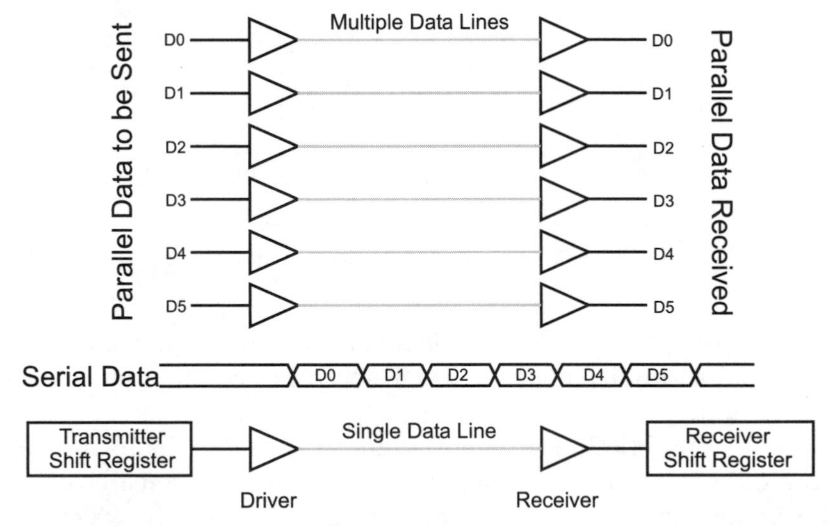

Fig. 9-9. Parallel vs. serial transmission hardware block diagrams.

into a set of parallel bits. Figure 9-8 shows this process with eight parallel data bits being converted into a bit stream and transmitted to a receiver, which "recreates" the eight bits back into their parallel data format.

The differences between serial and parallel data transfers are shown in Fig. 9-9. To send six bits in parallel, a half dozen transmitting "drivers" and an equal number of "receivers" are required. To send six bits serially, just a single driver and receiver is required, but the sending circuit must have a "shift register transmitter" and the receiving circuit must have a "shift register receiver". The parallel data can be sent in the time required for just one bit while the serial data requires enough time to send each of the six bits individually.

It probably looks like transmitting data serially requires a lot of overhead and it slows down the data transfer. There are a number of factors to consider before making this assumption. The first is that most chips are not made out of individual logic gates as the simple chips presented here so far; they are

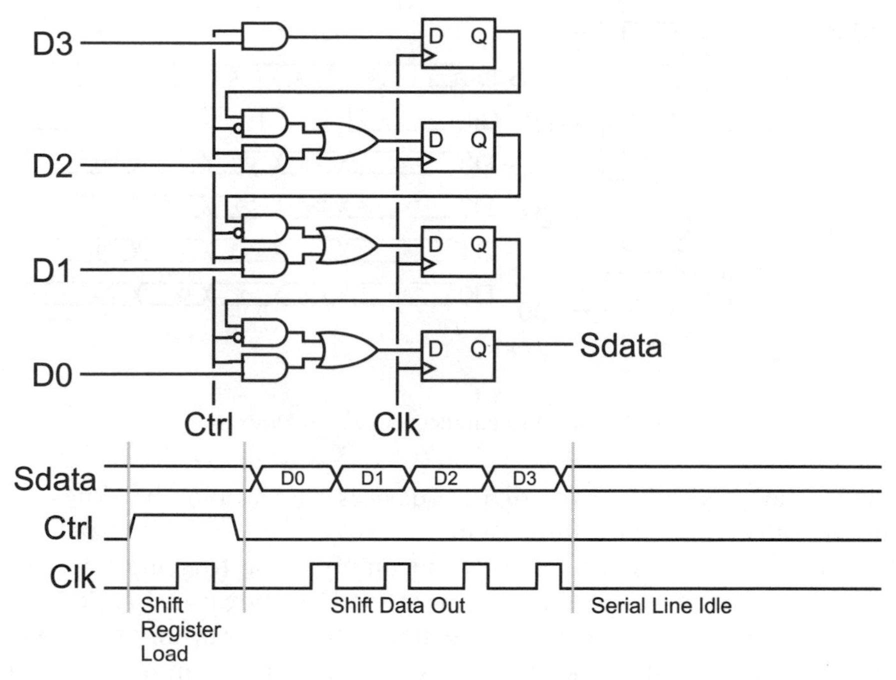

Fig. 9-10. Parallel to serial conversion hardware.

usually very dense circuits consisting of thousands of gates, with the impact of adding serial shift registers being very minimal. Another issue to consider is that it can be very difficult to synchronize all the parallel bits to "arrive" at the receiver at the same time in high-speed circuits. Finally, multiple wires can take up a lot of space and be quite expensive; if chips or subsystems could have shift registers built into them, then it often makes sense (both practical and economic) that data be transferred serially.

The circuit that converts the parallel data into the serial stream is quite simple. Figure 9-10 shows a circuit along with a waveform showing how the circuit works. Four bits are first loaded in parallel into a series of four flip flops. These four flip flops can be driven with data either from an external source or from the next significant bit depending on the "Ctrl" bit state. If "Ctrl" is high, when the "Clk" ("clock") is cycled, the data in the D3:0 bits are stored in the four flip flops. If "Ctrl" is low, when "Clk" is cycled, each bit is updated with its next significant bit and data is shifted out, least significant bit first.

The process of each bit of data "passing" through each of the flip flops is known as "shifting". As can be seen in Fig. 9-10 that four data bits are "shifted" out on the "Sdata" line in ascending order, with the "Clk" line specifying when a new bit is to be shifted out. If this method was used to

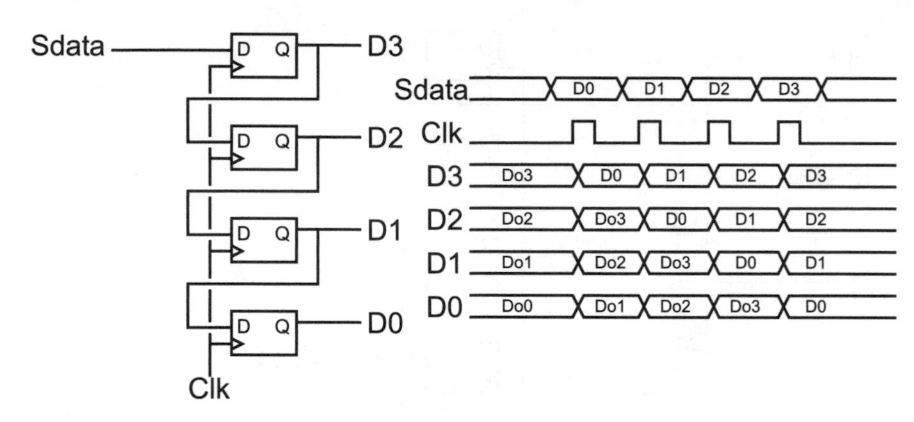

Fig. 9-11. Serial to parallel conversion hardware.

transmit data between two digital devices, it would be known as "synchronous serial data transmission".

Receiving "Sdata" is accomplished by simply using four flip flops wired with their outputs wired to the next input, as I've shown in Fig. 9-11. The same clock that is used to shift out the data from the transmitter should be used to shift in the data in the receiver. Along with the circuit used to shift in the data, I have included a waveform diagram for you to take a look at in Fig. 9-11. One potentially confusing aspect of the waveforms is my use of the "DoX" convention to indicate the previous values within the receiver. These bits will be shifted out in a similar manner as to how the data was shifted in.

There are a number of very common synchronous data protocols that are used in computer systems to provide simple interfaces to common peripherals. These interfaces, which include "Microwire", "SPI" and "I2C", are very easy and relatively fast ways of adding peripherals such as analog to digital converters and external memory to microcontrollers and complete computer systems. In fact, your PC has an I2C processor peripheral bus for controlling power supplies and monitoring the processor's chip temperature.

Linear Feedback Shift Registers

One of the most interesting logic devices you can work with is the "linear feedback shift register" ("LFSR"). It is built from a shift register along with two or more XOR gates modifying the contents of the register as shown in Fig. 9-12. This circuit can be used to "pseudo-randomize" data, encrypt and

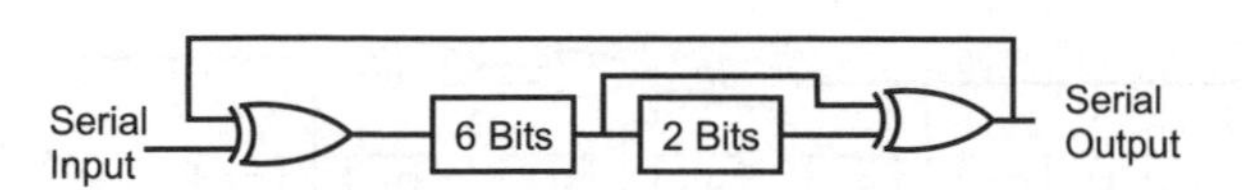

Fig. 9-12. Basic linear feedback shift register (LFSR).

decrypt serial data and provide very good serial data integrity checking. You may have heard the term "cyclical redundancy check ("CRC") when applied to data transmission; this is a type of linear feedback shift register. Linear feedback shift registers can also be implemented fairly easily in software with a microcontroller or microprocessor, although it is in hardware where the device is the most efficient.

The simple LFSR illustrated in Fig. 9-12 feeds back bits 5 and 7 of the shift register through XOR gates to the input. This changes the bit values in the shift register according to the formula:

$$\texttt{Bit}_0 = \texttt{Bit}_{\texttt{in}}\ \text{XOR}\ (\texttt{Bit}_5\text{XOR}\ \texttt{Bit}_7)$$

The LFSR is typically used for three purposes:

1. Creating a "checksum" value known as a cyclical redundancy check (CRC), which is a unique value or "signature" for a string of bits. Both the transmitter and receiver will pass the data through LFSRs and, at the end of the process, the CRC produced by the transmitter will be compared to the CRC produced by the receiver. If there is a difference in the CRCs, then the receiver will request that the transmitter resend the data.
2. Encrypting a string of bits. LFSRs can be used as an encryption/decryption tool with part of the encryption being the initial value in the LFSR. The value output from the LFSR is dependent on the initial value loaded into the LFSR. Decrypting data is also accomplished by using an LFSR, but configured as the complementary function.
3. Producing "pseudo-random" numbers. One of the most challenging computer tasks that you will be given is to come up with a series of random numbers. Computers are designed to be "deterministic", which means that what they are doing at any given time can be calculated mathematically. This property is important for most applications (nobody wants a computer to boot differently each time or to have a word processing program that responds randomly to keystrokes), but it is a problem for many applications which rely on the pseudo-random numbers for animated displays or "lifelike" responses to user input.

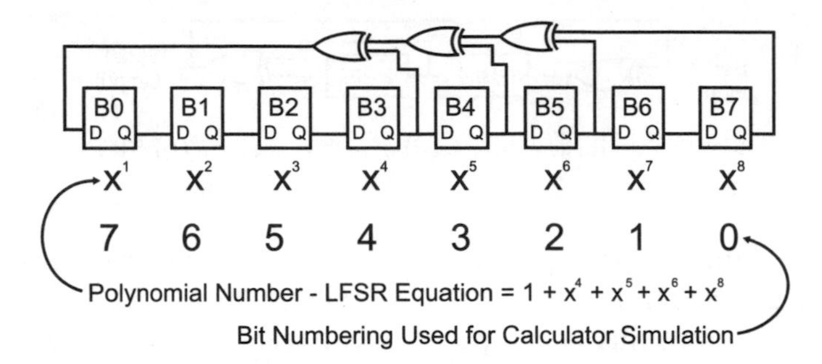

Fig. 9-13. Eight bit LFSR with defining polynomial expression.

In all of these applications, the LFSR is an ideal choice as a solution because it can be built very simply from just a few gates (meaning low cost and fast operation). The LFSR can also be implemented in software, as I will show below.

If you were going to express this LFSR to somebody else, you could send a graphic something like Fig. 9-13, or you could express it in the "polynomial" format like:

$$f(x) = 1 + x^4 + x^5 + x^6 + x^8$$

The polynomial format is the traditional way of expressing how an LFSR works and is used by mathematicians to evaluate an LFSR operation.

There are a few important facts about LFSRs that you should be aware of:

1. The LFSR can *never* have the value zero in it. If it contains zero, then none of the internal bits will ever become set.
2. The ideal LFSR implementation will be able to produce $2^n - 1$ different values. It should be obvious that the one value that cannot be produced is zero.
3. A poorly specified LFSR may have the situation where it ends out with a value of zero.

The operation of a single shift of the 8 bit LFSR in Fig. 9-13 can be modeled using the "C" function:

```
int SingleLFSRShift (intx)
{                                 //Shift the "CurByte" Value
//  with the polynomial 1 + x**4 + x**5 + x**6 + x**8
int LowBit;                       //XOR'd Low Bit
```

```
LowBit = x >> 7;                       //"x**8" Term
LowBit = LowBit ^ (x >> 5);            //"x**6" Term
LowBit = LowBit ^ (x >> 4);            //"x**5" Term
LowBit = LowBit ^ ((x >> 3)&1);        //"x**4" Term
return((x << 1) + LowBit) &0x0FF;      Return shifted 8 Bit value
} //  End SingleLFSRShift
```

Hardware State Machines

The hardware state machine circuit was originally designed to allow designers to create a complex application using a simple, single ROM, a few register bits and some basic logic gates instead of a complex sequential circuit design or a processor-based solution. State machines are not widely used in modern applications because the costs of the parts needed to make up the circuit can very easily exceed that of a microcontroller. Almost ironically, hardware state machines are used as the control mechanism for most modern computer systems because they are fairly easy to design, program and debug. The use of hardware state machines (which are typically referred to as just "state machines") as the control mechanism for computer processors has given a new importance to the understanding of state machines.

The typical "state machine" is shown in Fig. 9-14. This circuit consists of an ROM (usually EPROM) which has part of its output data fed back as a "state address". Other address lines are used as circuit inputs and the state machine changes its state address based on these inputs.

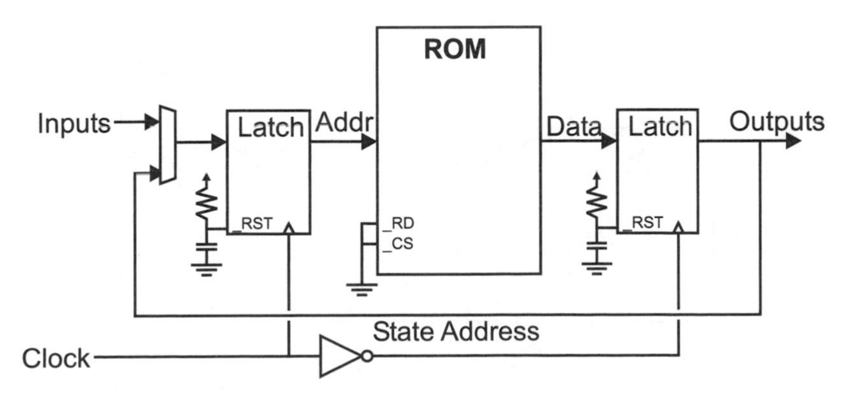

Fig. 9-14. General case hardware state machine.

The clock is used to pass the new address to the ROM and then pass the output from the ROM to the output and input state circuits. The two latches are operated 180° out of phase to prevent "glitches" from the ROM changing state from invalidly affecting any output circuits. A single edge triggered register is not typically used with the state machine because toggling inputs while the ROM is being accessed could result in invalid data being passed into the latches.

As few output bits are used as the "state address" as possible. The reason for this is to maximize the number of outputs and minimize the number of states which have to be programmed. Each state requires two to the number of inputs to function. Each state responds differently according to the inputs it receives.

A typical application for state machines is a traffic light. If a press-button crossing light, as shown in Fig. 9-15, is considered, a state machine circuit, like that shown in Fig. 9-16 could be used.

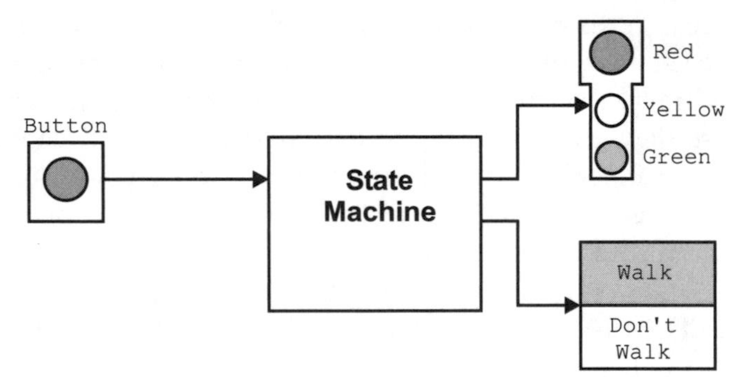

Fig. 9-15. Traffic light state machine block diagram.

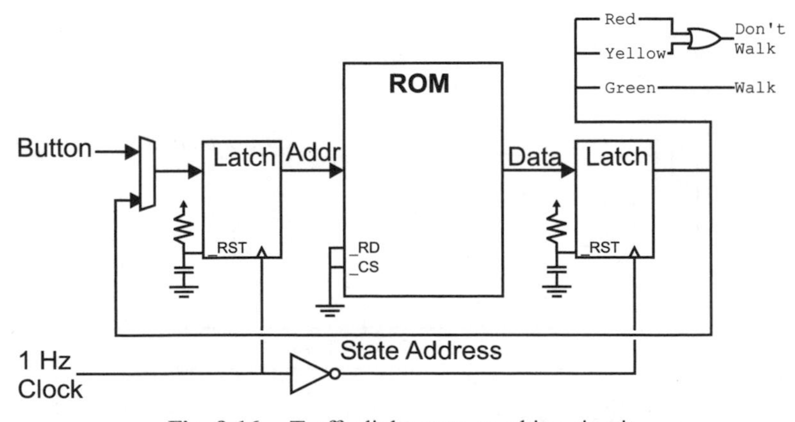

Fig. 9-16. Traffic light state machine circuit.

In normal operation (which is known as "state 0"), the green light is on and the button is not pressed. If the button is pressed, then execution jumps to state 1, which turns on the yellow light for 5 seconds (states 2, 3, 4 and 5), after which the red light is put on for 26 seconds (states 6–31). If the button is pressed during states 7–31, then execution jumps to state 6 to reset the timer.

Table 9-1 ROM programming for simple traffic light state machine.

State	Button	New state	Green	Yellow	Red	Comments
B'00000'	1	B'00000'	1	0	0	Power up
B'00000'	0	B'00001'	0	1	0	Power up/button press
B'00001'	x	B'00010'	0	1	0	Yellow LED on
B'00010'	x	B'00011'	0	1	0	Yellow LED on
B'00011'	x	B'00100'	0	1	0	Yellow LED on
B'00100'	x	B'00101'	0	1	0	Yellow LED on
B'00101'	x	B'00110'	0	1	0	Yellow LED on
B'00110'	x	B'00111'	0	1	0	Yellow LED on
B'00111'	x	B'01000'	0	0	1	Red LED on
B'01000'	1	B'01001'	0	0	1	Red LED on
B'01000'	0	B'00111'	0	0	1	Red LED on/reset Ctr
B'01001'	1	B'01010'	0	0	1	Red LED on
B'01001'	0	B'00111'	0	0	1	Red LED on/reset Ctr
⋮		⋮				⋮
B'11111'	1	B'00000'	0	0	1	Return to green
B'11111'	0	B'00111'	0	0	1	Red LED on/reset Ctr

To keep the circuit simple, I want to use an eight bit data bus ROM with six inputs (five state, one button). This means that 2**6 (or 64) states are required in the ROM. These states are listed in Table 9-1. The reset on the input address latch is used to reset the state to 0 on the power up. The button is assumed to be "pressed" if a "0" is returned.

Table 9-1 would then be converted into bits and burned into the ROM. An "x" means both input states have the same result on outputs.

This application is reasonable to code and build, but a problem arises with very complex state machines (ones that require tens of inputs and hundreds of different states). These state machines are normally hard coded into a custom chip rather than built out of discrete parts like I have shown for this application. The reason for placing it within a chip is to give more outputs as well as more states in a custom application. The depth and the width of the data in "real" applications is better suited to custom chips which can have non-custom memories added much more easily than in the situation where only commercial chips are used.

In the example above, I have used a state machine with a one second clock. Obviously in this situation there can be problems (such as the missed input if the button is pressed for less than 1 second and it isn't released after it is pressed). This function makes state machines unattractive for rapidly changing inputs and any kind of sophisticated real-time processing of inputs is simply not economical to do with the state machine. When I say "not economical", I am thinking in terms of the memory and properly programming the many states.

Quiz

1. In the sequential circuit block diagram where is the clock signal passed to?
   (a) To the "state memory" and "output formatter" blocks
   (b) To just the "state memory" block
   (c) To the "input formatter" block
   (d) To the "reset control circuitry" block
2. Asynchronous digital logic design is being pursued because:
   (a) It will result in simpler chip designs
   (b) Circuitry designed under this philosophy will be easier to interface to

(c) The end of performance gains using traditional design methodologies is in sight
(d) It offers faster operations with less power usage

3. Why are edge triggered registers used for counters instead of latches?
(a) It will result in simpler circuit designs
(b) Circuitry designed under this philosophy will be easier to interface to
(c) Less power is required
(d) It offers faster operations with less power usage

4. Ripple counters are:
(a) Always the fastest way to implement counters
(b) Usually more complex electronically than other counter designs
(c) Always the slowest way to implement counters
(d) Similar to ripple adders in operation

5. What are advantages of serial data transmission over parallel data transmission?
(a) Reduced number of drivers and receivers
(b) Faster data transmission
(c) Lower product costs
(d) Higher product quality

6. Where is serial data transmission not used?
(a) The internet
(b) Broadcasting stations to TVs/radios
(c) Keyboard to PC interface
(d) PCI bus interfaces

7. Linear feedback shift registers are built from:
(a) The system architectural drawings
(b) The high-speed circuits to support communications
(c) Shift registers and XOR gates
(d) The basic system serial interface

8. When the value of a linear feedback shift register equals zero:
(a) The operation has completed
(b) Either the initial and input values are zero or there is a problem with the LFSR design
(c) There was an error in encrypting a message
(d) Power has been removed from the circuit

9. Hardware state machines are rarely used except in:
   (a) Computer processors
   (b) Military and space applications
   (c) High-performance custom logic applications
   (d) Situations where old ROMs are easily available

10. State machines are normally built:
    (a) Out of discrete chips and ROM chips
    (b) In complex custom chips
    (c) On specially designed carrier PCBs
    (d) With the checksum of the ROM printed on them

Circuit Interfaces

For a sequential digital electronic circuit to be effective, it has to interface with something. This something could be a person or it could be other digital electronic circuits. If you were to look at different interfaces for either case (human or machine), you will discover that as the function of the circuit increases in sophistication, so does the interface. The reasons for this increase in interface complexity can be attributed to an increased amount of data to present as well as an increased number of operating parameters to choose from and select. The challenge is to come up with a way of adding these user and device interfaces simply, effectively and not affect the operation of the central sequential circuit.

Simple logic level switches and individual LEDs for each bit are perfect examples of the types of interfaces that I am talking about; to add these devices to your application, you generally don't require any types of busses nor do you need to have any special communications protocols for communicating with the devices. These interfaces are simple to add and modify to an application.

The problem with simple logic level switches and LEDs connected to each bit is that they cannot be very descriptive; nor are they very efficient methods of transferring data. An eight bit system is quite manageable, but it becomes very difficult when there are tens, hundreds or even thousands of bits to control and monitor. Early computers started out using simple switches and lights for input and output, respectively, but quickly outgrew them and began using printers, teletypes and punch cards to get state information from the computer. Today's computer systems have very sophisticated input and output capabilities, requiring the power of a processor that would have been identified as a "supercomputer" 10 years ago or less.

An example of a complex interface that you would be hard pressed (if it were possible at all) to create digital electronics for is the Hitachi 44780 based LCD module (Fig. 10-1). The controller hardware is fairly complex and must be accurately timed. The LCD module works like a "teletype" or a single line TV display – as you write characters to it, a "cursor" will move to the right, to prepare for the next character. The character interface consists of the eight data bits and three I/O pins listed in Table 10-1.

Fig. 10-1. Sample LCD output.

Table 10-1 Hitachi 44780 pin interface.

Pins	Description/function
1	Ground
2	Vcc
3	Contrast voltage
4	"RS" – _instruction/register select
5	"RW" – _write/read select
6	"E" clock
7–14	Data I/O pins

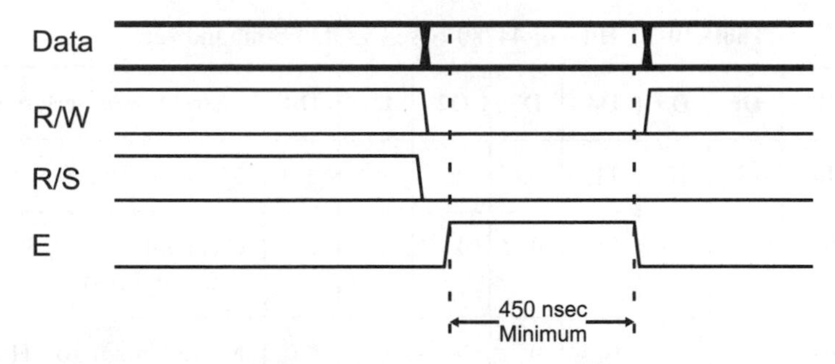

Fig. 10-2. LCD write waveform.

I typically attach a series of pins to the 14 connector pins so that the LCD can be easily mounted on a breadboard. In some LCDs, you may discover that there are 16 connector holes with the extra two holes used for backlighting. Some other LCD modules have two rows of seven or eight pins. For the ease of creating the experiments in this book and wiring them to the breadboard you should just use LCD modules that have a single row of pins.

Wiring the LCD to the hardware is quite straightforward as you will see in the waveform diagram (Fig. 10-2). The only unexpected aspect of the interface circuit is a potentiometer used to set the "contrast voltage" used by the LCD. The potentiometer is wired as a voltage divider, with the contrast voltage pin connected to the wiper of the potentiometer. Depending on the type of LCD that you are using, you will find that the voltage producing the best contrast will either be high or low, depending on the technology used in the LCD.

To communicate with the LCD, you will have to send the data words listed in Table 10-2 via the LCD interface. These bytes are commands that set the operating mode of the LCD or command it to perform some other operation. In Table 10-2, I have listed the different commands, along with the "RS" and "RW" lines that are used to control them. To clock in the command, the "E" bit must have a high value ("1") written to it and then a low value ("0").

Data displayed on the LCD is, for the most part, ASCII and you can pass ASCII characters directly from the hardware to the LCD. I say that the LCD can display ASCII "for the most part" because you will find that some characters are not supported (such as the backslash, "\") and if you go outside the normal ASCII character limits, you will see Japanese characters on the display. If you were to send a carriage return, line feed or any of the other ASCII terminal command characters, you would discover that they

Table 10-2 Hitachi 44780 based LCD command set.

RS	RW	D7	D6	D5	D4	D3	D2	D1	D0	Instruction/description
4	5	14	13	12	11	10	9	8	7	LCD I/O Pins
0	0	0	0	0	0	0	0	0	1	Clear Display (Takes up to 5 ms)
0	0	0	0	0	0	0	0	1	*	Move Cursor to "Home" (5 ms)
0	0	0	0	0	0	0	1	ID	S	ID = 1, Increment Cursor after Write S = 1, Shift Display after Write
0	0	0	0	0	0	1	D	C	B	D = 1, Turn on Display C = 1, Cursor On B = 1, Cursor Blink
0	0	0	0	0	1	SC	RL	*	*	SC = 1, Shift Display after Write RL = 1, Shift Display to Right
0	0	0	0	1	DL	N	F	*	*	Reset the 44780 Interface Length DL = 1,8 Bits/ DL = 0,4 Bits N = 1, Two Display Lines F = 1, 5x10 Font (Normally 0)
0	0	0	1	A	A	A	A	A	A	Move Cursor to Graphic RAM Address B'AAAAAA'
0	0	1	A	A	A	A	A	A	A	Move Cursor to LCD Position B'AAAAAAA'
0	1	BF	*	*	*	*	*	*	*	Poll LCD "Busy Flag" (Active "1")
1	0	D	D	D	D	D	D	D	D	Write Data to the LCD
1	1	D	D	D	D	D	D	D	D	Read Data from LCD at Current Cursor

*, "Don't Care".

result in a strange character being displayed. If you want to provide more "terminal"-like functions to the 44780 based LCD you will have to write them yourself and add them to your application.

Most commands execute in 160 μs or less with the display clear and move cursor to home commands can take up to 5 ms. The initialization process for the LCD is:

1. Wait more than 15 ms after power is applied.
2. Write 0x030 to LCD and wait 5 ms for the instruction to complete.
3. Write 0x030 to LCD and wait 160 μs for instruction to complete.
4. Write 0x030 AGAIN to LCD and wait 160 μs or Poll the Busy Flag.
5. Set the Operating Characteristics of the LCD:
    - Write "Set Interface Length"
    - Write 0x010 to disable display shifting
    - Write 0x001 to clear the display
    - Write "Set Cursor Move Direction" setting cursor behavior bits
    - Write "Enable Display/Cursor" & "Enable display and optional cursor".

The LCD could be controlled by a state machine, but there would be a significant amount of work to do this (and the state machine would be quite large). Along with the eight bit interface, the LCD can also be controlled by a four bit interface; each character and eight bit instruction is passed in four bit blocks through the D7:4 pins, but this interface would probably be even more difficult to create for the LCD module.

Address and Data Decoders

When you have decided upon the interfaces to your application, you will probably have to determine the best method of selecting which device is active at any time. The method that would make the most sense is to use the same method that a microprocessor uses: output a bit value, selecting the device and one control bit to activate the interface device. Depending on the resources available, the section bits may consist of a number of bits, each one passed to a different interface device, or a binary value, which is decoded into a specific control bit.

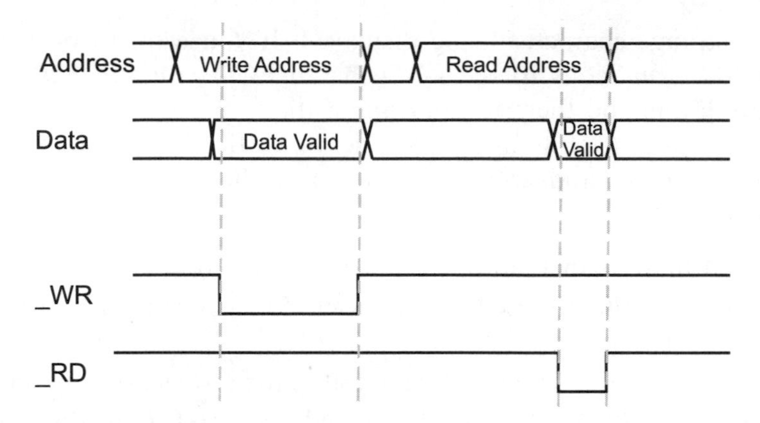

Fig. 10-3. Memory bus reads and writes.

Ideally, the signal being passed to the interface device would look something like Fig. 10-3 – an "Address" value is passed to the device and after a data "set up" time, a Read ("_RD") or Write ("_WR") line becomes active. A "Read" action polls the interface device and returns the value to the sequential circuit. A "Write" action does the exact opposite: it sends a value from the sequential circuit to the interface device.

You should notice that the timing of the read and write operations are quite a bit different. The short "Read" pulse is indicative of the expected operation of the device being accessed; once it receives the "Read Address" which selects the device, it takes some time to prepare the data before it can be read out. Similarly, when writing data, the _WR line is active for a surprisingly long period of time to allow the interface device to pass the data internally and prepare the interface circuitry to correctly store the data.

The interface read and write operations are good examples of situations where the latches rather than registers are used. When the _RD and _WR signals become active, data should be passed through them as quickly as possible rather than being held on a rising or falling value of the signal edge. For most applications, this need for taking advantage of every possible picosecond of time for data transfer is not needed, but you will find that it's a good idea to work with a standard design interface that will work in all situations.

Following this philosophy, rather than providing an individual bit to each interface device, how about a binary "address" that can be decoded to an individual address using a "decoder" like the 74139 (Fig. 10-4) that converts a two bit value into four individual active low outputs. The 74139 contains

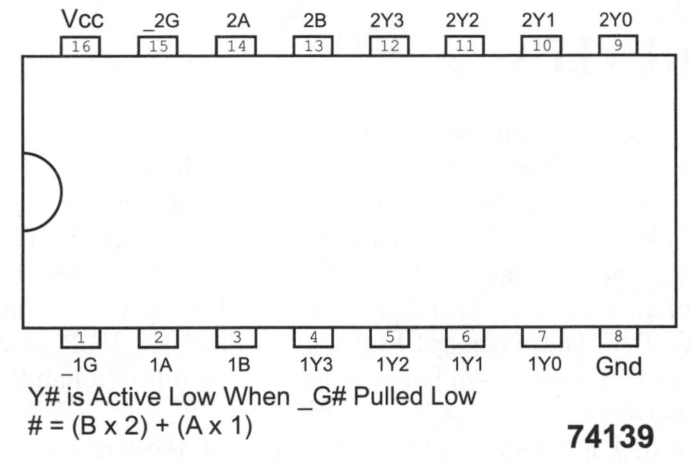

Fig. 10-4. 74139 dual 2 to 4 decoder chip pinout.

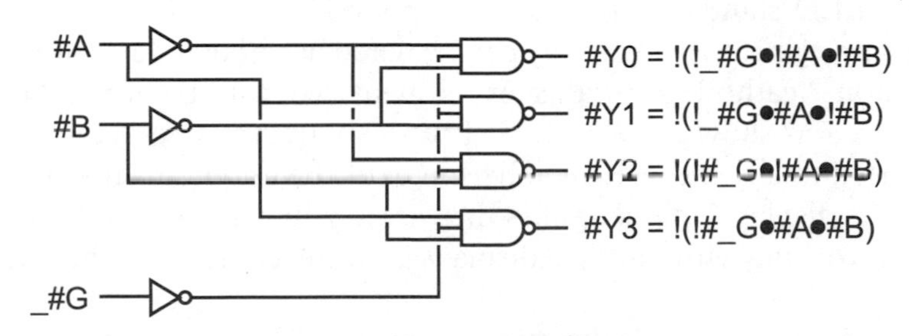

Fig. 10-5. 2 to 4 decoder logic diagram.

two "two to four" decoders like the ones shown in Fig. 10-5. Decoders (also known as "demultiplexors") convert binary values into individual output lines and are primarily used to decode memory addresses to individual chips.

The resulting "interface" is essentially a computer "bus" and will allow you to add standard computer interfaces to your circuit, probably with no modification. You might have been wondering why I went through the effort of providing an interface select function along with the _RD and _WR signals – you may be thinking that a single pin that both selects the device as well as initiate the read or write would be enough; the philosophy of interfacing like in a computer is only so useful. This is true if you are going to only work with custom-designed interfaces, but there are a lot of standard interfaces (including the various ones presented in this chapter) designed for being accessed by computer systems that you will want to take advantage of.

Multi-Segment LEDs

I think you would be hard pressed to find somebody in the industrialized world that has never seen a "seven-segment" LED display (Fig. 10-6) before. It first became popular in the 1970s and is used in almost literally *everything* from digital clocks to car instruments. Seven-segment LEDs can be found virtually everywhere, being used not only in digital watches but also in kitchen appliances, cars, instruments and, of course, in videocassette recorders (VCRs). The flashing "12:00" on a clock or VCR created using seven-segment LEDs is the symbol of a person's inability to handle the latest in technology.

In Fig. 10-6, I have shown the appearance of the seven-segment LED display – it can be put in the same "footprint" as a 0.300 inch" wide 14 pin DIP package, but some of the pins ("N/C" for "no connect") are not present. The "DP" LED stands for the "decimal point".

The seven-segment LED display can be wired as either a "common anode" or "common "cathode"; in this experiment we will be using "common anode", wired as shown in Fig. 10-7. For this part, the two "common" pins are connected to all (and occasionally some) of the anodes of the eight LEDs built into the display. This simplifies the wiring you will have to do somewhat and makes working with multiple displays a bit easier, as I will show in a later experiment.

Despite its commonality, the seven-segment LED display is not trivial to work with. There are a number of chips on the market that make the component easier to work with in some applications, but when you are working with your own sequential circuits, you will find that these "canned" functions never quite do what you hope for.

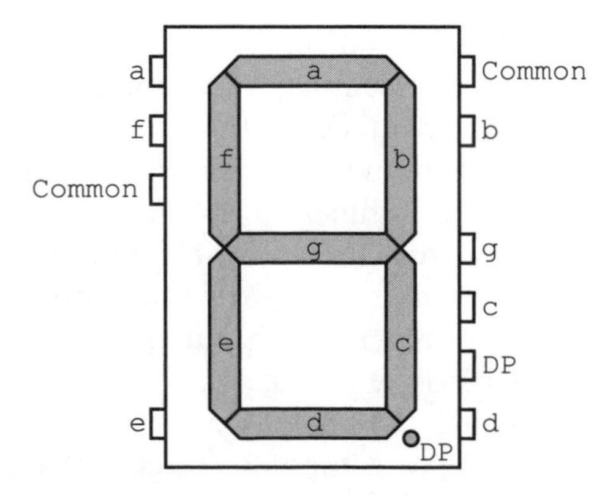

Fig. 10-6. Seven-segment LED display with pinout.

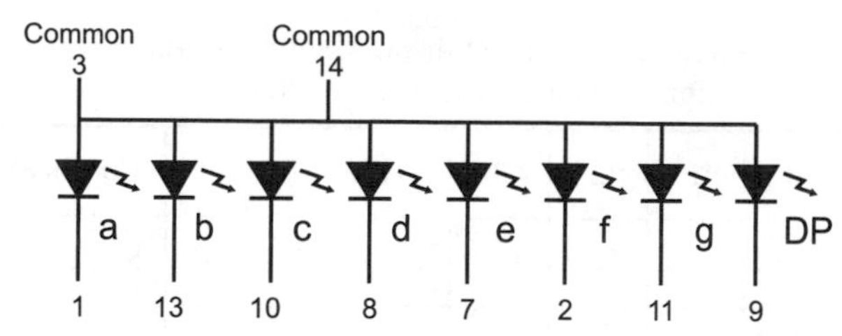

Fig. 10-7. Common anode seven-segment LED display internal wiring.

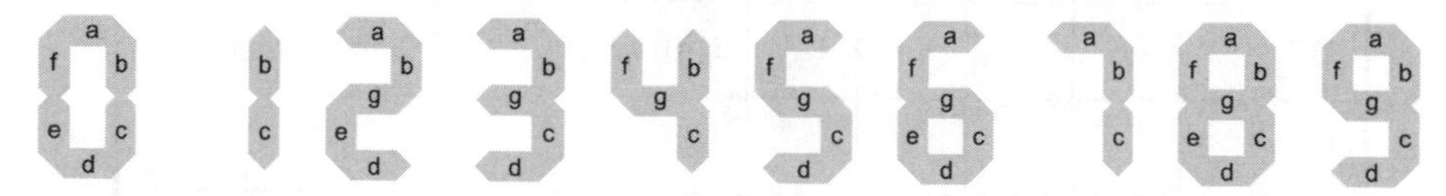

Fig. 10-8. Lit seven-segment LED display elements for different numbers.

As you are probably aware, by turning on each of the different LEDs with their unique values individually, you can create a multiple-digit display. Figure 10.8 shows how by turning on different LED segments, the display can be used to display the 10 numeric characters. Along with the 10 numbers, there are a number of letters that can be displayed, although only a few of them look exactly like the characters they are supposed to represent. If you want to display letters as well as numbers, then you will have to use a LED with more segments – these are available as either 16-segment displays or as matrixes of LEDs that display the character as a multi-dot "font" like on your computer screen.

Each LED in the display can be wired conventionally to control whether or not they are turned on or off. Controlling individual LEDs in a single display is quite easy. It gets quite a bit more difficult when you have to display different values on multiple LED displays. To convert incoming bits to meaningful characters on the display, you will have to pass the bit values through a combinatorial circuit, like the one shown in Table 10-3 for the first four decimal digits.

In Table 10-3 in the "Comments" column, you can see that I have noted any commonalities between different equations and noted that segment "B" is always active for all four digits. I should point out that coming up with the equations for each of the segments is good practice for working with Boolean arithmetic equations, but it is much easier and simpler to buy a seven-segment LED driver. The 7447 chip is commonly used for decoding the incoming bits and driving the LEDs and is an excellent solution when there is

Table 10-3 Seven-segment LED display combinatorial specification for the first four decimal characters.

Segment	"0"	"1"	"2"	"3"	Terms	Comments
A	1		1	1	(!A0 · !A1) + A1	Same as "d"
B	1	1	1	1	1	Always on
C	1	1		1	!A1 + (A0 · A1)	
D	1		1	1	(!A0 · !A1) + A1	Same as "a"
E	1		1		!A0	
F	1				(!A0 · !A1)	Uses AND from "a" & "d'
G			1	1	A1	

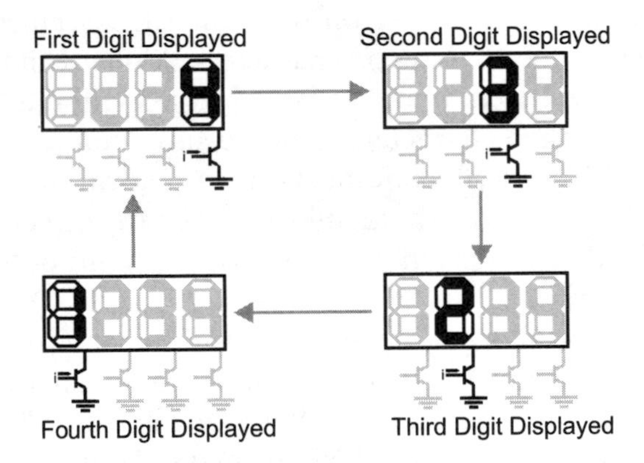

Fig. 10-9. Scanning through four seven-segment LED displays.

only one seven-segment LED display outputting information from the application.

When multiple LED displays are required, instead of providing multiple-digit drivers, a single-digit driver is used and different values are passed to it for different displays very quickly. Figure 10-9 shows a four-digit LED display with each digit having a different value. Each digit is turned on momentarily to display its value and then switched off for the next digit. The eye's visual persistence ignores the flickering if the sequencing is done fast

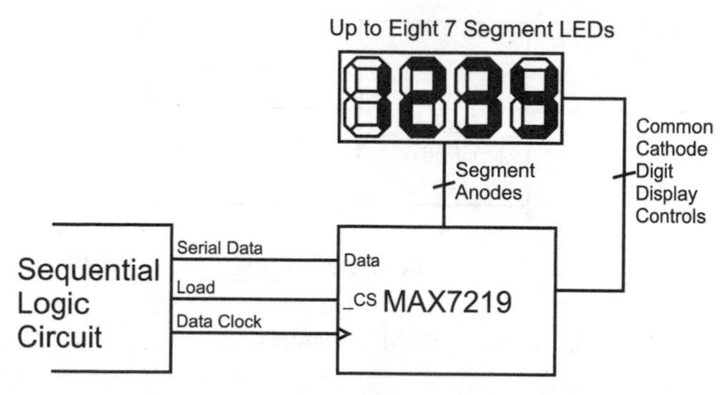

Fig. 10-10. Using a Maxim MAX7219 seven-segment LED controller chip.

enough and it appears that all the digits are on simultaneously, even though they are displaying different values.

As a rule of thumb, each display should be active 50 or more times per second. The slower each display is flashed on and off, the more likely the human eye will pick up the flashing. A flashing multi-character display is not attractive and could cause headaches in some people (especially if the displays are very bright). The time each display is turned on must be as equal as possible. If one display is on for a longer period of time than the others, then it will appear brighter and, conversely, a display active for a shorter period of time will appear dimmer. When working with multiple displays, in order to meet the 50 times per second guideline, you are actually going to have to loop through your individual display action 50 times per second multiplied by the number of displays. So, for a four-digit display, you will have to loop 200 times per second and each digit will be on for 5 ms at a time. There are some chips, such as the very popular Maxim MAX7219 (Fig. 10-10), which can control multiple seven-segment LED displays. This chip takes care of all the driving and timing requirements for the displays; the only catch is that you must shift in the desired value for the display.

Pulse Width Modulation

Despite showing how logic gates and other digital devices are built from simple analog components, they do not handle working with analog voltages very well. There are some circuits that will produce a valid analog (an arbitrary voltage, not just logic "high" and "low") voltage but they do not work very well if the circuit has to drive a high current device. Instead of varying the voltage level to provide varying levels of power, I produce a

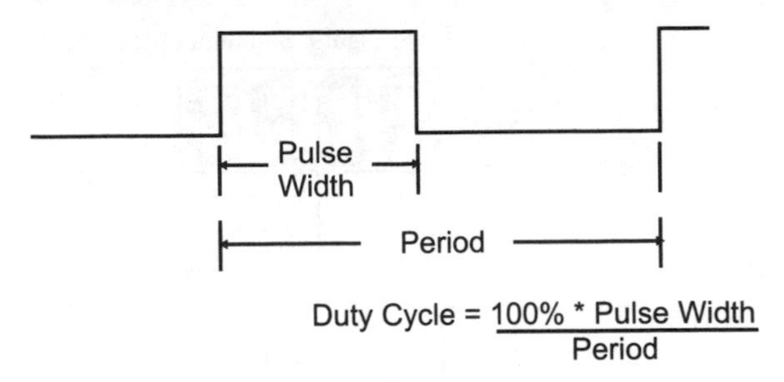

Fig. 10-11. PWM waveform features.

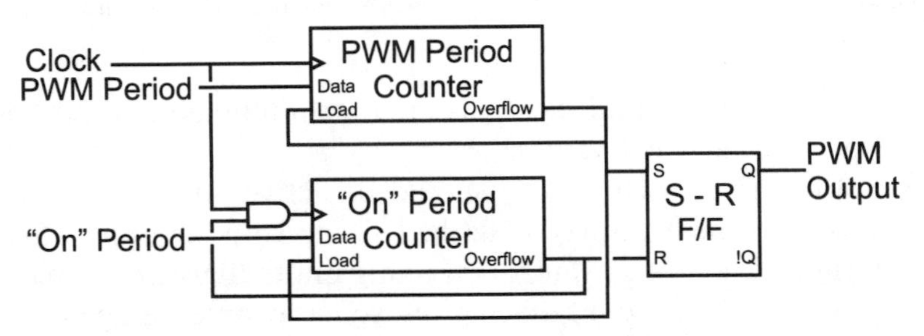

Fig. 10-12. PWM generator block diagram.

string of timed pulses known as a "pulse width modulated" ("PWM") signal (Fig. 10-11). A PWM signal is a repeating signal that is "on" for a set period of time that is proportional to the voltage being output. I call the "on time" the "pulse width" in Fig. 10-11 and the "duty cycle" is the percentage of time the "on time" is relative to the PWM signal's "period".

To output a PWM signal, there are several possible methods. One way is to use two counters that have a common clock. When one counter overflows, it resets itself and the second counter. Until the second counter overflows, the output of the circuit is set to "1". When the second counter overflows, the output of the circuit is reset until the first counter overflows and the process is repeated. Figure 10-12 shows how this type of circuit could be implemented.

This PWM generator circuit uses counters that are reloaded (from the "Data" pins) upon an "Overflow" positive pulse. The "PWM Period Counter" (the "first counter") runs continuously and when it overflows (reaches the final count), it resets and reloads the count value for not only itself but also for the second counter (the '"On" Period Counter').

When the PWM Period Counter resets, it "Sets" the S-R flip flop, driving the "PWM Output" high for the start of the PWM signal output. The "On" Period Counter is reset and reloaded by the PWM Period Counter and runs until it overflows. When the "On" Period Counter Overflows, the PWM

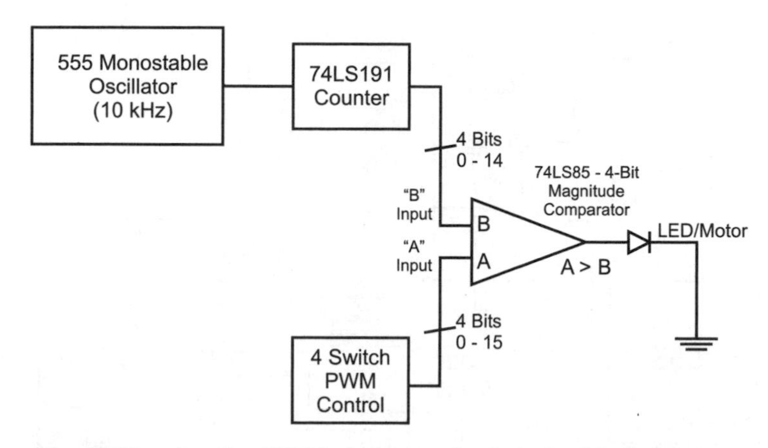

Fig. 10-13. Another PWM generator circuit design block diagram.

Output is halted and it also stops running until the PWM Period Counter reloads it, which resets the "Overflow" output and allows the Counter to drive the "On" Period Counter once more.

Another type of PWM generator is shown in Fig. 10-13. The counter output will be continuously compared against a bit value and when the bit value is greater than the counter value, a "1" will be output. The block diagram for the circuit that I envisioned is shown in Fig. 10-13 and can be built quite easily as I show in this section.

When you study Fig. 10-13, there will probably be one point that won't make sense to you: I show that the counter ranges from 0 to 14 and not 0 to 15, as you would expect for the typical four-bit counter. I wanted the counter to reset itself at 14 rather than 15 so that when the binary values were compared, a 100% duty cycle could be produced as well as a 0% duty cycle by outputting a "1" when the set value was greater than the counter value. If the counter ran from 0 to 15 then the circuit would not be able to produce a PWM with a 100% duty cycle.

To produce the bit range from 0 to 14, I used the 74×191 chip counting down and tying the "_LOAD" pin to the "_RIPPLE" pin and driving the inputs to 14. The "_R" ("Ripple" Output) pin becomes active when the chip is "rolling over" from one extreme to another and the "_LD" pin moves the value at the input pins into the counter's latches when it is active. Normally, when a four-bit counter is "rolling over" as it counts down it goes from 0 to 15, but by tying the "_R" pin to the "_LD" (negative active "Load") pin of the 74×191, you can load in a new value when the counter reaches 0 and is about to roll over. This feature is ideal for this application as it ensures the count stays in the range of 0 to 14.

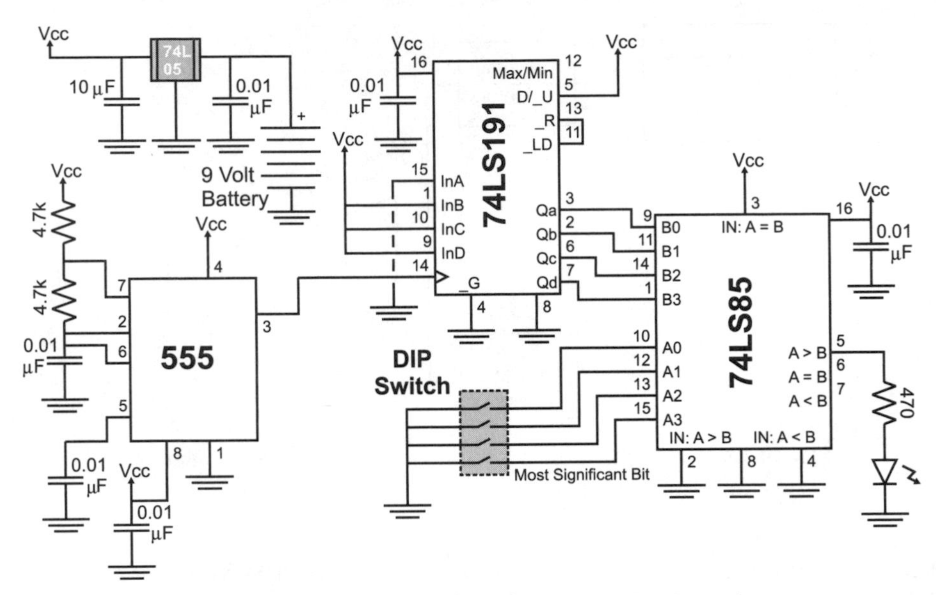

Fig. 10-14. Sample PWM generator circuit schematic.

Converting the block diagram to a schematic one is very straightforward (Fig. 10-14) and wiring it onto the PCB's breadboard is tight but not really a challenge (Fig. 10-15). The PWM output value is specified by the four-position DIP switch. I placed 0.01 µF decoupling capacitors on all of the power inputs of each of the chips. These decoupling capacitors are very important when working with the standard (not CMOS) 555 because it can place large transients on the power line.

I used TTL chips (powered by 5 volts from the 78L05 regulator) rather than CMOS chips because I found that it is difficult locating 74C85 chips. An advantage of using TTL instead of CMOS for this circuit was that I could simply pull the comparator inputs to ground without the need of a pull up resistor. If you build this circuit with CMOS chips, make sure that you have 10 k pull up resistors on the DIP switch to ensure a high voltage is passed to the comparator.

Once you have built the circuit, you will find that the LED's brightness will be dependent on the value on the DIP switch. It will be confusing, as the value on the DIP switch will seem to be the opposite to the behavior of the PWM. When all the switches are "on", the LED will be off and, for what seems to be a "large" value, the LED will be dim. When all the switches are "off" the LED will be full on. This confusion is a result of the "on" marking indicating when the switches are closed, not when the signal is a "1" or "high" (which is often extrapolated to being "on") – when the switches

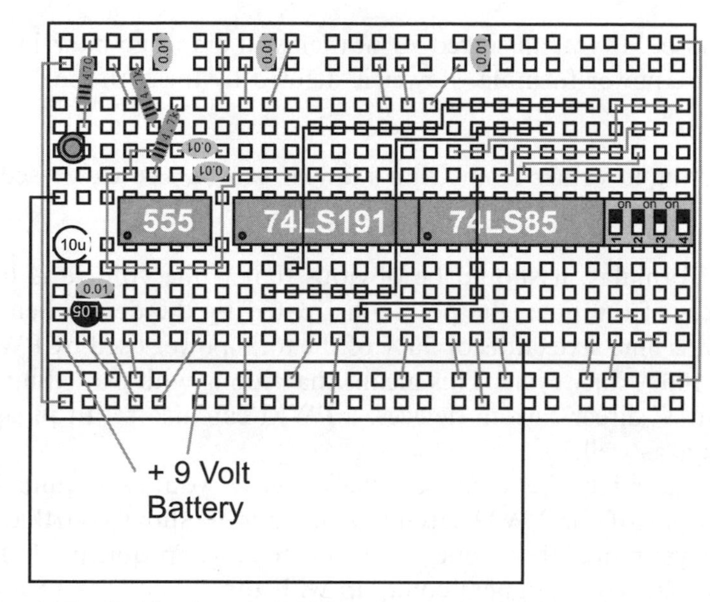

Fig. 10-15. Sample PWM generator circuit wiring diagram.

are closed ("on"), the comparator input is pulled to ground and has the value "0".

For all PWM circuits (not just the two I've shown here), you must remember that the effective frequency is the input clock frequency divided by the counter value. For the example circuit shown here, the 10 kHz signal is divided by 15 (how many cycles the 74LS191 counts before resetting) so the resulting output signal frequency is 667 Hz, which is still faster than the human eye can perceive a flashing LED, but much lower than required for some DC motors. PWMs are commonly used to control the speed of electric motors and if the PWM frequency is within the audible range of human hearing, you will hear a definite "whine" from the motors. The solution to this problem is to either run the PWM at frequencies above human hearing (greater than 18 kHz) or below the range of human hearing (60 Hz or below).

The lower PWM frequencies should not be an issue to produce, but the higher ones can be a challenge, especially if more bits are used in the counter. For example, to create a 20 kHz PWM output signal, you will have to provide a 300 kHz clock for a 15-value PWM and 5.1 MHz for a 255-value PWM! You may find that to get a practical circuit, you will have to find a compromise between the number of bits used in the PWM for the signal level and the speed of the oscillator that is going to be used with it.

An interesting feature of a PWM is how it can save you power. If you were to run the PWM with a 75% duty cycle, what do you think the average power output would be? If you answered 75%, then you didn't go back in the book to look up the power formula. Power is defined by the formula:

$$P = V \times i$$

and substituting in values from Ohm's law, it can be also expressed as:

$$P = V^2/R = i^2 \times R$$

From these formulas, it should be obvious that if the voltage is high only three-quarters of the time, the power dissipated by the device being driven by the PWM is nine-sixteenths or 56% of the total power used by PWMs running with a 100% duty cycle. This means that, along with providing the ability to "throttle" direct current devices, a PWM can also result in significant power savings as well.

Finally, you might be confused that I gave you two quite different implementations of the PWM circuit; I did this to show you that there is almost always more than one solution to any problem. I normally recommend that new designers come up with three solutions to a problem before going ahead and implementing *something*. Having three solutions to choose from will allow you to compare features and drawbacks and choose the solution that is best for the application.

Button "Debouncing"

I consider the issue of debouncing switches and buttons to be one of the most important and vexing problems that you will have to deal with when you are developing applications that work with operator input. Most people think that electrical connections happen instantaneously; you might be surprised to discover that the contacts within a switch actually bounce a few times before the switch makes a constant contact. This is shown in the oscilloscope picture in Fig. 10-16.

Earlier in the book, I showed you a simple method of debouncing a switch input by creating a small memory device from two inverters. A major drawback of this circuit is that it "backdrives" the outputs of one of the inverters, but this problem can be eliminated through the use of CMOS inverters and a 10 k current limiting resistor. Even with this fix in place, there is another problem to consider when deciding whether or not to use this circuit – finding double throw push buttons can be difficult. This circuit is well suited for double throw switches but, from the practical difficulty of finding double throw buttons, it becomes impractical.

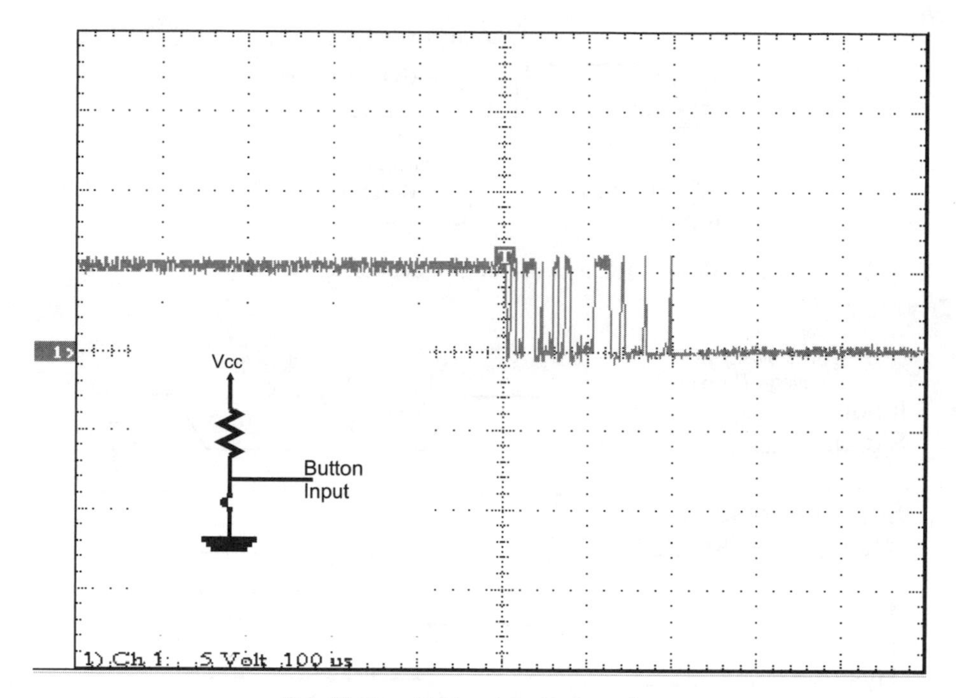

Fig. 10-16. Button bounce waveform.

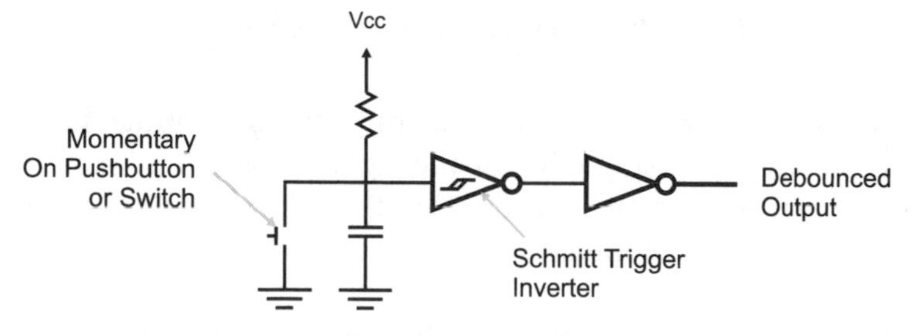

Fig. 10-17. Simple button debounce circuit.

The debounce circuit that I recommend you use is shown in Fig. 10-17. This circuit consists of a resistor–capacitor network that charges over a given amount of time or discharges quickly through a closed switch or button. Figure 10-18 shows the filtering of the bouncing; it is not perfect, but it is much better than what we started with.

The inverter with the funny symbol in Fig. 10-17 is called a "Schmitt Trigger Input Inverter" and provides an extra measure of filtering of the button input. Schmitt trigger inputs are designed to change state on the rising or falling edge of a signal with "hysteresis", as shown in Fig. 10-19.

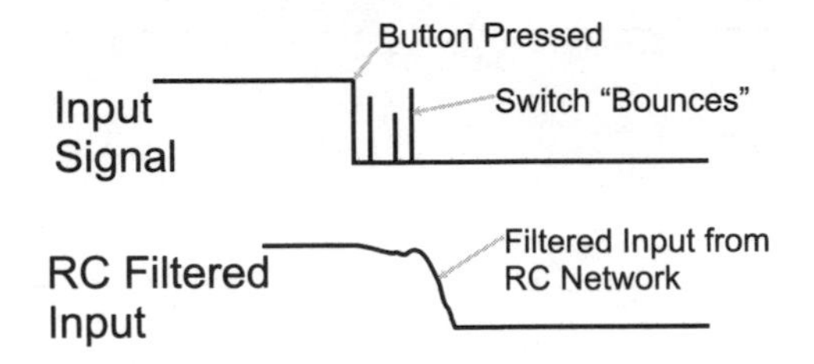

Fig. 10-18. RC network button debounce operation.

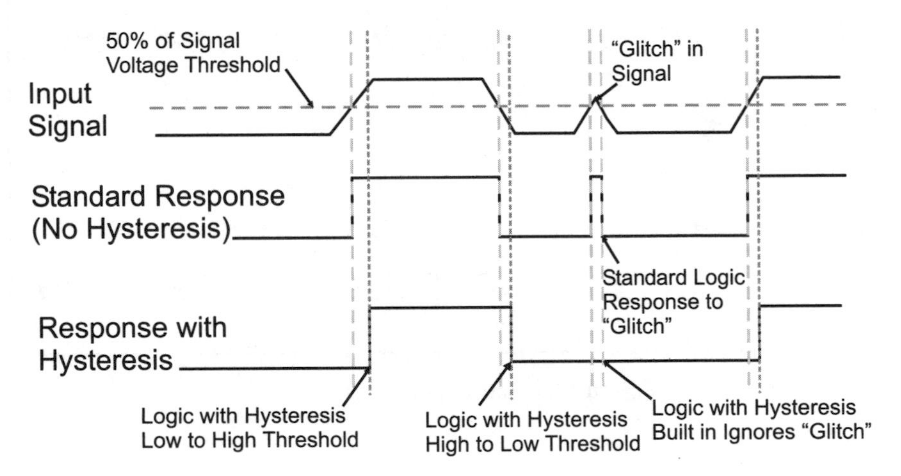

Fig. 10-19. Schmitt trigger input operation.

"Hysteresis" is the property of the Schmitt trigger inputs in which the threshold point for the rising edge of the signal is different than the falling edge. Looking at Fig. 10-19, you can see that the rising edge threshold is above the "normal gate voltage threshold", while the falling edge threshold is less.

These changing threshold values are the reason for the strange symbol on the inverters, indicating Schmitt trigger inputs. Figure 10.20 shows the input versus the gate response on an "X-Y" chart. The "X" axis is the input voltage with rising voltages to the right and the "Y" axis represents the response of the Schmitt trigger input. By following the numbers, you can see the response of the input and that it forms the same symbol that I put on the inverter gates. For comparison, a traditional logic gate does not use this symbol – the response threshold is the same for rising and falling edge signals.

Another method of debouncing button inputs is to use a 555 or monostable circuit. In Fig. 10-21, I show a 555 wired as a monostable, driving out a pulse from a button press. The internal waveforms of the circuit

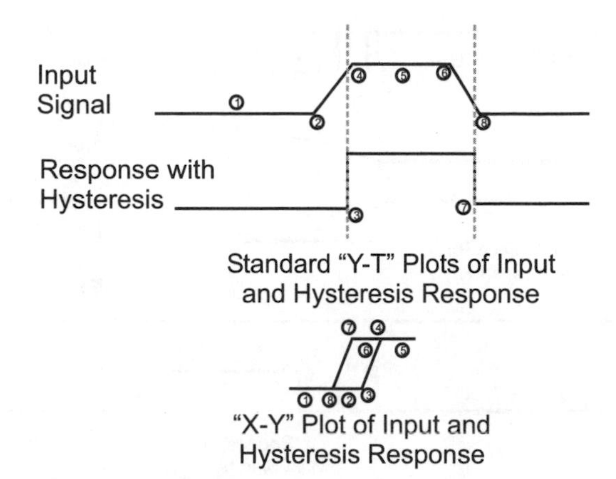

Fig. 10-20. Schmitt input hysteresis.

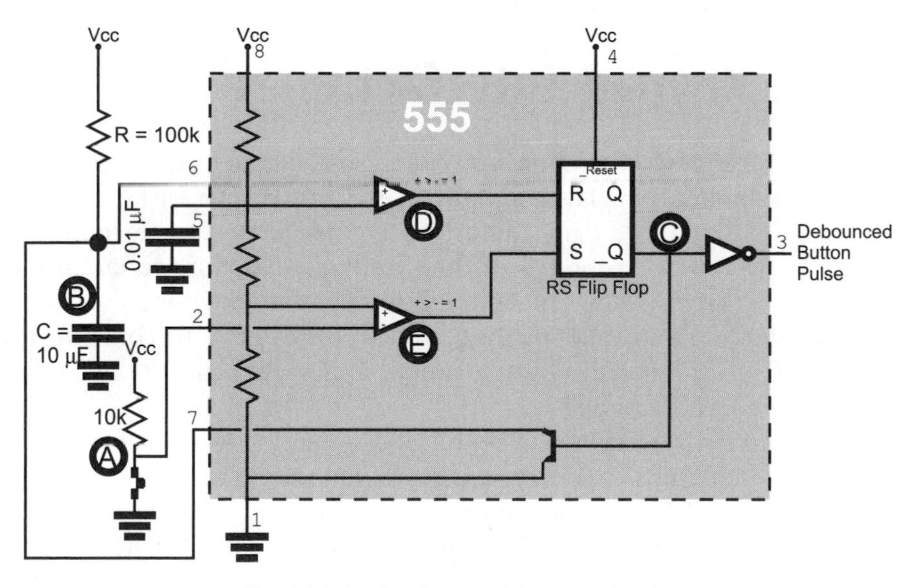

Fig. 10-21. 555 button debounce circuit.

are shown in Fig. 10-22, which shows that any subsequent bouncing of the button after it makes its first connection are ignored by the circuit as the pulse is being output. If you work out the pulse time from R and C, you'll discover that the pulse time is roughly 1 second in length. This should be long enough for a single button press to be registered and the user to remove his fingers. Obviously, this delay is too long to implement multiple buttons or even any kind of data entry functions in the circuit. To do this, you should consider the next section.

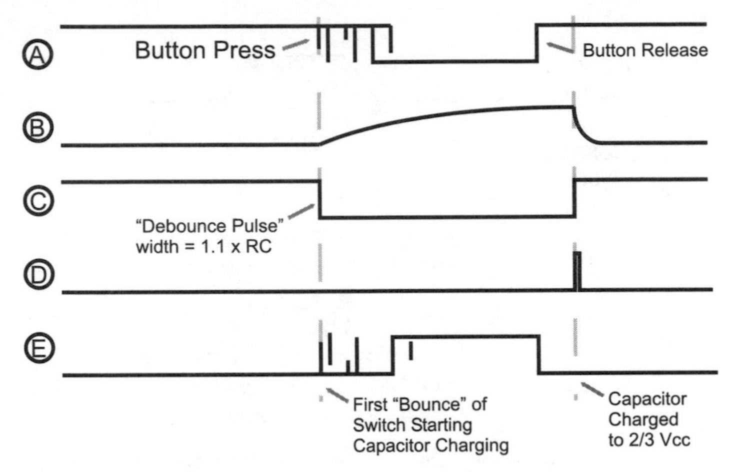

Fig. 10-22. 555 button debounce circuit waveform.

Switch Matrix Keypad Interfacing

As I ended off the previous section, you cannot use simple button debouncing techniques to implement a large number of buttons or even a keyboard for data entry. Just so there's no confusion, I consider a "large number of buttons" to be four or more; providing individual debounce circuits for anything more than a couple of buttons is expensive and time consuming. Along with the cost and time involved, you will also have to come up with some way of prioritizing the button inputs and recognizing non-standard keys like "shift" and "control".

The keys and buttons in PC keyboards and numeric keypads are arranged in "rows" and "columns" and they can be drawn out in such a way that they look like a "matrix". A "momentary on" switch is placed at the intersection of each row and column, as shown in Fig. 10-23. This "switch matrix" provides the ability to "scan" a large number of button inputs with a relatively small number of lines. Your PC's 104/105 keyboard usually has a 22 by 7 matrix connection to a microcontroller, which scans through the keys and reports any key presses using the algorithms presented in this section. Keyboards with a 100 keys or more are an extension of the four-button key matrix shown in Fig. 10-23 and have the same concerns and issues to watch out for.

You probably cannot see immediately how the individual keys or buttons of the switch matrix shown in Fig. 10-23 can be polled, but the operation will probably become clearer when you see the resistors and transistors I've added to the switch matrix in Fig. 10-24.

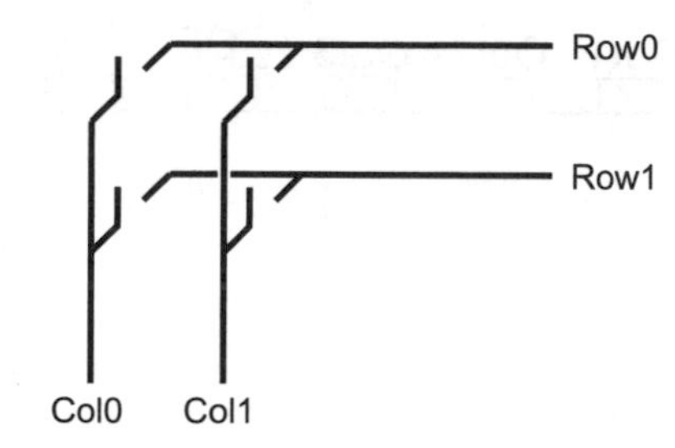

Fig. 10-23. Simple switch matrix.

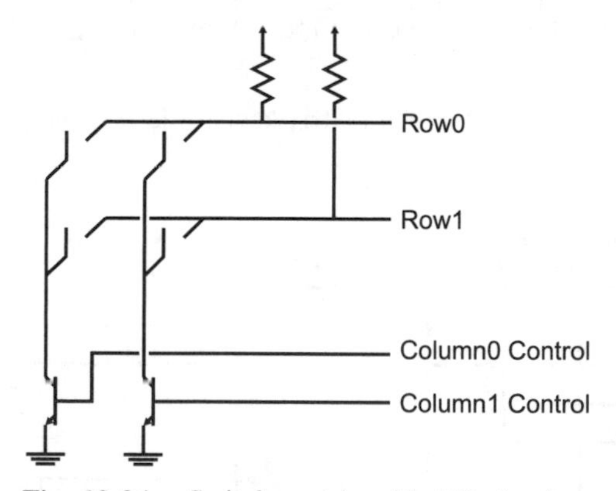

Fig. 10-24. Switch matrix with I/O circuitry.

In this case, by connecting one of the columns to ground, if a switch is closed, the pull down on the row will connect the line to ground. When the row is polled by an I/O pin, a "0" or low voltage will be returned instead of a "1" (which is what will be returned if the switch in the row that is connected to the ground is open due to the pull up on it). To scan the keyboard, the column transistors are turned on, one at a time, and while the column transistor is on and the column is pulled to ground, the rows are compared to a logic level of "0", which would indicate that the button is pressed.

This methodology for handling switch matrix keypad scans I've outlined here probably seems pretty simple. Depending on your familiarity with programming and different microprocessors and microcontrollers, you will probably realize that implementing these functions could be done even simply in assembly language programming or "C". You should also realize that this code would be quite difficult to implement just using logic chips.

To avoid the complexities of trying to develop TTL logic that will carry out the functions described in the pseudo-code presented above, I normally

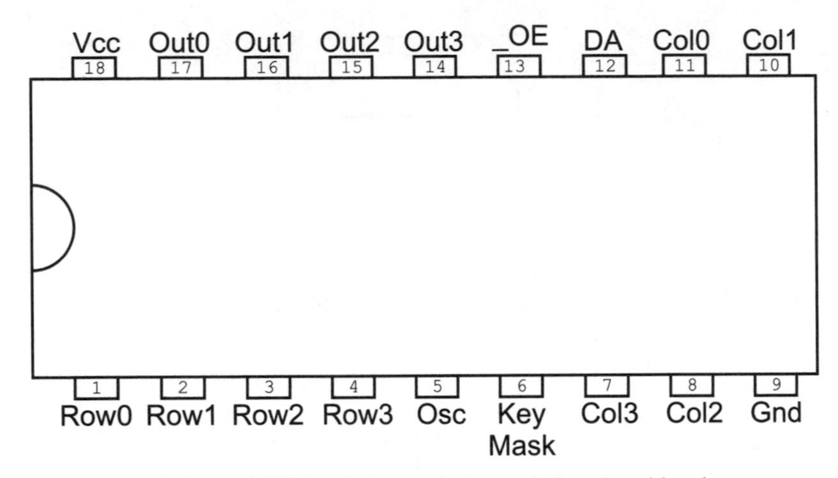

Fig. 10-25. 74C922 switch matrix keypad decoder chip pinout.

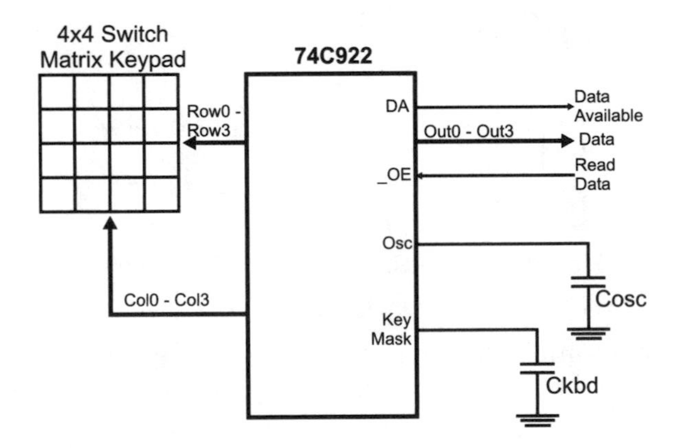

Fig. 10-26. Basic 74C922 wiring.

use the 74C922 keypad decoder chip (Fig. 10-25). This chip can be used to debounce and encode up to 32 buttons (although 16 is the normal maximum) and carries out button debouncing internally as well as keeping track of two currently held-down keys when new keys are pressed. The 74C922 is quite easy to wire to a four by four (16 button) switch matrix keypad, as shown in Fig. 10-26. By "doubling up" rows of sensors of the 74C922, you can add a number of additional keys to the application. In the next section, I will show how this is done to create a 20 button input device (with up to 32 possible).

The two capacitors are used to create a relaxation oscillator within the chip that is used to "scan" through the buttons as well as provide a "debounce" delay count for the application. The two capacitor values are calculated as:

$$\text{Cosc} = \text{Scan Rate}/10\ \text{K}$$
$$\text{Ckbd} = 10 * \text{Cosc}$$

I like a debounce interval of 20 ms: plugging this into the formulas above, I get a value of 2 μF and 20 μF. When I build my own applications, I tend to have a lot of 10 μF (for power filtering) and 1 μF (for MAX232 RS-232 level converters) electrolytic capacitors on hand. I have not found any problems with using these components and I would recommend that you use them as well to avoid having to stock multiple capacitor values for different applications.

Quiz

1. What are ideal sequential circuit interfaces?
   (a) LCDs
   (b) Individual switches and LEDs
   (c) Keyboards
   (d) USB flash disks
2. What is the suggested digital electronics interface to an Hitachi 44780 controlled LCD display?
   (a) Hardware state machine
   (b) Microcontroller
   (c) Sequential circuit
   (d) Combinatorial circuit
3. Is a bus "Read" or "Write" faster?
   (a) Write is faster
   (b) Read is faster
   (c) Using a synchronous clocked circuit, they take the same amount of time
   (d) Read is slower due to the need to retrieve data from the interface device
4. Seven-segment LED displays have a common:
   (a) LED anode or cathode
   (b) Segment pins

(c) LED drivers built into the package
(d) Pin interface that is used by all devices, regardless of the number of digits in the package

5. Multiple seven-segment LED displays show different values by using:
   (a) Linear feedback shift registers that have encoded the bit patterns
   (b) Turning on each individual digit with its unique value periodically
   (c) Multiple LED driver circuits that drive the value for its respective digit to the LED display
   (d) Multiple bits of memory, one for each segment, which are loaded according to the display value

6. For a PWM circuit running its logic at 5 volts and a duty cycle of 67%, what is the "on" voltage level of the output signal?
   (a) 0.67 volts
   (b) 5 volts
   (c) 2/3 volts
   (d) 3.35 volts

7. The power dissipated by a PWM running with a 20% duty cycle will be what fraction of a 100% duty cycle?
   (a) 0.04
   (b) 40%
   (c) 0.4
   (d) 400%

8. A 555 monostable with $R = 100\,k$ and C of $4.7\,\mu F$ will output a pulse of:
   (a) Approximately 0.5 s
   (b) Approximately 4.7 s
   (c) Approximately 1.1 s
   (d) Insufficient data given to determine the pulse width

9. Rows and columns in a switch matrix keypad have what connected to them?
   (a) The rows have a transistor connected to ground and the columns have a capacitor
   (b) The columns have transistors connected to ground and the rows are left open

(c) The columns are left open and the rows have a transistor connected to ground
(d) The columns have transistors connected to ground and the rows have pull up resistors

10. The 74C922 reads a switch matrix keypad by:
(a) Pulling the columns of a switch matrix keypad to ground and scanning the rows for pulled down bits
(b) Driving a 1 kHz square wave on the rows and polling the columns for the signal
(c) Using an internal microprocessor
(d) Measuring the capacitances of individual lines and looking for changes

Reading Datasheets

I've never understood why college and university courses do not give an introductory course in reading digital electronic device datasheets. Despite how prepared you are for them, you will feel quite overwhelmed the first time you have to look through a number of datasheets trying to find a part that meets your requirements. When I first started working with electronics, datasheets were generally quite poor, with only a few standout companies providing good documentation for their chips. Fortunately, this has changed over the past 10 years, the Internet and the capability of downloading good-quality datasheets being almost a marketing tool to help engineers select the parts they are going to use in their designs.

Personally, I find it more daunting to look at datasheets over the Internet because they are generally encoded as Adobe Acrobat pdfs that take a while to load and you can never flip the pages on the screen as fast as you would like. To make matters worse, it can be very difficult to put multiple datasheets up on a computer display to allow you to compare the features of the different chips. This difficulty gives rise to the most important recommendation that I can make about downloading datasheets from the Internet – print them out! I have several binders of printed out datasheets for parts that I often use. By printing them out, I have immediate access to them and I can flip back and forth between pages effortlessly. It is my opinion that documentation shouldn't be "paperless".

The first sheet of the datasheet is usually a one page description of the part. It normally contains:

1. Part number
2. High-level part description
3. Part pinout
4. Common/related/pin compatible parts
5. Important chip features
6. Basic operations truth table

When looking at a datasheet, you should first check out the part number of the datasheet versus what you are interested in. This means that you should be checking not only the numeric code for the device but also the high-level identifier and the technology identifier. For example, if you were looking for a low-power TTL dual input NAND gate and looked up the datasheet based on a web search for "TTL dual input NAND", you could see such diverse part numbers as:

```
74LS00
74C00
54LS00
74W00
74ALS03
```

with the question being: Which is the one that you want?

For these parts the high-level identifier is the "74" or "54". In this book, I have focused on the 74 series of logic – but if you look at "54" series logic you will see that its operation is identical and may decide to go ahead and order the parts. This could be a big problem because "54" series parts are military-grade chips and they tend to cost 10 times that of standard "74" series logic and do not necessarily have the same pinout as "74" series chips. The technology identifier is the letter code between the high-level identifier and the part number. In the list of five chips above, I have presented traditional TTL low-power logic ("LS"), CMOS logic ("C"), advanced Shottkey low power ("ALS") and single gate CMOS ("W"). The danger of not reading the datasheet's part number is that you could end up ordering the wrong part number, resulting in higher than expected costs and lost time looking up and reordering the correct part.

Always read through the datasheet's first page high-level part number description. This can range from a single sentence to four or five bulleted items. Like the part number check, this should just be a filter operation, resulting in you making sure the part will do essentially what you want it to do.

The part pinout is something that is critical to know when you are wiring a circuit. Except for the wiring experiments presented in this book, I have listed only a few part pinouts because a part's pinout may vary between manufacturers of the same part and they may vary according to the packaging type. The part pinout may also change according to packaging technology. It isn't unusual to see a pin through hole (PTH) packaged chip with a specific pinout but its surface mount technology (SMT) sibling having extra pins or different connections to different pin numbers. I'm sure that both of these statements are a bit hard to understand; you might be thinking that the part numbers are standard. I wish I could tell you how many times I have been bitten by these two little traps. Circuit design systems also make assumptions about part pinouts based on the pinouts from specific manufacturers and don't bother checking the pinouts from others.

The important chip features listed on the front page of the datasheet will not list the features of the chip to the lowest possible level, but it will give you some ideas about how the chip works and if there are any issues that could be a problem with you using the chip in your application.

A lot of times you will discover that a part will not have exactly the functions that you want but, by checking the datasheet, it may list related parts that provide a similar function that you can take a look at. Finally, for very simple chips, the front page of the datasheet will present you with truth tables describing the operation of the chip or the different parts of the chip.

The front page of a chip's datasheet can be incredibly useful to you and by spending a few minutes familiarizing yourself with it, you can decide whether or not the chip is appropriate for your application without having to delve into the minutia of the following pages.

Chip Operating Characteristics

An important feature of the datasheet is the "operating characteristics" for the chip. This section of the datasheet explains such operating parameters as:

1. Input pin voltage thresholds and currents
2. Output voltages along with current source and sinking capabilities

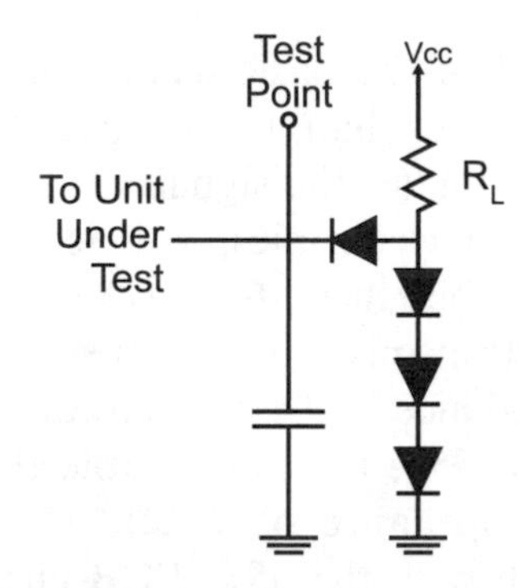

Fig. 11-1. Sample chip test circuit.

3. Gate delay timing
4. Expected input and output pin line impedances
5. Miscellaneous operating information.

Each chip datasheet lists the logic thresholds, along with their characteristics when subjected to different parameters. Often you will see a two-axis graph, with a curve showing the chip characteristic response to the changing input parameters. This part of the drawing should not be difficult to understand but what can be confusing is the small schematic marked as a "test circuit" that often accompanies the graph. An example of such a schematic is shown in Fig. 11-1; it shows the circuit that was connected to the pin while the test was taking place. These test circuits simulate other circuits connected to the chip, helping to ensure that the chip is operating as it would in a typical application.

The output voltage and current characteristics are really a function of the logic technology used and not unique to the individual chip's pins. If you were to read other datasheets of chips built from the same technology you would discover that the output parameters are the same between the two chips and, by extrapolation, all the chips built from this technology. If you were to search the manufacturer's web site, you would discover that this information has been published for all parts in the technology family and the information in the chip's datasheet is really redundant. The reason why the information is repeated in the individual chip's datasheets is to minimize the amount of cross-checking that you will have to do.

The previous comment could be made about gate delay timing but there is a wrinkle in the specification in the datasheet. Is the quoted "gate delay" for the chip function or for the individual basic technology gate (i.e. the "NAND" gate for TTL)? Normally, the datasheet will list the chip function gate delay instead of the basic technology gate delay because the actual gate delay is probably less than the product of the basic gate delay time multiplied by the number of gates the signal has to pass through.

As you learn more about electronics, you learn that not only do wire connections have resistance but they also have capacitance and impedance. All these factors affect the transmission of data signals and are known by the term "characteristic impedance". Printed circuit boards (PCBs) have a characteristic impedance of 55 Ω (the coax cable that sends signals to your TV has a characteristic impedance of 75 Ω). The input and output pins must be designed to match with the 55 Ω PCB characteristic impedance to ensure that signals pass between pins as efficiently as possible.

All these chip characteristics and any miscellaneous data that the chip manufacturer feels important enough to include should be read through and understood in order to best wire a chip into your application circuit.

IEEE Logic Symbols

When you look at some datasheets, you will see the function of the chip described using a graphical system that is different from the one that I have used in this book. Instead of unique shapes for each gate, they are represented as a rectangular block like the one in Fig. 11-2. These blocks are part of the "IEEE Standard Graphic Symbols for Logic Functions". This standard is often used to describe the operation of a chip instead of the graphical symbols that I have used in the book.

The IEEE gate definition contains a single character to indicate what the function is. Table 11-1 lists the basic characters and their functions. For negated outputs and inputs, the gate pin modifiers presented in Fig. 11-3 are used. Note in Table 11-1, only the four unique gate functions are listed – NAND and NOR gates are represented with the gate modifiers shown in Fig. 11-3.

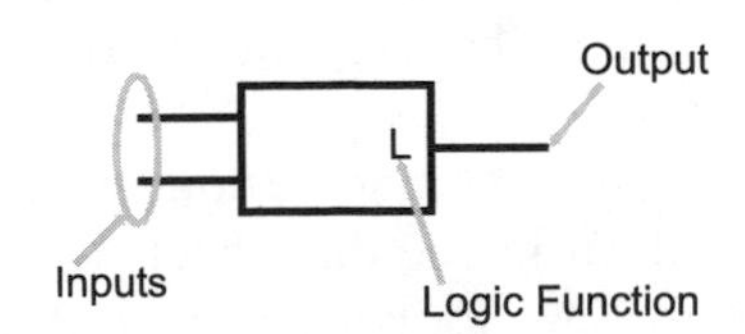

Fig. 11-2. Basic IEEE logic gate symbol.

Table 11-1 Four basic IEEE logic functions.

IEEE gate symbol	Gate type
&	AND
|	OR
!	NOT
^	XOR

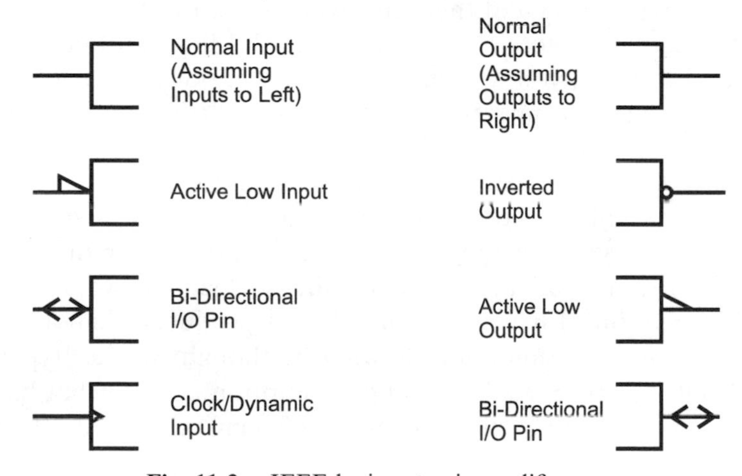

Fig. 11-3. IEEE logic gate pin modifiers.

For the complete IEEE logic symbol definition, I suggest that you download and printout ANSI/IEEE Std 91a-1991 from:

http://www.ee.ic.ac.uk/pcheung/teaching/ee1_digital/logic symbols.pdf

This document outlines the conventions used to identify each of the different functions used to describe different logic functions in the IEEE standard format. This method of presenting chip functions may seem to be rather difficult to decode when you first see it, but after you've worked with it a while, it will become second nature to you.

Having said this, I would suggest that you avoid working with these symbols until you are very familiar with working with digital electronics. The standard graphic symbols have been well thought out and are immediately recognizable when you are first learning to work with digital electronics and logic gates. The IEEE symbols can be difficult to distinguish when you are first starting out and you can very easily get yourself into trouble if you misread a symbol or forget the purpose of a pin modifier.

Power Usage and Fanouts

An important consideration for selecting a chip is the amount of power dissipated. This information is necessary not only for the individual gate but also for the complete application. The sum of all the power is the total power needed by the application and this value will dictate the power methodology used as well as the cooling requirements for the final product. While not explicitly a correlation, you will find that the more power a logic technology uses, the more external inputs that can be driven (this is the technology's "fanout").

When you look at a chip's current (which is related to its power) consumption, remember to look at not only the current required to power the gate but also at maximum input sinking and output sourcing or sinking. These currents should all be added together to get the worst-case power consumed by the chip. I have seen a number of products where the designer expressed the power consumption by what he thought was a "typical case" and found out that the actual current consumption is somewhat higher and the specified power supply did not have sufficient margin for the product to work reliably.

Along with the current consumption, the datasheet should also specify the number of input pins the chip's output pins can drive. It is important to note that the number of input pins quoted is the *same technology* as the chip. When you are mixing technology, you will have to understand the input current requirements of the input pins and, as "a rule of thumb", make sure that the total current drawn by the input pins does not exceed 50% of the total sinking current capability of the output pin. This will ensure that the logic functions will be at the correct levels regardless of the circumstances.

Actually, I would recommend that for your first applications, you *never* drive more than three inputs from a single output and strive to drive no more than two outputs in the design. Marginal signals due to overloaded output pins are very difficult to recognize from the failure symptoms and difficult to confirm when the problem is suspected.

Quiz

1. What isn't on the first page of the datasheet?
   (a) Part number
   (b) Part pinout
   (c) Chip cost
   (d) Important chip features
2. What is the technology identifier in the part number "74S174"?
   (a) 174
   (b) 74
   (c) 74S
   (d) S
3. What parameter isn't a chip operating characteristic?
   (a) Gate delay timing
   (b) Chip logic function
   (c) Input pin voltage thresholds and currents
   (d) Expected input and output pin line impedances
4. The chip gate delay specification
   (a) Is for ideal conditions
   (b) Is for the basic technology gate delay
   (c) Is for the chip function
   (d) Is for the NAND gate delay
5. IEEE symbols
   (a) Will replace the standard graphical symbols
   (b) Represent negative output functions by placing a symbol on the output pin
   (c) Is used to define all chips
   (d) Are only used for basic logic gates
6. Starting out using IEEE symbols
   (a) Is a bad idea as the symbols are not immediately recognizable
   (b) Is the recommended way to learn about digital electronics
   (c) Will help you design highly optimized digital electronics circuits
   (d) Will encourage you to buy from manufacturers that properly document their products

7. When planning for the current consumption of a product, which current specification should be ignored?
   (a) Standby current
   (b) Output low current sink
   (c) Input low current drain
   (d) None

8. The maximum number of inputs a single output can drive is:
   (a) Determined by the total current drawn by the inputs
   (b) Three in all cases
   (c) The output sink current specification divided by the average input current drain
   (d) Infinite

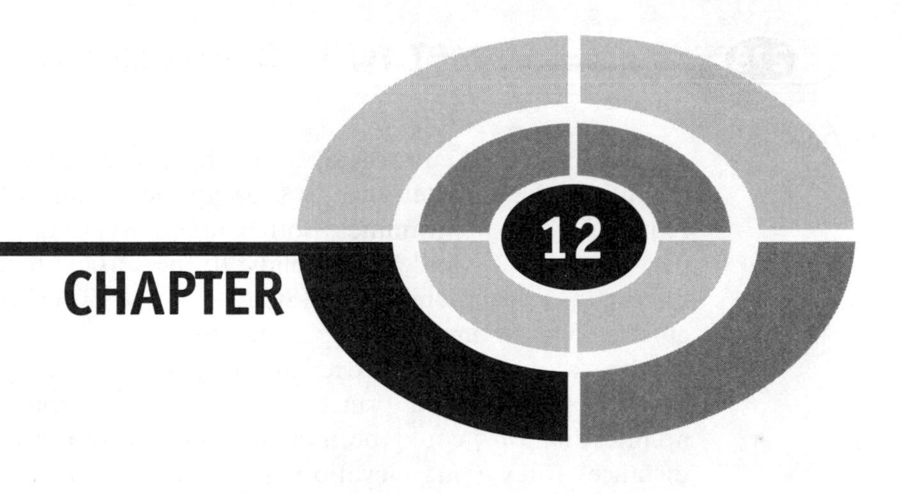

Computer Processors and Support

I'm sure that you realize that computer processors are really just a great big sequential circuit, but I'm sure that you have no idea where to start understanding how they work. Traditional computer processors are designed using a selection of six or so basic design philosophies that give them different characteristics. In this chapter, I will introduce you to the different issues that have to be confronted in designing computer processors, along with some of the technologies that have been developed to support them.

From a high level, computer processor architects choose from making the processors "RISC" ("Reduced Instruction Set Computers" – pronounced "risk") based or "CISC" ("Complex Instruction Set Computers") based. CISC processors tend to have a large number of instructions, each carrying out a different permutation of the same operation (accessing data directly, through index registers, etc.) with instructions perceived to be useful by the processor's designer while RISC systems minimize the instruction set, but give them as much flexibility and access as much of the memory in the system as possible. CISC processors also have the same requirement, but by definition, they are designed to simplify the amount of manipulation that is required by the programmer. Both computer types have their advantages and disadvantages – the RISC tends to be easier to design and executes instructions faster while the CISC tends to be easier to program but may be cumbersome in implementing some functions.

The second option processor designs have came from a competition between Harvard and Princeton universities to come up with a computer architecture that could be used to compute tables of naval artillery shell distances for varying elevations and environmental conditions. Princeton's response was for a computer that had common memory for storing the control program as well as variables and other data structures. It was best known by the chief scientist's name "John Von Neumann". Figure 12-1 is a block diagram of the Princeton architecture. The "Memory Interface Unit" is responsible for arbitrating access to the memory space between reading instructions (based upon the current Program Counter) and passing data back and forth with the processor and its internal registers. In contrast, Harvard's response (Fig. 12-2) was a design that used separate memory banks for program storage, the processor stack and variable RAM. By separating the data and program memories and avoiding the need to arbitrate data movements between them, there was an opportunity for programs to execute faster in Harvard's computer.

It may at first seem that the Memory Interface Unit of the Princeton architecture is a bottleneck between the processor and the variable/RAM space – especially with the requirement for fetching instructions at the same time. In many Princeton architected processors, this is not the case because of the time required to execute an instruction is normally used

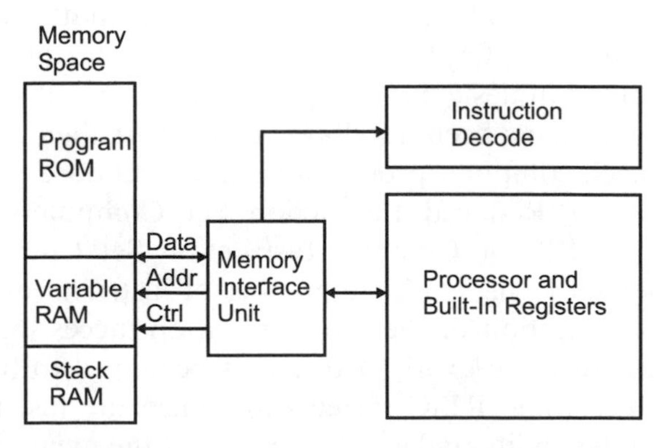

Fig. 12-1. Princeton processor architecture.

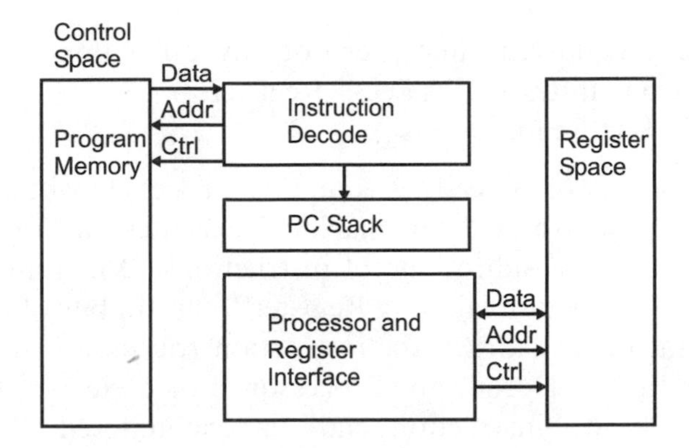

Fig. 12-2. Harvard processor architecture.

to fetch the next instruction (this is known as "pre-fetching"). Other processors (most notably the processor in your PC) have separate program and data "caches" that can be accessed directly while other address accesses are taking place.

The Princeton architecture won the competition because it was better suited to the technology of the time. Using one memory was preferable because of the unreliability of then current electronics (this was before transistors were in widespread general use): a single memory and associated interface would have fewer things that could fail. The Harvard architecture is really best for processor applications that do not process large amounts of memory from different sources (which is what the Von Neumann architecture is best at) and be able to access this small amount of memory very quickly.

Once the processor's instruction set philosophy and architecture have been decided upon, the design of the processor is then passed to the engineers responsible for implementing the design in silicon. Most of these details are left "under the covers" and do not affect how the application designer interfaces with the application. There is one detail that can have a big effect on how applications execute, and that is whether or not the processor is a "hardcoded" or "microcoded" device. Each processor instruction is in fact a series of instructions that are executed to carry out the instruction. For example, to load the accumulator in a processor, the following steps could be taken:

1. Output Address in Instruction to the Data Memory Address Bus Drivers.
2. Configure Internal Bus for Data Memory value to be stored in Accumulator.
3. Enable Bus Read.

4. Compare Data read in to zero or any other important conditions and set bits in the "STATUS" Register.
5. Disable Bus Read.

A microcoded processor is really a computer processor within a processor. In a microcoded processor, a "state machine" executes each different instruction as the address to a subroutine of instructions. When an instruction is loaded into the "Instruction Holding Register", certain bits of the instruction are used to point to the start of the instruction routine (or microcode) and the "uCode Instruction Decode and Processor" Logic executes the microcode instructions until an "instruction end" is encountered. This is shown in Fig. 12-3.

A "hardwired" processor uses the bit pattern of the instruction to access specific logic gates (possibly unique to the instruction) which are executed as a combinatorial circuit to carry out the instruction. Figure 12-4 shows how

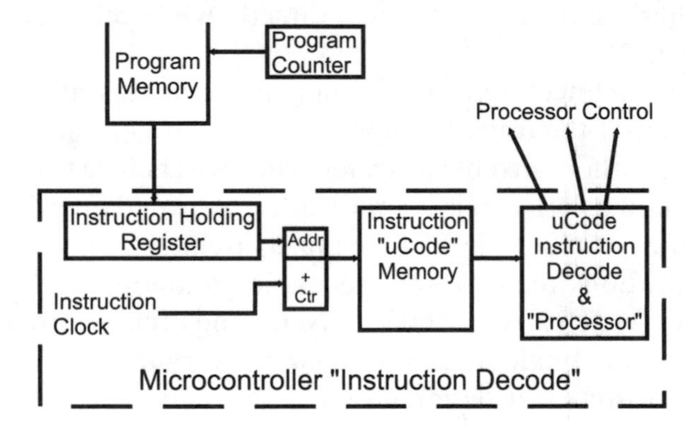

Fig. 12-3. Microcoded or state machine processor instruction decode circuitry.

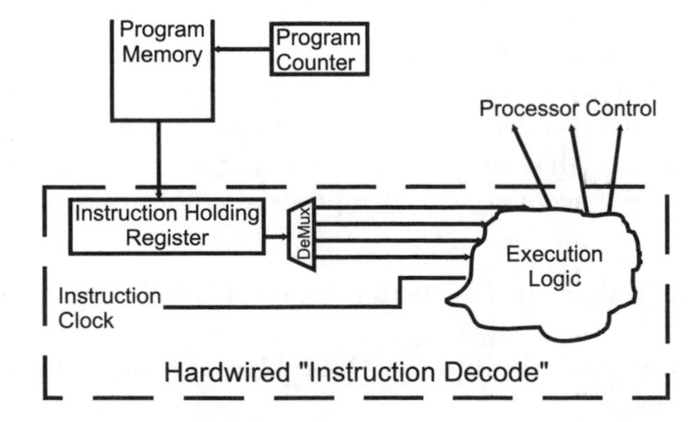

Fig. 12-4. Hardwired processor instruction decode circuitry.

the instruction loaded into the Instruction Holding Register is used to initiate a specific portion of the "Execution Logic" which carries out all the functions of the instruction.

Each of the two methods offers advantages over the other. A microcoded processor is usually simpler than a hardwired one to design and can be implemented faster with less chance of having problems at specific conditions. If problems are found, revised "steppings" of the silicon can be made with a relatively small amount of design effort. The hardwired processor tends to execute instructions much faster but is much harder to modify.

IEEE754 Floating Point Numbers

When I introduced binary numbers earlier in the book, I discussed binary integers, but I did not discuss how binary "real" numbers were produced or how they were manipulated in workstation processors. It should not be a surprise to discover that binary floating point numbers are analogous to decimal floating point numbers.

For example, if you were going to convert decimal 7.80 to binary, you would first convert the value equal to or greater than one to binary. Decimal 7 becomes B'0111', leaving decimal 0.80 to convert. This is accomplished by knowing that decimal fraction digits are multiplied by *negative* exponents of the base 10. The same methodology can be used for binary numbers.

To convert decimal 0.80 to a binary fraction, I will start with the exponent "−1" which is equal to 0.5 decimal and test to see if it can be removed from the fraction. Since it can, my binary number becomes B'0111.1' with a remainder of 0.30. Going to the next negative exponent ("−2"), I discover that I can subtract this value, giving me the binary value B'0111.11'. Continuing this on for another four bits, the binary value is B'0111.110011'. It's interesting to see that the binary number is irrational; the bit pattern will change the smaller the fraction that is calculated even though the decimal number ends at the first digit after the decimal point.

This method can be expressed as the "C" function, which converts the floating point number to a binary string:

```
BinaryFraction(float Value, char* BinaryValue)
{                        // Convert the Floating Point Value to
                         // a string pointed to by BinaryValue
```

```
int i, k;
int j = 0;
      BinaryValue(j ++) = `B´;      BinaryValue(j ++) = `\´´;
                                    // Initialie Binary String
   for(i = 0; (2<<i) < = Value; i ++);
                                    //Find Most Significant Bit
                                    // of Value´s Integer Value
   for (; i ! = -1; i --)
     if ((Value - (2<<i))> = 0)
        {                           // Take Away Binary Value
      Value = Value - (2<<i);
        BinaryValue (j ++) = `1´;
      } else                        // Can´t take anything away
      BinaryValue (j ++) = `0´;
        BinaryValue (j ++) = `.´;   // Find fractional value
                                    // to 4 digits
   for (I = 1, k = 0; k<4; i ++, k ++)
     if (Value - (1 / (2<<i)) > = 0)
     {                              // Can take away binary fraction
       Value = Value - (1 / (2<<i));
       BinaryValue (j ++) = `1´;
     } else
       BinaryValue(j ++) = `0´;
   BinaryValue(j ++) = `\´´;   //Close Binary String
   BinaryValue(j ++) = `\0´;
} // End BinaryFraction
```

This operation can be performed within high-performance processors (like the Intel Pentium), but instead of producing a string of characters representing binary data, they generally put them into the IEEE754 format, which stores the floating binary value in a format which is similar to that of "Scientific Notation":

```
(Sign) Mantissa × 2** (Exponent Sign) Exponent
```

The "Mantissa" is multiplied by the signed exponent to get values less than or greater than one.

Table 12-1 IEEE754 data sizes.

Data format	Data range	Data size
Long word	+/− 9.2(10**18)	8 Bytes
Packed BCD	+/− 10**17	8 Bytes
Single precision real	+/− 10**38 to +/− 10**−38	4 Bytes
Double precision real	+/− 10**308 to +/− 10**−308	8 Bytes
Extended precision real	+/− 10**4932 to +/− 10**−4932	10 Bytes

Table 12-1 lists the data formats supported by the Intel Pentium. These different formats give you a lot of flexibility to work with a wide range of numbers in different applications. All the number formats can be processed together, with the final result being in the most accurate format (i.e. a "word" and "single precision" combined together will have a result as a single precision number).

Memory Types

A number of different memory types are currently available. In this introduction, I will first show you three different technologies and discuss where (and why) they are used in a computer system. The boot up, non-volatile memory used in a computer system is based on ultraviolet light "Erasable PROM" ("EPROM") program memory (Fig. 12-5) and was first introduced in the late 1960s. An EPROM memory cell consists of a transistor

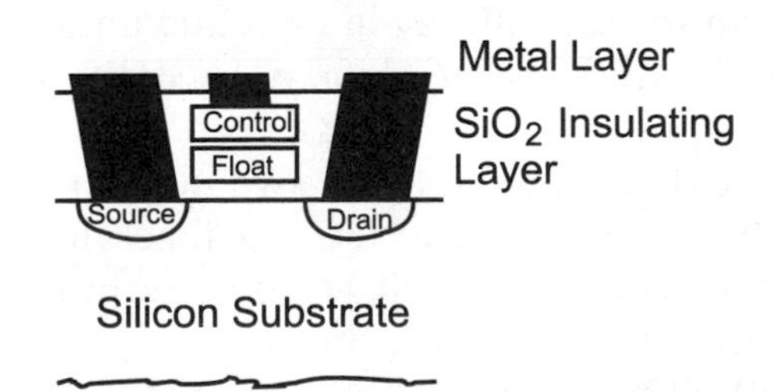

Fig. 12-5. MOSFET EPROM memory cell.

that can be set to be always "on" or "off". Figure 12-5 shows the side view of the EPROM transistor.

The EPROM transistor is a MOSFET-like transistor with a "floating" gate surrounded by silicon dioxide above the substrate of the device. "Silicon dioxide" is best known as "glass" and is a very good insulator. To program the floating gate, the "Control" gate above the floating gate is raised to a high enough voltage potential to have the silicon dioxide surrounding it to "break down" and allow a charge to pass into the floating gate. With a charge in the floating gate, the transistor is turned "on" at all times, until the charge escapes (which will take a very long time that is usually measured in tens of years).

An improvement over UV erasable EPROM technology is "Electrically Erasable PROM" ("EEPROM"). This non-volatile memory is built with the same technology as EPROM, but the floating gate's charge can be removed by circuits on the chip and no UV light is required. There are two types of EEPROM available. The first type is simply known as "EEPROM" and allows each bit (and byte) in the program memory array to be reprogrammed without affecting any other cells in the array. This type of memory first became available in the early 1980s.

In the late 1980s, Intel introduced a modification to EEPROM that was called "Flash". The difference between Flash and EEPROM is Flash's use of a bussed circuit for erasing the cells' floating gates rather than making each cell independent. This reduced the cost of the EEPROM memory and speeded up the time required to program a device (rather than having to erase each cell in the EEPROM individually, in Flash the erase cycle, which takes as long for one byte, erases all the memory in the array).

For high-speed storage, data is saved in "Static Random Access Memory" ("SRAM") which will retain the current contents as long as power is applied to it and is known as "volatile" memory. This is in contrast to the "EPROM" or Flash, which does not loose its contents when power is taken away but cannot have its contents changed as easily as "SRAM". Each bit in a SRAM memory array is made up of the six transistor memory cell, as shown in Fig. 12-6. This memory cell will stay in one state until the "Write Enable" transistor is enabled and the write data is used to set the state of the SRAM cell.

The SRAM cell could be modeled as the two inverters shown in Fig. 12-7. Once a value has been set in the inverters' feedback loop it will stay there until changed. Reading data is accomplished by asserting the read enable line and inverting the value output (because the "read" side contains the inverted "write" side's data). The driver to the SRAM cell must be able to "overpower" the output of the inverter in order for it to change state.

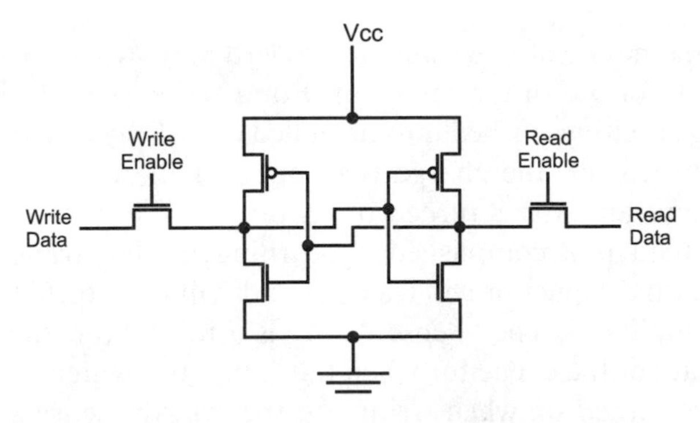

Fig. 12-6. MOSFET SRAM cell.

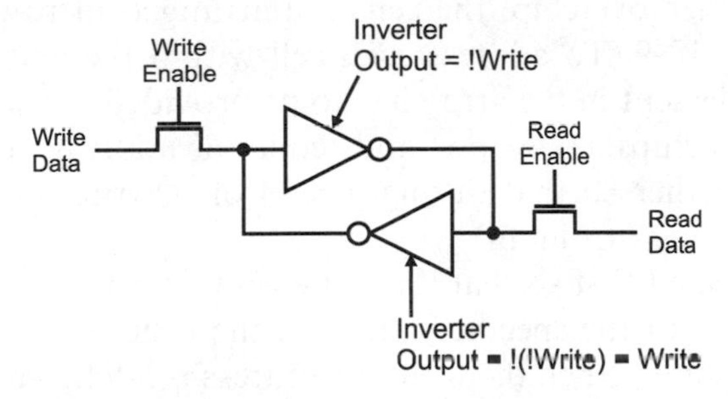

Fig. 12-7. Equivalent circuit to SRAM cell.

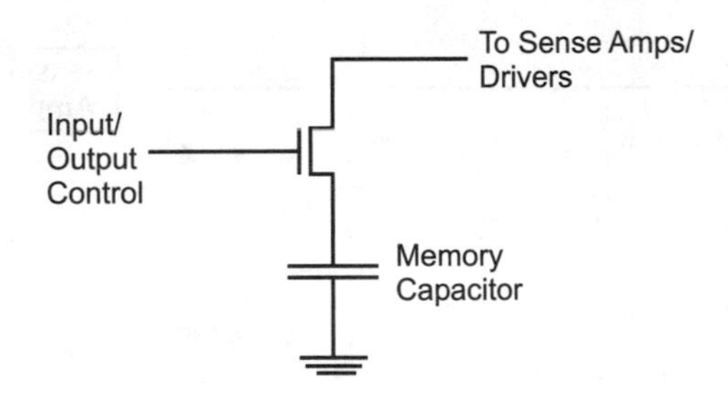

Fig. 12-8. DRAM memory cell.

The outputs of the inverters are usually current limited to avoid any backdriving concerns.

Large, inexpensive and reasonably high-speed memory can be built from "Dynamic Random Access Memory" ("DRAM") cells. You may have heard the term "Single Transistor Memory Cells" for descriptions of DRAM and that's actually a pretty good description of each cell, which is shown in Fig. 12-8.

In DRAM memory cell, the transistor is used as a switch to allow a charge to be moved into or out of the capacitor. For a write, the transistor is turned on and a charge is either pushed into or pulled out of the capacitor. When the transistor is turned off, the charge is trapped in the capacitor and cannot change until the transistor is turned on again.

A DRAM read is accomplished by turning on the transistor and any charge that is in the capacitor will leak out and will be detected and amplified by a "Sense Amplifier". The "Sense Amp" is a metastable flip flop that will be set to the state of the capacitor when the transistor switch is closed. Before the transistor is turned on when writing to the cell, the sense amp will be set to a specific state to load the correct charge into the capacitor.

In a DRAM memory chip, the cells are arranged in rows and columns, as shown in Fig. 12-9. To address each cell within the chip, a row/column address for the element in the array has to be provided. Usually, to save pins on the DRAM chips, the row and column address lines are shared (multiplexed) together so that during a read or a write, first the "Row" is selected and then the "Column".

The row is selected first so that if a write is taking place the sense amp for the row can be set to the specific value. All the other sense amps are set in their metastable state. When the Column address is latched in, the transistors for the array row are turned on. Next, when the row address is available, the "Input"/"Output Control" transistor is turned on, then the read/write takes

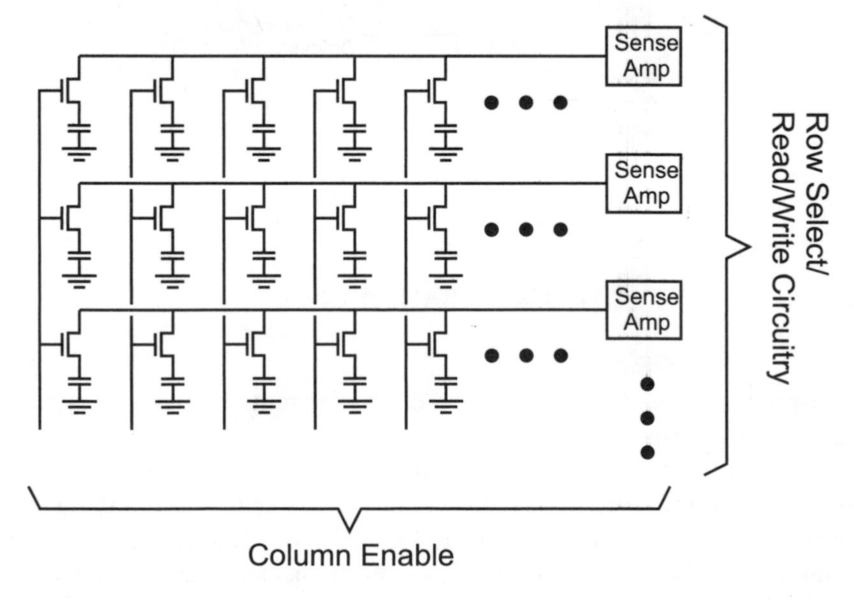

Fig. 12-9. DRAM cell array.

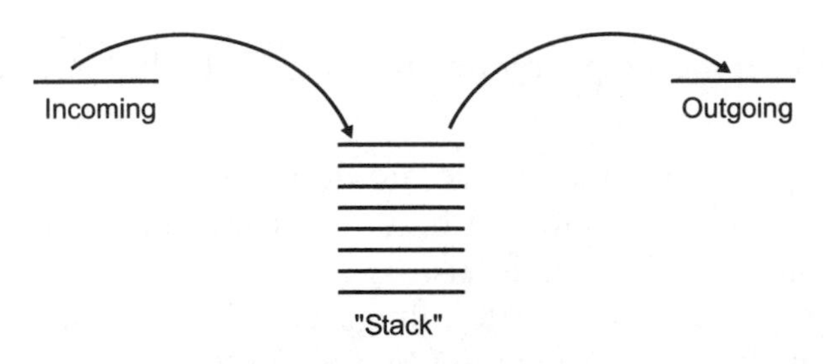

Fig. 12-10. Data stack operation.

place. For the cells not being written to, the sense amps will not only read what the charge is in the capacitor but will also "refresh" it as well.

This is a very good thing in DRAM because the capacitor is actually a MOS transistor built into the chip, acting as a capacitor. Over time, any charge in this capacitor will leak away into the silicon substrate. By periodically "refreshing" the charge by performing a read (which will cause the sense amps to amplify the charge), the contents of the memory will never be lost.

Refreshing is typically done by enabling all the transistors in a column (without first specifying a row) and letting the sense amps do their thing. A read of an incrementing column address is usually implemented in the DRAM support hardware and is known as a "CAS ("Column Address Strobe") Only Refresh". In the original PC, 5% of the processor bandwidth was lost due to DRAM refresh requirements. To overcome these potential deficiencies, PC designers have come up with a few hardware features.

Along with being arranged as simple, single-dimensional arrays of data, memory can also be built into "stacks" (Fig. 12-10). Processor stacks are a simple and fast way of saving data during program execution. Stacks save data in a processor the same way you save papers on your desk and is known as "last in/first out" ("LIFO") memory. As you are working, the work piles up in front of you and you do the task that is at the top of the pile.

Power Supplies

It is surprising to many people that you can add a simple voltage regulator to power your projects for just a few dollars; cheaper than a set of rechargeable batteries. Voltage regulators, powered by an AC/DC "Wall wart" power converter, will convert one DC voltage to another that can be used by the electronics in your circuit and, more importantly, will be tolerant of changes in the AC supply and the current load. In this section, I will introduce you to some simple power supply circuits that have the following characteristics:

1. They are safe for their users and designers.
2. They are relatively efficient in terms of the amount of power that is lost converting voltage levels.
3. They provide very accurate voltage levels, independent of the voltage input or the current required by the application.
4. They are inexpensive.
5. Their design can be optimized for the application that they are providing power for.
6. These supplies source up to 1 amp of current.

The power supply ideas presented here are very appropriate for the simple circuits discussed in this book; the 250 watt power supply used for your PC requires methodologies and circuits for producing this much power that are quite a bit different than what is required for the simple power supplies presented here. Advanced degrees are normally required for properly designing high current power supplies that work at high efficiencies.

There are some semiconductor-based circuits, like Zener diode power supplies (Fig. 12-11) that do lend themselves to being modeled using water analogs. The Zener diode power supply works as a shunt regulator – applying a specified amount of current to a circuit at a rated voltage and shunting the rest away as wasted power.

When the term "shunt" is used, it is simply saying that excess voltage and current is turned away from the circuit. This concept can be illustrated with a water pressure regulator created from a catch basin with a hole at the bottom; water coming out of the hole is at a pressure which is determined by the depth of water in the basin. To maintain this depth (and bottom pressure), even though water is being drawn from the hole at the bottom, "source" water is continually poured into the basin. More water is pouring in than is expected to exit through the hole in the bottom, with the excess leaking out over the side. This is exactly how the Zener diode works, except

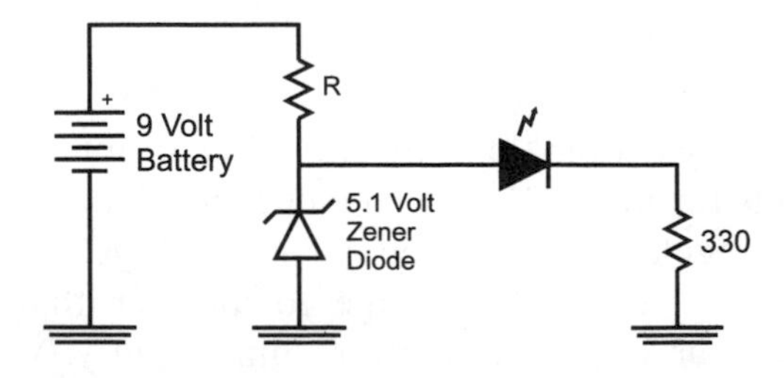

Fig. 12-11. Zener 5.1 volt regulated power supply to LED.

that extra current does not "leak out over the side" but is passed (or "shunted") through the diode. The diode itself is expected to be reverse biased when it is wired into the circuit and it will pass current through it to maintain a set voltage level at its anode (positive terminal). This property is known as "breakdown" and it is not unique to the Zener diode. All diodes will "breakdown" when a high enough reverse bias voltage is applied to them. The breakdown voltage for a Zener diode is usually specified to be in the range of 1.5–25 volts where the breakdown voltage for a typical diode (say the 1N4148/1N914 that I usually use) is 75–100 volts.

Specifying a Zener diode for use as a power supply in an application isn't very difficult but it will require you to understand what your incoming power specifications are as well as what the required current is for the circuit being powered. The powered circuit's voltage should be the same as the rating of the Zener diode. For 5 volt circuits, I use a Zener diode rated at 5.1 volts. Specifying the resistor that is to be used with the Zener diode as well as the Zener diode's power rating can be somewhat complex. Care must be taken to ensure that the circuit has enough current to be powered in all circumstances, including if the input power "sags" (if it is powered by a battery that is discharging). To do this, some kind of "margins" must be designed into the circuit.

For this experiment, I would like to use a 5.1 volt Zener diode to act as a power supply for a LED circuit requiring approximately 10 mA to light the LED. The circuit is shown in Fig. 12-11 and, before it can be assembled, the value for the Zener diode's current limiting resistor "R" must be determined. For a Zener diode power supply to be 100% efficient in terms of current (no current is shunted through the Zener diode), "R" must be chosen so that the voltage drop through it will allow the same amount of current as the powered circuit uses to pass through it. In this application, I am going to assume that the LED has a 2 volt drop, so using the basic electrical formulas, I can determine the current through the LED:

$$\begin{aligned} i &= V/R \\ &= (5.1\,V - 2\,V)/330\,\Omega \\ &= 9.39\text{ mA} \end{aligned}$$

Assuming that the battery produces an even 9 V, the value of R can be calculated:

$$\begin{aligned} R &= V/i \\ &= (9\,V - 5.1\,V)/9.39\text{ mA} \\ &= 415\,\Omega \end{aligned}$$

There are no standard 415 Ω resistors available, but I can make a 420 Ω resistor using a 200 Ω and a 220 Ω in series. This will result in a current of 9.29 mA (a difference of about 1% from the targeted value).

When I described the Zener diode regulator as acting like a basin of water in which the unused current was simply lost, I'm sure that many people grimaced because they knew of devices which are much better at regulating fluid pressure. If this book was written in the 1980s (or earlier) just about everybody would know about the commonly used fluid regulator that is used in older cars called a carburetor (Fig. 12-12). Virtually all cars built in the past 15 years have utilized some form of computer-controlled "fuel injection" which relies on active, rather than passive, control of the fuel being passed to the engine.

The carburetor is a very clever device that only provides fuel on demand. In Fig. 12-12, I have drawn the situation where no fuel is being drawn from the carburetor – a "float" is connected to a simple valve that closes when the fuel in the bowl that the float is in is full. When fuel is drawn from the bowl, the fuel level within the bowl drops (along with the float) and the valve opens, allowing more fuel into the bowl (Fig. 12-13). The carburetor is quite efficient and very simple in operation.

The carburetor acts as a regulator, just providing the volume of fuel (current) as required and the shallow bowl will result in lower pressure (pressure regulation) than what was available from the high-pressure source (the fuel pump). An electrical version of the carburetor would look

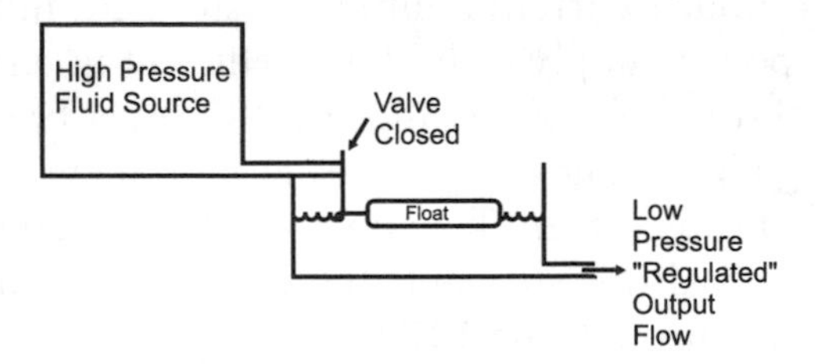

Fig. 12-12. Carburetor with no fluid being drawn from the bowl.

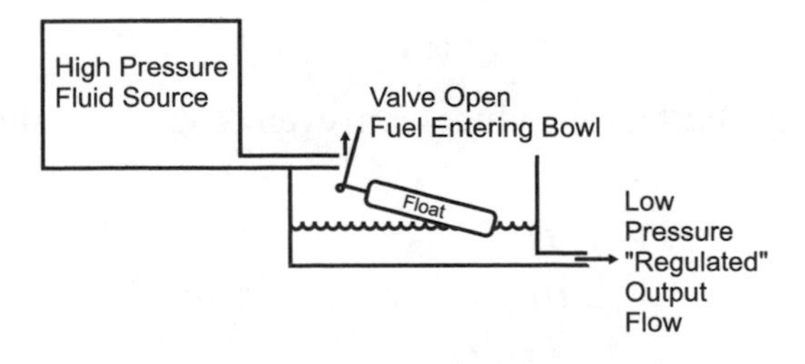

Fig. 12-13. Carburetor in operation.

like Fig. 12-14; current from the high voltage source is switched through a PNP bipolar transistor, with the control of the transistor being the output of the comparator. The comparator's inputs are the current voltage level of the regulator's output and the specific "output" voltage which comes from some kind of voltage reference. The voltage reference is usually a Zener diode that has a miniscule amount of current passing through it; the comparator does not need a lot of current to operate.

Adding the current and temperature "crowbar" sensors is implemented something like in Fig. 12-15. When either the current output or temperature exceeds the preset limits, the reference voltage is pulled to ground using an NPN transistor (remember that the voltage reference is very low current so this can be done safely). In some regulators, if the current or temperature parameters are exceeded, they "latch" the failing state until power is removed and the crowbar conditions are reset. The need for the current sense and shut down should be pretty obvious to you; if the current drawn exceeds the

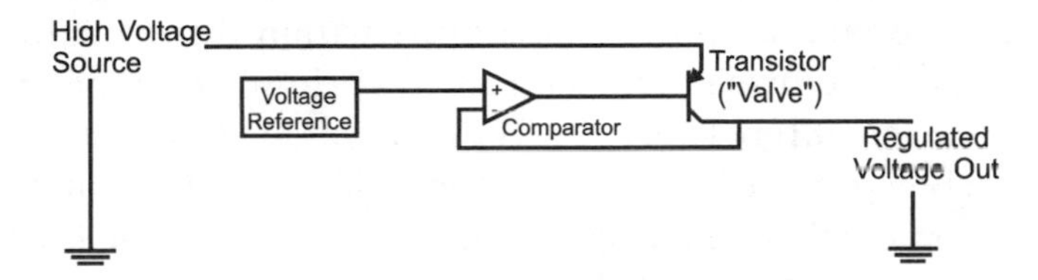

Fig. 12-14. Linear voltage regulator block diagram.

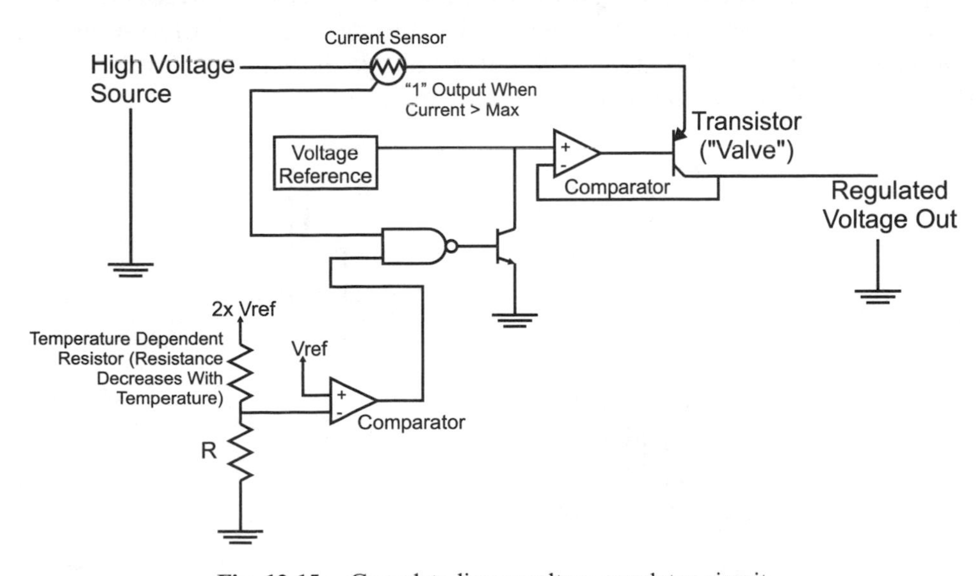

Fig. 12-15. Complete linear voltage regulator circuit.

maximum rating for the PNP transistor, it could be damaged. The temperature sensor may be a bit more unexpected but shouldn't be surprising when you consider what is happening in the regulator when it is transforming a high voltage into a lower one. The difference between the input voltage and the regulated voltage multiplied by the current being drawn by the circuit being regulated is the power dissipated by the regulator. For example, if you had a 12 volt voltage source and a 5 volt regulator providing 200 mA current, the power being dissipated by the regulator would be 1.4 watts. This level of power dissipation could damage the internal circuitry of the regulator or, at the very least, raise the temperature of the part so that it does not work as designed.

The most popular linear voltage regulators that provide the crowbar features are the 78xx and 78Lxx series. The 78xx (or the LM2940 series of regulators which have the same pinout and package) shown in Fig. 12-16 ("xx" standing for the voltage, so a 5 volt regulator is a "7805") can normally source up to 500 mA and up to 1 A with heat sinking. The heat sink is used to dissipate the power and keep the temperature within the regulator less than 125°C, which is the crowbar temperature. For lower current applications (up to 100 mA), the 78Lxx (Fig. 12-17) can be used. For either device, the input voltage should be at least 2 volts above the regulated output voltage. When wiring the regulator in circuit, you should include at least 10 μF of capacitance on the input and a 0.1 μF capacitor on the output.

While the Zener diode and linear power supplies presented so far in this chapter are useful and easy to work with, they do have two concerns that can make them problematic when they are being used in a battery-powered application. First off, they require a higher voltage than the regulated output; this can be an issue when you want to use very simple power like two AA cells

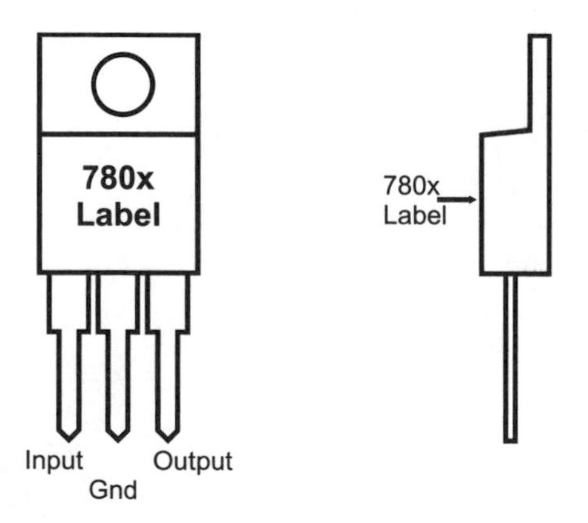

Fig. 12-16. 780x voltage regulator chip pinouts.

for a digital electronics circuit. Secondly, they are not terribly efficient. It isn't unusual for 80% or more of the power input to the Zener diode power supply to be lost and 40% or more lost in the linear power supply. What is required is a power supply circuit that is very efficient and will "step up" voltages.

While these two requirements seem impossible, they can actually be achieved very easily through the use of the "switch mode power supply" (SMPS). The basic SMPS circuit (Fig. 12-18) is quite simple and relies on the energy storing characteristic of the inductor or "coil". While the capacitor stores energy in the form of charge, the coil stores energy in the form of a magnetic field which is maintained by current running through the coil. When this current is shut off, the magnetic field produces a voltage "spike" (which I called "kickback" when discussing magnetic devices) that can be used as the basis for an output voltage.

Using the circled letters in Fig. 12-18, I have drawn the waveforms (Fig. 12-19) that you can expect to see in the SMPS. The "Control" signal is a PWM produced by a "voltage controlled oscillator" (VCO). A voltage controlled oscillator oscillates at a different frequency based on the voltage at an input. The input to the VCO used in the SMPS is the output voltage

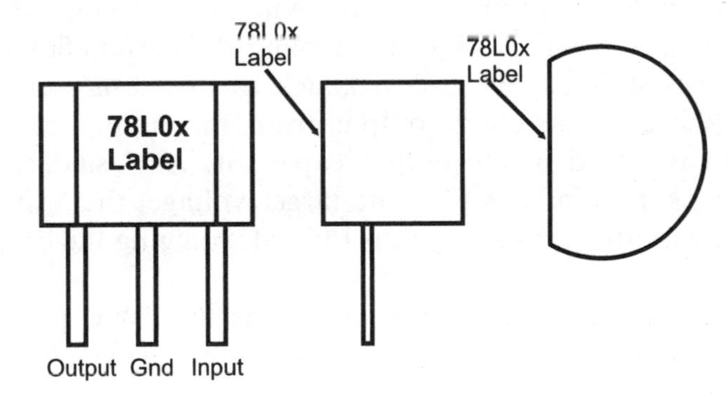

Fig. 12-17. 78L0x voltage regulator chip pinouts.

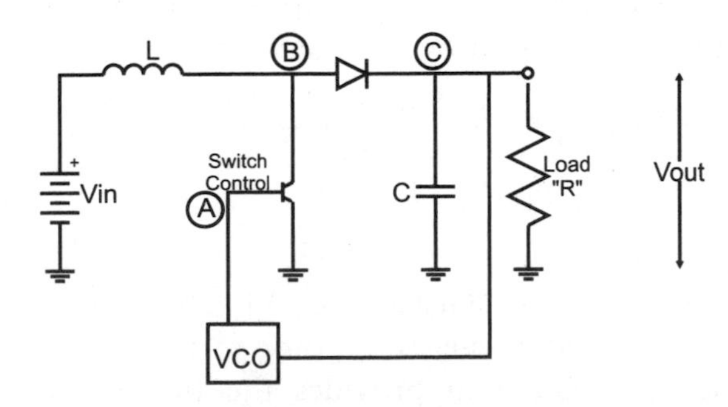

Fig. 12-18. Basic switch mode power supply (SMPS).

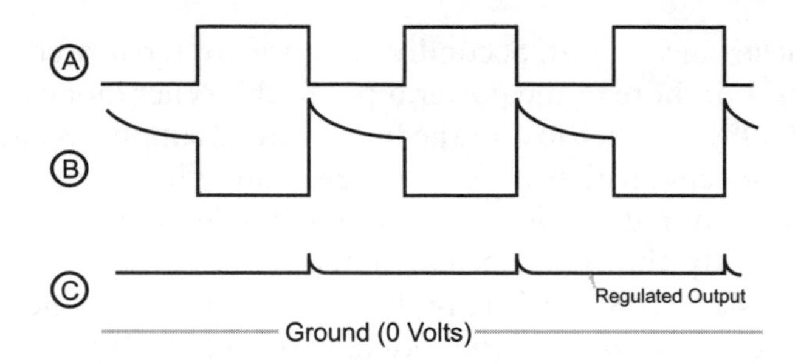

Fig. 12-19. Switch mode power supply operating waveforms.

of the power supply; the VCO frequency will change according to the power supply output to ensure the output stays as stable as possible at the required voltage. The output of the VCO is the base of a transistor that periodically pulls one side of the coil to ground, allowing current to flow through it. When the transistor connected to the coil is turned off, current flow through the coil stops and the magnetic field "kicks back", producing a higher voltage.

The operation of the VCO PWM output along with the coil's response and the output voltage is shown in Fig. 12-19. When the VCO is turning on the transistor, the coil (symbol "L") is tied to ground and current flows through it. When the transistor is off, the coil kickback can be seen and any voltage greater than the current voltage output from the supply passes through the diode and is stored in the output capacitor. As I said above, if the output voltage is more or less than the target voltage, the VCO frequency changes along with the transistor control PWM, bringing the output voltage into line.

To determine the correct coil value as well as the PWM parameters, the following three formulas are used once the output voltage ("V_{out}") is known along with the expected output current draw ("I_{out}") and the input voltage ("V_{in}"). These formulas are used repeatedly until the values for "L" (the coil value), "T_{on}" (time the transistor is on) and "T_{off}" (time the transistor is off) are values that can be produced by reasonable hardware.

$$I_{peak} = 2 \times I_{out} \times (V_{out}/V_{in})$$
$$T_{off} = L \times I_{peak}/(V_{out} - V_{in})$$
$$T_{on} = (V_{out}/V_{in}) - 1$$

Designing an SMPS is not a trivial exercise. While you may think you can do it using something like a 555 timer, I'm going to recommend that you use a commercially available chip that provides the function for you, like the LT1173-5. This chip can be used to create 5 volts (neccesary TTL and

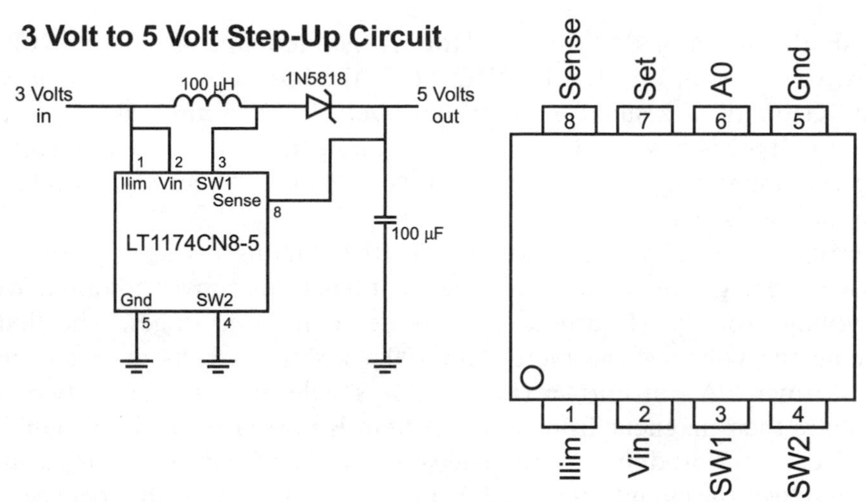

Fig. 12-20. LT1174CN8-5 switch mode power supply controller chip pinout and sample circuit.

many CMOS logic chips) from 3 volts, as shown in the basic circuit in Fig. 12-20.

With the appropriate regulator selected, you now have to find a source of DC current to power the application and regulator. By far the most popular way of providing power to an electronic device is by simply plugging it into a wall socket.

> I must caution you that the power coming out of your wall socket can conceivably destroy your application, cause a fire or hurt you (e.g. burns or electrocution). Despite the fact that it is commonly used for appliances, light and electronic devices in the home, electricity is not to be trifled with.
>
> The circuits provided below may not be appropriate for where you live. The information provided here is strictly "rule of thumb" and is primarily written for use in North America. If you are going to design a power supply for a specific country's use, make sure you understand what are the characteristics of the local power supply, along with any laws or regulations that are appropriate when connecting to it.

Power coming from your wall sockets ("the mains"), comes in as either a 110 or 220 volts "peak-to-peak" as a "sine wave" with a frequency of 50 or 60 cycles per second (or "hertz" ("Hz")). In North America, power is provided at 110–120 volts peak-to-peak voltage (typically 115 volts) at 60 Hz. Different countries around the world will use different peak-to-peak voltage levels and operating frequencies.

This power coming in is normally provided by a "socket", which is built into your walls. Figure 12-21 shows the layout of the socket and labels the individual connections. "Live" or "Hot" is the incoming alternating voltage sine wave shown in Fig. 12-21. "Neutral" is the return path for this current, while "Ground" is a shunt to "earth ground" if the circuit is damaged and the live voltage is passed to the neutral connection. If these three signals are being wired manually by convention, "Live" is black, "Neutral" is white and "Ground" is Green.

Because the AC voltage coming from the "mains" is so high and has positive and negative voltage components, it has to be converted into a lower DC voltage for the electronics. This is done in three stages. The first is reducing the voltage from more than 100 volts to 15 volts or less using a "transformer". A transformer (Fig. 12-22) is a device made up of two coils that share their magnetic field. When current is passed through one coil, the second coil will produce an "inducted" voltage and current, which can be used to power the circuit. Figure 12-22 also gives the relationship between the voltage and current on the secondary side coil based on the number of turns for each coil.

Note that the current is inversely proportional to the turns ratio. In North America (which has 110 volts AC), an 8:1 transformer is often used. This means that with 110 volts in, there will be 14 volts out. For 220 volts, a 16:1 transformer should be used for the same voltage output.

While the voltage has been lowered, it is still "AC" and it is still going positive and negative. This voltage has to be "rectified" into a straight

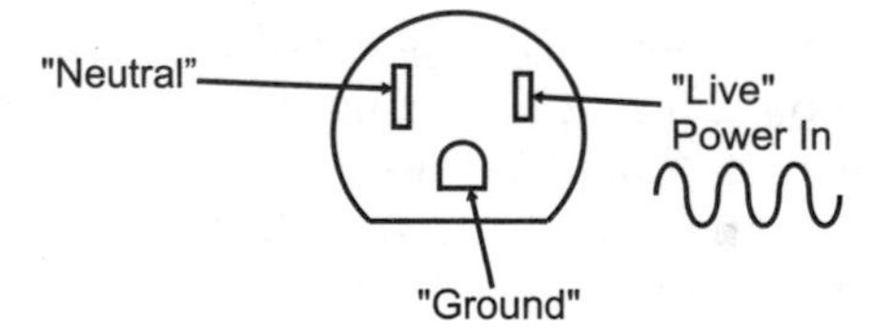

Fig. 12-21. North American wall plug.

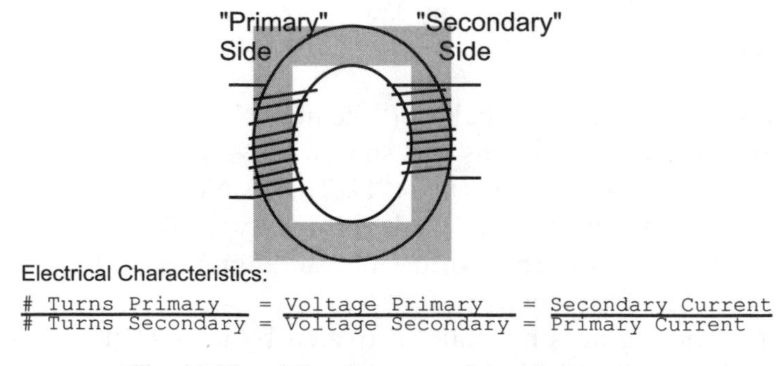

Fig. 12-22. AC voltage transformer operation.

DC voltage. This is done using diodes in either a "half wave" or "full wave" rectifier. Full wave rectifiers transform the positive and negative "lobes" of the AC circuit into a positive voltage, whereas the half wave rectifier "clips" the negative wave (providing half the total power available to the circuit). Inputting the rectified signal directly from the diodes into a voltage regulator should not be attempted; instead, a filtering electrolytic capacitor or a few tens of μF should be used. The filtered signal output from the full wave rectifier was shown in Chapter 3.

As long as the rectified signal does not drop below the minimum voltage of the Voltage Regulator, the regulated DC voltage output will be constant. The filtering cap should be a minimum of 10 μF with a good rule of thumb being that for digital circuits; a 20 μF capacitor is required for each Amp of current drawn. For DC electric motors, this value increases to 100 μF per amp drawn to help prevent inductive "kickback" "spikes" from being driven back through the transformer to the mains circuit.

Using the transformer, full wave rectifier, an electrolytic filter capacitor and a 7805 voltage regulator, a +5 volt 0.5 amp power supply for digital logic applications could be created, as shown in Fig. 12-23. The *voltage regulator* converts the rectified transformer-reduced AC voltage into a voltage that can be used by the digital logic.

There are a few things to note in Fig. 12-23. The first is that the mains ground is connected to the case and not to the "digital ground". In any DC-powered circuit, the negative terminal of the full wave rectifier can be called "digital ground", but should be left "floating" relative to "earth ground", which is provided by the AC plug. In this case, "digital ground" is simply a common negative terminal for the circuit. I have put a "fuse" in the power line, which will cut out in high current draw situations (like short circuits).

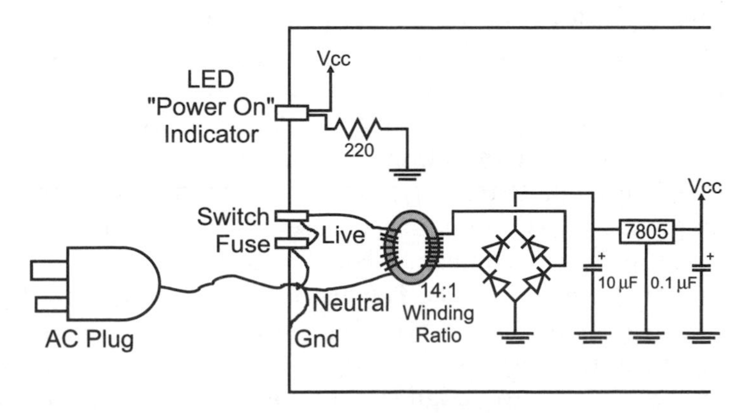

Fig. 12-23. Sample AC to 5 volt DC power supply for digital electronics.

It is rated at 0.1 amps, which may seem low, but remember that current output is inversely proportional to the turns ratio of the transformer: 0.1 amps at 110 volts translates into 1.4 amps at 8 volts at the output. Without this fuse, very large (and very dangerous) currents could build up inside the circuit. For example, 2 amps at 110 volts translates to 28 amps at 8 volts or 224 watts of power. Along with the fuse, the "Switch" in the circuit should be one that is certified for switching AC voltages. AC switches usually have a mechanical assembly inside them that "snaps" the switch contacts on and off. This minimizes "arcing" within the switch. This may seem hard to believe, but if you look inside an AC switch while it is opening or closing, you will see a blue spark and sometimes hear a "pop". This is caused by high inductive voltages produced by the transformer coils that "kickback" when the AC power is shut off.

If you do build mains power supply circuits, like this one, I recommend that you use 14-gauge stranded wire for all connections. Connections should consist of soldered connections (not household "Marette" connectors) for safety. "Heat shrink tubing" should be placed over all solder joints and bare wire. As well, only UL/CSA (or the local country testing organization) approved plugs, wires, switches, fuse holders and transformers should be used in a properly grounded metal case.

If any of these terms are unfamiliar to you or you doubt your ability to build the circuit safely, then don't build it!

Programmable Logic Devices

Programmable logic devices ("PLDs") are chips which have logic gates and flip flops built in, but are not interconnected. The application designer will specify how the gates and flip flops are interconnected in order to create a portion of the application's circuit. Most people feel that programmable logic devices are a relatively new invention, but they have been around for many years. It has only been quite recently (in the last 10 years or so) that reusable chip technology (i.e. EPROM and flash)-based PLDs have been available at prices hobbyists and small companies could afford.

There are several types of PLDs. The first is the simple array of logic gates and devices that are built of this type are known as "PALs" and "GALs" (I generically refer to them as "PALs"). The chips themselves are quite simple and relatively easy to design circuits for. These circuits are normally arranged as a "sum of products" in which signals on the chip can be easily

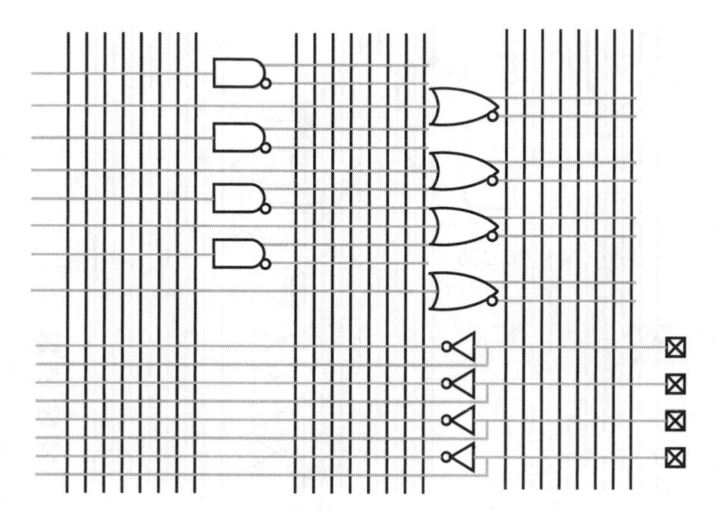

Fig. 12-24. Simple unprogrammed programmable logic device (PLD).

interconnected to form more complex logic functions. The chips are normally blocked out as a series of inputs and outputs, as shown in Fig. 12-24.

The vertical lines or "busses" in Fig. 12-24 are referenced to the gates and I/O pins they are connected to.

To form logic functions, the "sum of products" is used. In Fig. 12-24, a simple 4 I/O, 12 gate PAL is shown. Every output is driven on a bus in both positive as well as negative format. Connections are made between the gates and the busses to create logic functions.

For example, the "XOR" gate which is characterized by:

```
A XOR B = A ^ B
        = _A AND B OR A AND _B
        = (_A * B) + (A *_ B)
```

which is not often available in standard logic. Taking Fig. 12-24 and connecting the busses to the different I/O pins and gates within the PAL, I can implement the XOR gate, as shown in Fig. 12-25.

Note in Fig. 12-25, that an I/O pin changes from an input to an output by simply connecting it directly to a gate output. This feature allows the pins to be used as either input or output.

Options for PALs include varying numbers of inputs to the internal AND and OR gates. For the PLD shown in Fig. 12-25, I have left open the option that any of the pins can be used for any purpose. This is a bit unusual and, normally in PALs, the number of inputs to a gate is restricted. Another

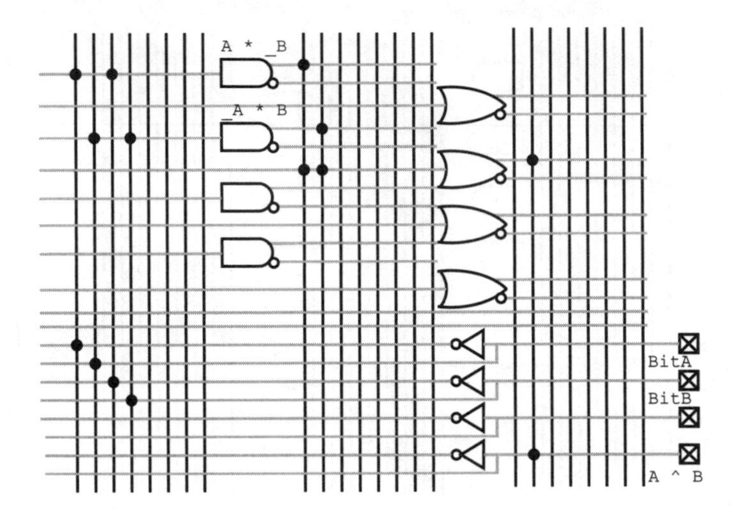

Fig. 12-25. PLD programmed as an XOR gate.

option is to include built in flip flops to store states and turn the PAL from a combinatorial circuit into a sequential one.

PALs may seem simple, but they can result in large decreases in the chip count for an application. In some cases, PALs may be more expensive than the chips they replace, but they reduce application power and board space chip requirements. These savings could result in all-over product savings. It is not unusual for 10 TTL chips to be replaced by a single PAL, resulting in huge PCB and power supply cost savings.

At the high end of the programmable logic device family range, some devices are virtually "ASICs" ("Application Specific Integrated Circuits") and use the same programming language ("VHDL") and development tools as ASICs. These complex parts generally have their functions broken up into "macros". An ASIC/PLD macro can be an AND, or XOR not, logic gate, flip flops or collections of functions (such as multiplexors and arithmetic logic units) which simplify the task of circuit development and eliminate the need for wiring individual gates into basic functions.

The high-end programmable logic device's programming information is often directly transferable to the technology. This allows initial production to use programmable logic devices that require little cost to program and, when the design is qualified, ASICs can be built at a chip foundry for reduced per unit costs.

Programmable logic devices have the advantage of being able to implement fast (less than 10 ns) logic switching, but they do not have the ability to store more than a few bits of data.

Often, programmable logic devices are used in proprietary circuits because their functions cannot be easily traced and decoded.

Programmable logic devices and ASIC development tools are generally function text based as opposed to graphically based applications (like a schematic drawing). This means that a text format, like the "XOR" definition above, must be used to define the functions. Most "compilers" for these statements are intelligent enough to pick the best gates within the device to work with and pick the best paths without your intervention. They are typically much more sophisticated (and expensive) than the compilers used for converting high-level program statements into instructions for a processor.

Quiz

1. The choices in processor design are:
   (a) Intel vs AMD vs PowerPC
   (b) CISC vs RISC, Princeton architecture vs Harvard architecture, hardcoded instruction execution vs state machine instruction execution
   (c) TTL vs CMOS logic
   (d) Speed vs minimal power consumption
2. Microcoded instructions are:
   (a) Short instructions which take less time to execute
   (b) Coded instructions that cannot be read by spies
   (c) Instructions that are specific to a microprocessor
   (d) State machine instructions outlining the steps needed to execute an instruction
3. Decimal 47.123 in binary is:
   (a) Invalid; you cannot perform this conversion
   (b) B'101111.00011111'
   (c) B'11111.101111'
   (d) 0x02F.1F7
4. Which statement is false? "Flash" memory cells:
   (a) Are designed from EPROM memory cells
   (b) Can be erased by applying an electrical voltage
   (c) Are built from flip flops
   (d) Are limited to 256 bits in size

5. DRAM Memory is:
   (a) Faster than SRAM memory
   (b) Less expensive per bit than SRAM memory
   (c) More reliable than SRAM memory due to "refreshing"
   (d) More expensive per bit than SRAM memory

6. What is not a feature of the DC/DC power regulators presented in the book?
   (a) They will convert AC to DC directly
   (b) They have current limiting capability
   (c) They have temperature limiting capability
   (d) They have voltage "Brown out Reset" capability

7. With 12 volts coming in, the current limiting resistor for a 5 volt, 200 mA Zener diode:
   (a) Insufficient information to calculate the resistor's parameters
   (b) 3.5 Ω, 10 watt
   (c) 350 Ω, 1 watt
   (d) 35 Ω, 2 watt

8. A switch mode regulator needs the following components to work:
   (a) Capacitors, diodes and inductor
   (b) Capacitor, diode and PWM driver
   (c) Capacitors, diode, transistor, PWM driver and inductor
   (d) Comparator, PWM driver, transistor and inductor

9. "PAL" is the acronym for:
   (a) Pound and lever
   (b) Peripheral aspect light
   (c) Partial AND logic
   (d) Programmable array logic

10. VHDL is used for:
    (a) Defining PLD electrical parameters
    (b) Defining PLD gate requirements
    (c) Defining PLD gate operations
    (d) Defining PLD speed parameters

CHAPTER 13

PC Interfacing Basics

From a practical point of view, chances are you will be designing an interface or enhancement to your PC. Most modern commercial devices utilize USB ports, but you can still do a lot of interesting projects with the "legacy" interfaces built into the PC. Along with this, a basic understanding of your PC will help you understand how commercial products are designed and may give you some ideas as to how you can design your own complex applications.

The PC "core" circuitry consists of the microprocessor, memory, the interrupt controller and a DMA controller, as shown in Fig. 13-1. This set of hardware can run any program or interface with hardware attached to this "local bus". While you may think of processor memory in terms of the megabytes that were advertised when you bought the PC, there are actually three different types of memory that are accessed.

The term "local memory" is kind of a loosely defined term that I use to describe memory on the PC's motherboard and not on external cards or subsystems. There are a number of different kinds of memory used on the motherboard, each with a different set of characteristic features. The first type, "ROM" ("Read Only Memory") is fairly slow and can only be read; it contains the PC start up ("boot") code. Next, there is what I will call

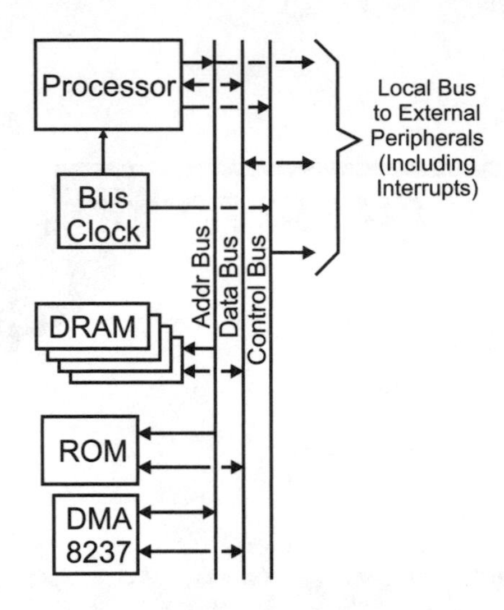

Fig. 13-1. PC major device block diagram.

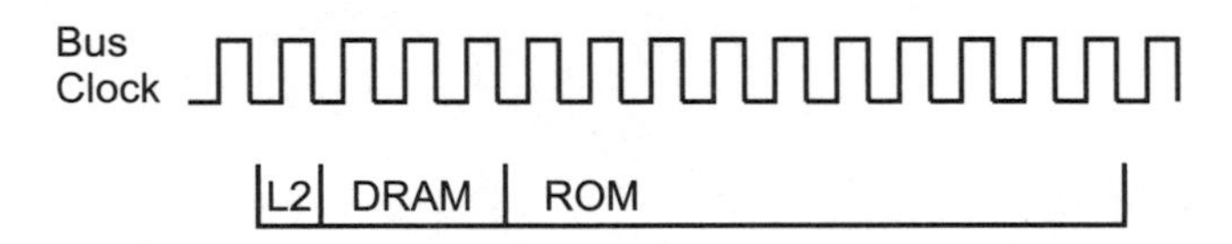

Fig. 13-2. PC memory access time comparison.

"main memory", which is measured in tens or hundreds of megabytes (and was prominently advertised when you bought your PC) and is moderately fast.

The PC's processor itself has "cache memory" which runs at the processor's speed and, ideally, all execution takes place from this area as this will help speed up the operation of the PC. Figure 13-2 shows the number of clock cycles for the different cases of the PC's processor reading from cache memory (at full local bus clock speed), main memory (which has 50 ns access time DRAMs) and ROM (150 ns access time). The importance of being able to run entire applications from cache memory should be obvious to you.

"DMA" stands for "Direct Memory Access" and consists of a hardware device that can be programmed to create addressing and control signals to move data between devices within the system without involving the processor. DMA is most typically used in the PC for moving data to and from the disk drives.

Interrupts are hardware events, passed to the "8237 interrupt controller" chip that requests the PC's processor to stop what it is currently doing and respond to the external request. This request does not have to be responded to immediately and it is up to the programmer's discretion to decide how to respond to the request.

All PC systems have multiple busses for system expansion and improved communications. Before going too much further, I should list the characteristics of what a computer bus is. A computer bus is defined as having:

1. A method for the controlling microprocessor to provide addresses bus hardware for both memory and I/O access.
2. The ability of the cards to request interrupt and DMA processing by the controlling microprocessor.
3. The ability to enhance the operation of the PC.

In a modern PC, there are usually three primary busses, each one delineated by the access speed they are capable of running at and the devices normally attached to them (see Fig. 13-3). The three busses have evolved over time to provide data rates consistent for the needs of the different devices. You might consider that there is a problem with cause and effect, but the busses and devices attached to them have sorted themselves out over the years.

The "Front Side Bus" (usually referred to by its acronym "FSB") runs at a very high speed specific to the processor. This is used by the processor to access DRAM memory directly as well as provide an interface to the system peripherals.

The "PCI" ("Personal Computer Interface") Bus. This bus is not only a staple of all modern PCs but is also available on many other system architectures as well. This allows PCI bus cards to be used across a number of

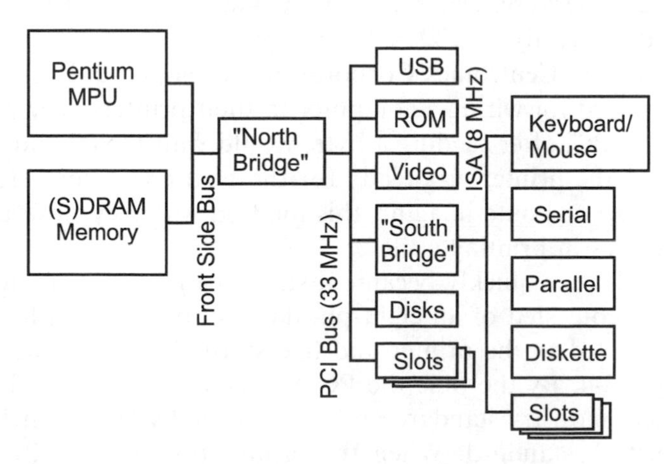

Fig. 13-3. Modern PC block diagram.

different systems and eliminates the need for designers and manufacturers to replicate their products for different platforms. PCI is somewhat of a "hybrid" bus, with some internal features of the PC (notably video and hard file controllers) using the PCI busses built into the copper traces on the motherboard as well as providing access to adapter cards in "slots". PCI is notoriously difficult to create expansion cards for. Along with the fairly high data rate speed (33 and 66 MHz), there are data transfer protocols which generally require an ASIC to decode and process bus requests.

The last bus interfaces the "legacy" interfaces of the PC together and is known as the "Industry Standard Architecture" ("ISA") bus. This bus typically has a data transfer speed of 125 ns – the same as the original PC/AT. When you are interfacing digital electronic devices to your PC, chances are the interface will be connected to this bus.

The Parallel (Printer) Port

When hardware is to be interfaced with the PC, often the first method chosen is the "Parallel" (Printer) port. If I was being introduced to the PC for the first time, I would probably look at this method first as well, but as I have learned more about the PC, using the parallel port would actually be one of the *last* methods that I would look at. The parallel port is really the most difficult interface in the PC to use because it is really device (printer) specific, has a limited number of I/O pins and is very difficult to time operations with accuracy. The only reason why I would see applications using a parallel port for interfacing is because, electronically, it does not need a level translator and can connect directly to TTL/CMOS logic.

In the 1970s, the Centronics Corporation developed a 25 pin "D-Shell" connector standard for wiring a computer to their printers. The early printer interfaces in the PC, after sending a byte, would wait for a handshaking line to indicate that the printer was ready for the next character before sending the next one. As you could imagine, this method was very slow and took up all the PC's cycles in printing a file.

This "pinout" very quickly became a standard in the computer industry. This connector consisted of an eight-pin data output port with strobe, four control lines passed to the printer and five status lines from the printer, as shown in Fig. 13-4. By the time the PC was being developed, this interface was the de facto industry standard and was chosen by IBM to help make the PC an "Industry" standard. When the printer port for the PC was being

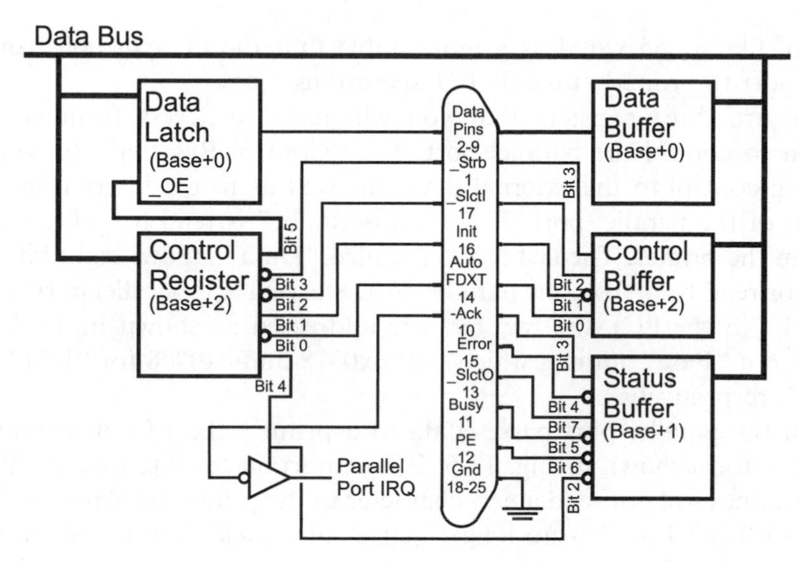

Fig. 13-4. PC parallel port wiring and pinout.

developed, little thought was given to enhancing the port beyond the then current standard level of functionality.

To improve the functionality of the printer port, IBM, when designing the PC/AT, changed the eight data bit output only circuitry to allow data bits to be output as well as read back. This was accomplished by using pulled up open collector outputs that were passed from the "Data Latch" of Fig. 13-4, to the 25 pin D-Shell connector and on to the "Data Buffer", which could read the eight bits and see if anything was changed.

When a read of the eight data bits takes place, the eight output bits should all be "High" to allow the parallel port's pull ups and the open collector outputs of the peripheral device to change the state of the pins. The bi-directional data capability in the PC/AT's parallel port allowed hardware to interface to the PC to be very easily designed. Further enhancing the ease in which devices could be designed to work with the parallel port was the standard timing provided by the PC/AT's ISA bus and BIOS.

Surprisingly, given this background, the biggest problem today with interfacing to the parallel port is timing signals properly. Depending on the different PCs that I have experimented with, I have seen parallel I/O port read and write timings from about 700 ns to less than 100 ns. When you design your interface to the IBM PC, you can either ignore the actual timings of the parallel port and execute signals that are many milliseconds long, "tune" the code to the specific PC you are working with or add a time delay function, like a 555 wired as a monostable that the PC can poll from the printer port to properly time its I/O operations.

There are three registers that you will have to access from your PC program to control the parallel port. The "Control Register", provides an operating control to the external device as well as to the internal interrupt function of the parallel port. The "Status Buffer" is read to poll the status bits from the printer. The last register, called "Data", allows eight bits to be written or read back from the parallel port. Starting at a parallel port (known as "LPT" in the PC), the registers are addressed as shown in Table 13-1 relative to a "Base" address which is at 0x0378 and 0x0278 for "LPT1" and "LPT2", respectively.

When the parallel port passes data to a printer, the I/O pins create the basic waveform shown in Fig. 13-5. It is important to note that the Printer BIOS routines will not send a new character to the printer until it is no longer "Busy". When "Busy" is no longer active, the "Ack" line is pulsed active, which can be used to request an interrupt to indicate to data output code that

Table 13-1 PC LPT port registers.

Register	Address	Function
Data	Base + 0	Read/Write 8 Data Bits
Status	Base + 1	Poll the Printer/Peripheral Status
Control	Base + 2	Write Control Bits to Printer/Peripheral and LPT Port Hardware

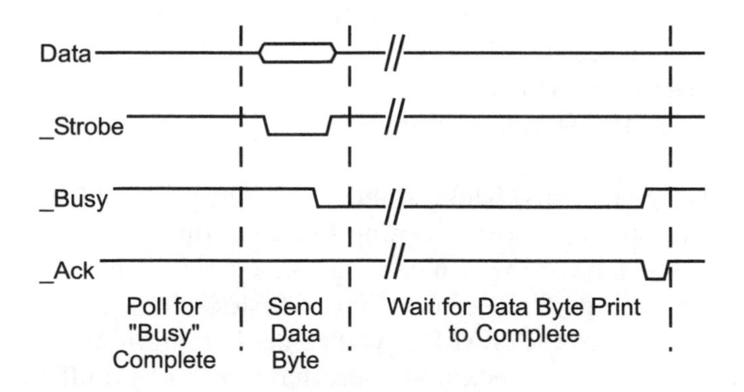

Fig. 13-5. PC parallel port write operation waveform.

the printer is ready to accept another character. The timing of the circuit for printer applications is quite simple, with 0.5 μs minimum delays needed between edges on the waveforms in Fig. 13-5.

Video Output

The first computer CRT display was the "vector" display in which the "X-Y" deflection plates moved the electron beam to a desired location on the screen and then drew a line to the next location. This was repeated until the entire image was drawn on the screen (at which time the process started over). This was popular in early computers because only a modest amount of processing power and video output hardware was needed to draw simple graphics. Because of the way the vector displays operated, a complex image was often darker than and could "flash" more than a "raster" (TV-like) display because a simple image would require fewer vector "strokes" and could be refreshed more often. Vector displays enjoyed some popularity in early computer video displays (including video games), but really haven't been used at all for over 10 years.

Today, a more popular method of outputting data from a computer is to use a "raster" display, in which an electron beam is drawn across the cathode ray tube in a regular, left to right, up and down, pattern. When some detail of the image is to be drawn on the screen, the intensity of the beam is increased when the beam passes a particular location on the screen, which causes the phosphors to glow more brightly, as shown in Fig. 13-6. If you have an old

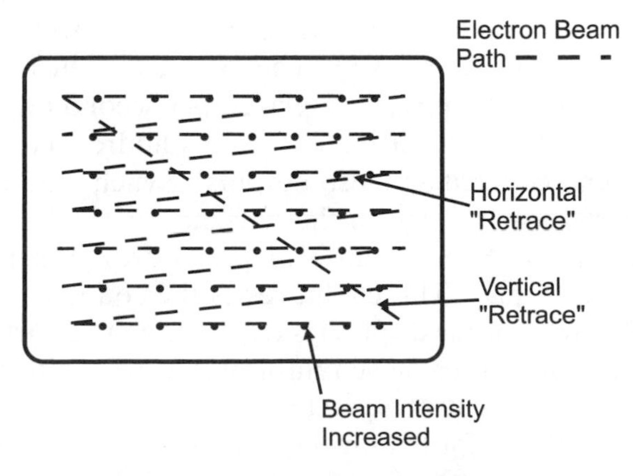

Fig. 13-6. Video raster.

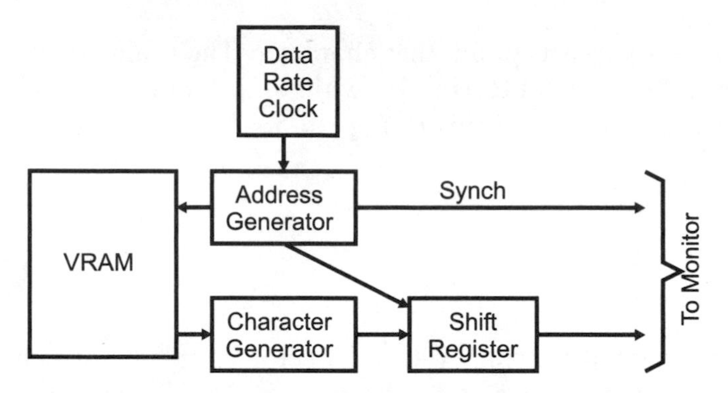

Fig. 13-7. Basic computer video driver circuit.

black and white TV (or monochrome computer monitor), chances are you can increase the "brightness"/"contrast" controls on a dark signal to see the different features I've shown above.

All raster computer output hardware consists of a shift register fed by data from a "Video RAM" ("VRAM"). In Fig. 13-7, the data to be output is read from the VRAM and then passed to a character generator, which converts the data from the Video RAM into a series of dots appropriate for the character. This series of dots is then shifted out to the hardware that drives the raster on the video monitor. If graphical data is output, then the "Character Generator" is not needed in the circuit and the output from the VRAM is passed directly to the shift register. The shift register may also be connected to a "DAC" ("Digital Analog Converter"), which converts the digital data into a series of analog voltages that display different intensities of color on the display.

The addresses for each byte's data to be transferred to the display are accomplished by using an "Address Generator", which controls the operation of the display (Fig. 13-8). The Address Generator divides the "data rate" (the number of "pixels" displayed per second on the screen) into character-sized "chunks" for shifting out, resets address counters when the end of the line is encountered and also outputs "Synch" information for the monitor to properly display data on the screen.

The circuit in Fig. 13-8 shows an example address generator for an 80 character by 25-row display. This circuit resets the counters when the end of the line and "field" (end of the display) is reached. At the start of the line and "field", horizontal and vertical synch information is sent to the monitor so the new field will line up with the previous one.

I have presented this information on standard video inputs to give you some background on how "NTSC" televisions work and what their input

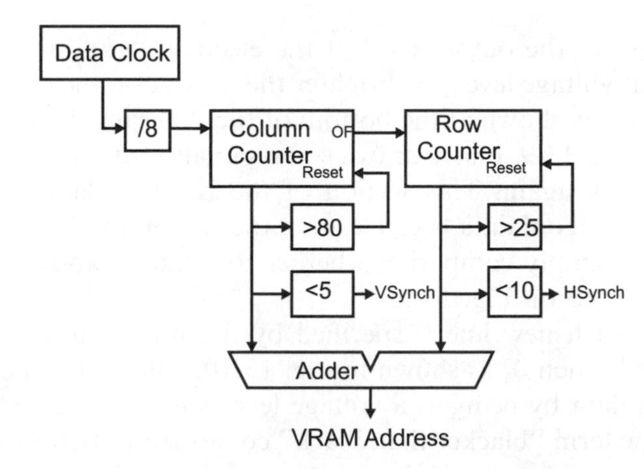

Fig. 13-8. Video RAM address generator.

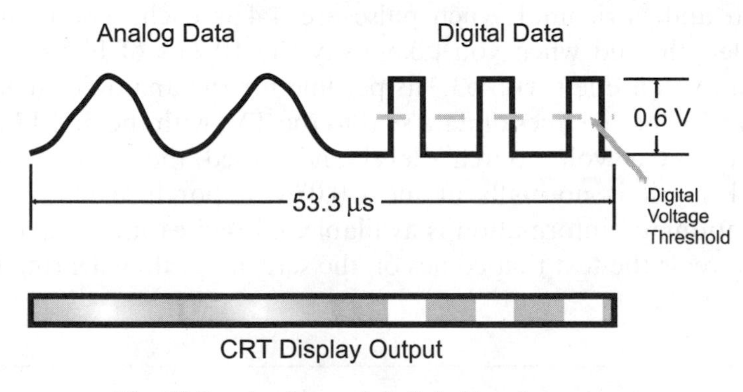

Fig. 13-9. Analog and digital video data output.

signals look like. "NTSC" is the television "composite video" standard for North America and is an acronym for the "National Television Standards Council", the governing body put in place for television in the United States in the 1940s (today, television broadcasting technical standards are controlled by the Federal Communications Commission or FCC).

"Composite NTSC" is the signal typically received by your TV set and consists of both the data input streams along with the formatting information (called synchronization or synch pulses) that are used by the TV set to recognize where the incoming signal is to be placed on the screen. The incoming data, shown in Fig. 13-9 consists of voltage levels (from 0 to 0.6 volts) varying over time. This data is wiped across the CRT from left to right as a single line raster.

On the left-hand side of the incoming serial data shown in Fig. 13-9, I have drawn the data as a typical, analog video stream. The analog voltage levels are used to control the output level of the electron "gun" in the CRT: the higher the input voltage level, the brighter the output on the current raster of the CRT display, as shown at the bottom of the diagram. To the right of the analog data of Fig. 13-9, I have drawn some signals that jump from 0 volts to 0.6 volts and back again. This section of the raster is labeled as "Digital Data". I call this digital data because it is either all on or all off and it can be read as data by simply comparing whether the data is above or below the "threshold" marked on Fig. 13-9.

The start of each new line is specified by the inclusion of a "horizontal synch pulse" ("hsynch"), as shown in Fig. 13-10. This pulse is differentiated from the serial data by being at a voltage level less than the "black" level – this is where the term "blacker than black" comes from. Before and after the synch pulse, there are two slightly negative voltage regions called the "front porch" and "back porch", respectively. The front porch is 1.4 μs long, the back porch and horizontal synch pulse are 4.4 μs each. The entire line is 63.5 μs in length, and when you take away the 10.2 μs of horizontal synch information, you are left with 53.3 μs per line for the analog or digital data.

There are 262 or 263 lines of data sent to the TV, with the first 12 normally being at the −0.08 volt "porch" level and called the "vertical blanking interval". I say it is normally at the −0.08 volt porch level, except when "closed captioning" information is available. Closed captioning, as I'm sure you are aware, is the text that comes on the screen to allow hearing impaired

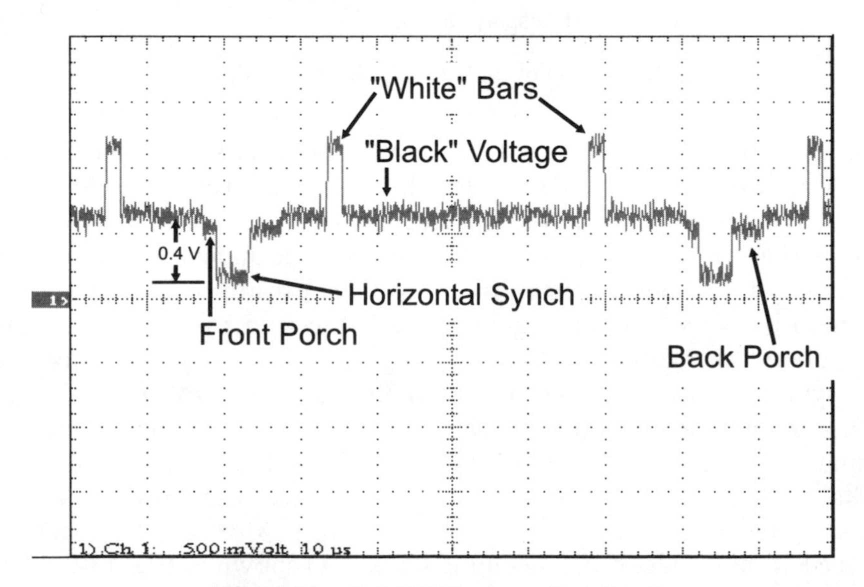

Fig. 13-10. Full NTSC composite video line.

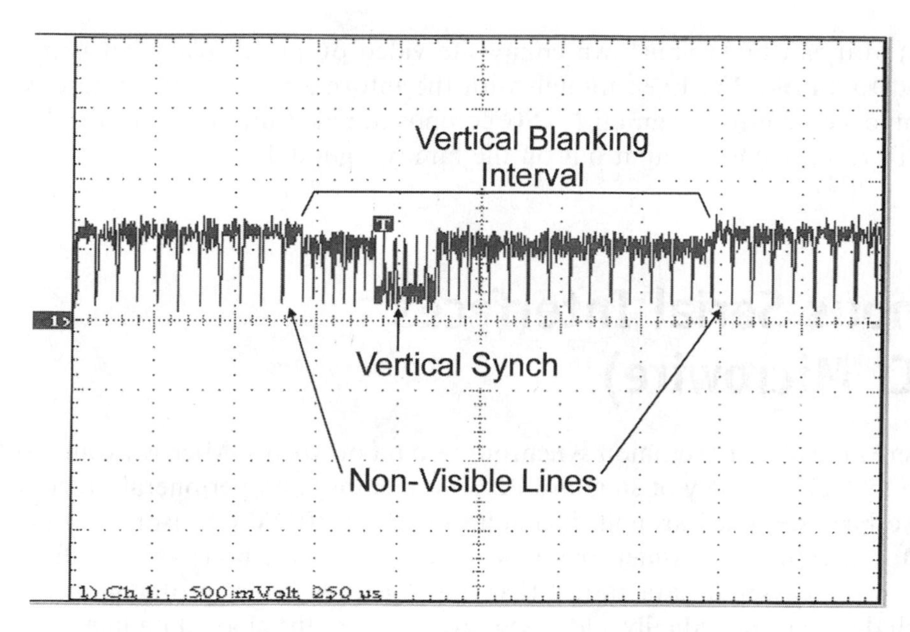

Fig. 13-11. Composite NTSC video vertical synch.

people to watch TV. When closed captioning data is included in the vertical blanking interval, digital pulses, similar to the ones shown in Fig. 13-11 are built into each line and decoded by circuitry built into the TV.

The TV set does not have a counter built into it to indicate when the CRT's gun should reset and go back to the top of the screen. Instead the incoming signal indicates when a vertical retrace must take place by its inclusion of a "vertical synchronization pulse" ("vsynch"). The oscilloscope picture (Fig. 13-11) shows what the vertical synch pulse looks like: it is actually a series of half-sized lines followed by a number of inverted half-sized lines.

So far, I have described NTSC video as having 262 or 263 lines, but if you look at the specifications for a TV set, you will see that it can accept 525 lines of NTSC data. A full NTSC screen consists of two "fields" of data: one 262 lines in size and the other 263 lines. The fields are offset by one half the width of a line – after the first field is displayed, the second fills in the spaces between the lines displayed by the first field. This is called "interlacing" the two fields together into a single "frame".

With the information contained within this section, you have enough information to create a monochrome video output circuit. I've done it a number of times with basic microcontrollers; as long as you keep your field width, vertical and horizontal synch pulses timed consistently, you will find that you can build your own composite video output display that you can send to an old TV. Even though with the information contained here, you should be able to design an NTSC composite video output perfectly, I still don't recommend trying it out on the family's good TV.

Synchronous Serial Interfaces (SPI, I2C, Microwire)

There are two very common synchronous data protocols, Microwire and SPI, from which a variety of standard devices (memory and peripheral functions) have been designed around, including serial EEPROMs, sensors and other I/O functions. I consider these protocols to be methods of transferring synchronous serial data rather than intelligent network protocols because each device is individually addressed (even though the clock/data lines can be common between multiple devices). If the chip select for the device is not asserted, the device ignores the clock and data lines. With these protocols, only a single "Master" can be on the bus. Normally, just eight bits of data are sent out at a time. For protocols like Microwire where a "Start bit" is initially sent, the "Start bit" is sent using direct reads and writes to the I/O pins. To receive data, a similar circuit would be used, but data would be shifted into the shift register and then read by the microcontroller.

The Microwire protocol is capable of transferring data at up to one megabit per second. Sixteen bits are only transferred at a time. After selecting a chip and sending a start bit, the clock strobes out an eight bit command byte, followed by (optionally) a 16 bit address word transmitted and then another 16 bit word either written or read by the microcontroller.

With a one megabit per second maximum speed, the clock is both high and low for 500 ns. Transmitted bits should be sent 100 ns before the rising edge of the clock. When reading a bit, it should be checked 100 ns before the falling edge of the clock. While these timings will work for most devices, you should make sure you understand the requirements of the device being interfaced to.

The SPI protocol is similar to Microwire, but with a few differences:

1. SPI is capable of up to 3 megabits per second data transfer rate.
2. The SPI data "word" size is eight bits.
3. SPI has a "Hold" which allows the transmitter to suspend data transfer.

4. Data in SPI can be transferred as multiple bytes known as "Blocks" or "Pages".

Like Microwire, SPI first sends a byte instruction to the receiving device. After the byte is sent, a 16 bit address is optionally sent followed by eight bits of I/O. As noted above, SPI does allow for multiple byte transfers. An SPI data transfer is shown in Fig. 13-12. The SPI clock is symmetrical (an equal low and high time). Output data should be available at least 30 ns before the clock line goes high and read 30 ns before the falling edge of the clock.

When wiring up a Microwire or SPI device, one trick that you can do to simplify the microcontroller connection is to combine the "DI" and "DO" lines into one pin, as was shown earlier in the book.

The most popular form of microcontroller network is "I2C", which stands for "Inter-Intercomputer Communications". This standard was originally developed by Philips in the late 1970s as a method to provide an interface between microprocessors and peripheral devices without wiring full address, data and control busses between devices. I2C also allows sharing of network resources between processors (which is known as "Multi-Mastering"). Your PC probably has several I2C busses built into it for controlling system control and monitoring functions.

The I2C bus consists of two lines: a clock line ("SCL") which is used to strobe data (from the "SDA" line) from or to the master that currently has control over the bus. Both these bus lines are pulled up (to allow multiple devices to drive them). A I2C controlled stereo system might be wired as shown in Fig. 13-13.

The two bus lines are used to indicate that a data transmission is about to begin as well as pass the data on the bus. To begin a data transfer, a "Master" puts a "Start Condition" on the bus. Normally, when the bus is in

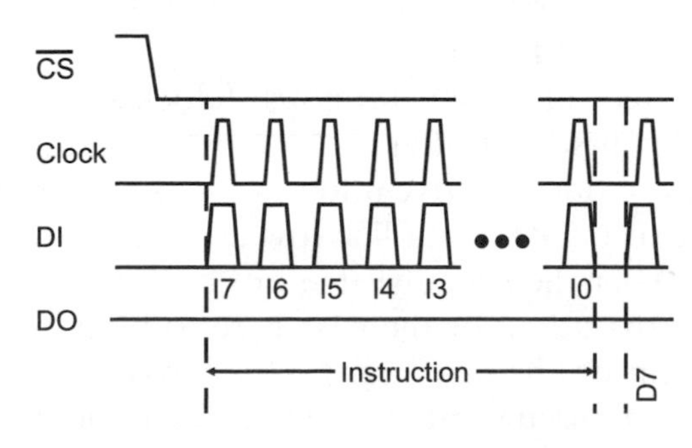

Fig. 13-12. SPI data transfer waveform.

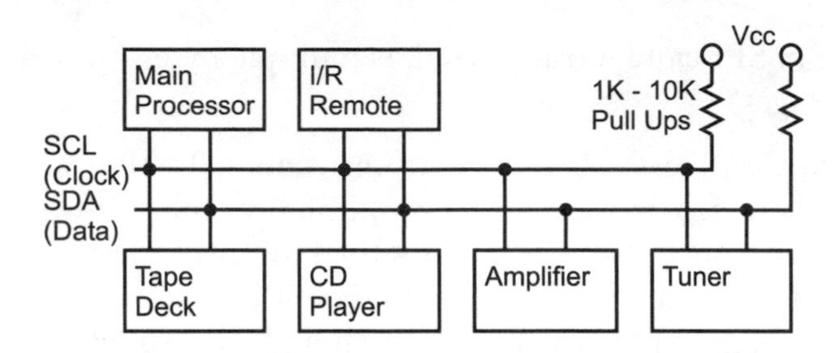

Fig. 13-13. Example I2C application wiring.

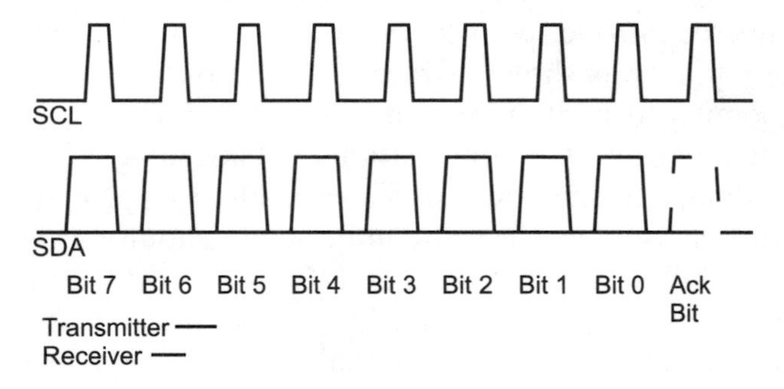

Fig. 13-14. I2C data packet waveform.

the "Idle State", both the clock and data lines are not being driven (and are pulled high). To initiate a data transfer, the master requesting the bus pulls down the SDA bus line followed by the SCL bus line. During data transmission this is an invalid condition (because the data line is changing while the clock line is active/high). Each bit is then transmitted to or from the "Slave" (the device the message is being communicated with by the "Master") with the negative clock edge being used to latch in the data. To end data transmission, the reverse is executed; the clock line is allowed to go high, which is followed by the data line.

Data is transmitted in a synchronous (clocked) fashion. The most significant bit is sent first and after eight bits are sent, the master allows the data line to float (it doesn't drive it low) while strobing the clock to allow the receiving device to pull the data line low as an acknowledgment that the data was received. After the acknowledge bit, both the clock and data lines are pulled low in preparation for the next byte to be transmitted or a Stop/Start Condition is put on the bus. Fig. 13-14 shows the data waveform.

Sometimes, the acknowledge bit will be allowed to float high, even though the data transfer has been completed successfully. This is done to indicate that the data transfer has completed and the receiver (which is usually a "slave device" or a "Master" which is unable to initiate data transfer) can prepare for the next data request.

There are two maximum speeds for I2C (because the clock is produced by a master, there really is no minimum speed). "Standard Mode" runs at up to 100 kbps and "Fast Mode" can transfer data at up to 400 kbps.

Asynchronous Serial Interfaces

Understanding how asynchronous serial communications works is easy; implementing them on a PC is frustrating and hard and used to be especially true for casual PC owners. This situation is improving with newer high-speed interface connection protocols (USB and Firewire), but actually sitting down and getting RS-232 communication between two devices working can be a pain. Much of the information I will present on RS-232 in this section will seem quite obvious, but I urge you to read through this material as there are some tricks and information contained here that could probably get you out of a bind or two.

Asynchronous long distance communications came about as a result of the Baudot "teletype". This device mechanically (and, later, electronically) sent a string of electrical signals (which we would call "bits"), like the one shown in Fig. 13-15, to a receiving printer. This data packet format is still used today for the electrical asynchronous transmission protocols described in the following sections. With the invention of the teletype, data could be sent and retrieved automatically without having an operator having to sit by the teletype all night unless an urgent message was expected.

The Baud Rate specified for the data transmission is the maximum number of possible data bit transitions per second (measured in "bits per second", abbreviated to "bps"). This includes the "Start", "Parity" and "Stop" bits at the ends of the data "packet" shown in Fig. 13-15 as well as the five data bits

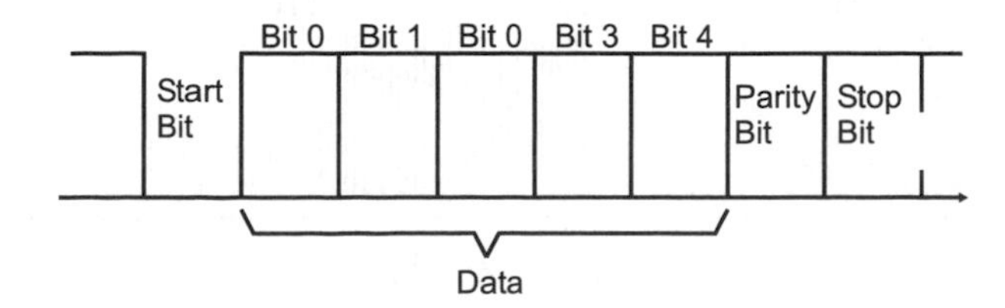

Fig. 13-15. Five-bit Baudot data packet.

in the middle. I use the term "packet" because we are including more than just data (there is also some additional information in there as well), so "character" or "byte" (if there were eight bits of data) are not appropriate terms. This means that for every five data bits transmitted, eight bits in total are transmitted (which means that nearly 40% of the data transmission bandwidth is lost in teletype asynchronous serial communications).

The "Data Rate" is the number of data bits that are transmitted per second. For this example, if you were transmitting at 110 baud, the actual data rate is 68.75 bits per second (or, assuming five bits per character, 13.75 characters per second).

With only five data bits, the Baudot code could only transmit up to 32 distinct characters. To handle a complete character set, a specific five-digit code was used to notify the receiving teletype that the next five-bit character would be an extended character. With the alphabet and most common punctuation characters in the "primary" 32, this second data packet wasn't required very often. In the data packet shown in Fig. 13-15, there are three control bits. The "Start Bit" is used to synchronize the receiver to the incoming data. In most hardware circuits designed to read an asynchronous serial packet, there is an overspeed clock running at 16 times the incoming bit speed which samples the incoming data and verifies whether or not the data is valid. When waiting for a character, the receiver hardware polls the line repeatedly at 1/16 bit period intervals until a "0" ("Space") is detected (down arrow at the left of Fig. 13-16). The receiver then waits half a cycle before

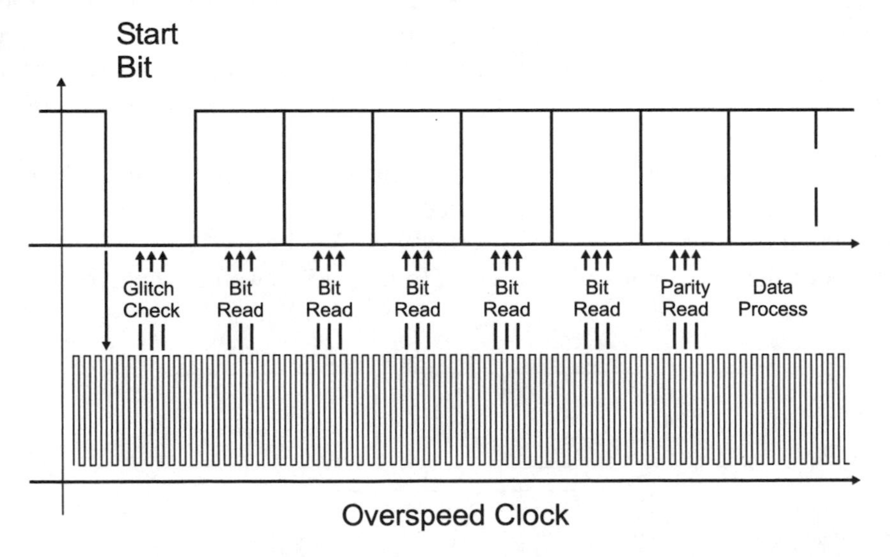

Fig. 13-16. Asynchronous serial data read.

polling the line again to see if a "glitch" was detected and not a Start bit. This polling takes place in the middle of each bit to avoid problems with bit transitions (or if the transmitter's clock is slightly different from the receivers, the chance of misreading a bit will be minimized).

Once the Start bit is validated, the receiver hardware polls the incoming data once every bit period multiple times (again to ensure that glitches are not read as incorrect data).

The "Stop" bit was originally provided to give both the receiver and the transmitter some time before the next packet is transferred (in early computers, the serial data stream was created and processed by the computers and not custom hardware, as in modern computers).

The "Parity" bit is a crude method of error detection that was first brought in with teletypes. The purpose of the parity bit is to indicate whether the data was received correctly. An "odd" parity meant that if all the data bits and parity bits set to a "Mark" were counted, then the result would be an odd number. "Even" parity is checking all the data and parity bits and seeing if the number of "Mark" bits is an odd number. Along with even and odd parity, there are "Mark", "Space" and "No" parity. "Mark" parity means that the parity bit is always set to a "1", "Space" parity is always having a "0" for the parity bit and "No" parity is eliminating the parity bit all together. I said that parity bits are a "crude" form of error detection because they can only detect one bit error (i.e. if two bits are in error, the parity check will not detect the problem). If you are working in a high induced noise environment, you may want to consider using a data protocol that can detect (and, ideally, correct) multiple bit errors.

The most common form of asynchronous serial data packet is "8-N-1", which means eight data bits, no parity and one stop bit. This reflects the capabilities of modern computers to handle the maximum amount of data with the minimum amount of overhead and with a very high degree of confidence that the data will be correct.

Having reviewed the data protocol for asynchronous serial communications, let's go on and look at the electrical specifications, starting with the connectors that you will find on the back of your PC (Fig. 13-17). Either a male 25 pin or male 9 pin connector is available on the back of the PC for each serial port – chances are you will have a 9 pin connector because it takes up the least amount of space.

Working with MS-DOS in early systems, only a maximum of four serial ports could be addressed by the PC, and of these probably only two were useable for connecting external devices to the PC due to conflicts with other hardware devices. In modern systems, which have "Plug and Play" capabilities and the Windows operating system, which can allocate resources

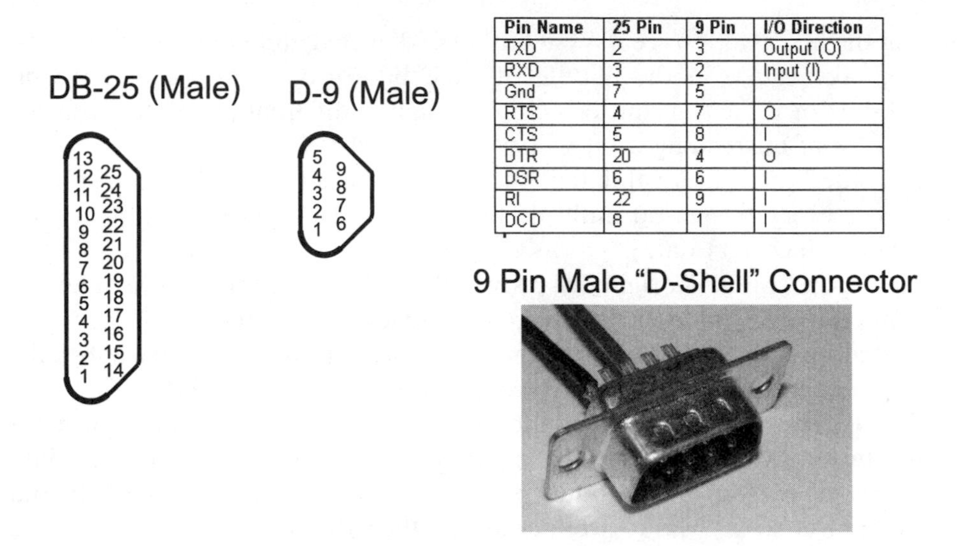

Pin Name	25 Pin	9 Pin	I/O Direction
TXD	2	3	Output (O)
RXD	3	2	Input (I)
Gnd	7	5	
RTS	4	7	O
CTS	5	8	I
DTR	20	4	O
DSR	6	6	I
RI	22	9	I
DCD	8	1	I

Fig. 13-17. PC asynchronous serial port connectors and pinouts.

throughout the system, the number of standard serial ports has been greatly expanded. My personal record is 64 serial devices for environmental chamber status test software.

The standard RS-232 data rates available to you in the PC are listed in Table 13-2. As an interesting exercise, I suggest that you find the reciprocal of the data rates listed in Table 13-2, multiply by 1,000,000 and then divide by 13 – what you will find is that the results for the data rates starting at 1200 bps and going higher will have a very small fraction. In fact, the error will be much less than 0.1%! This is a good trick to keep in your hip pocket when you have to implement an RS-232 interface and you don't have any crystals that have been cut specifically to provide a multiple of these data rates.

Asynchronous communications based on the Baudot teletype/RS-232 is known as "Non-Return to Zero" ("NRZ") asynchronous communications, because at the end of each data packet the serial line is high. There are other methods of sending asynchronous serial data, with one of the most popular being "Manchester encoding". In this type of data transfer, each bit is synchronized to a "start" bit and the following "Data Bits" are read with the "space" dependent on the value of the bit.

Manchester encoding is unique in that the "Start Bit" of a packet is quantitatively different from a "1" or a "0" (shown in Fig. 13-18). This allows a receiver to determine whether or not the data packet being received is actually at the start of the packet or somewhere in the middle (and should be ignored until a start bit is encountered).

Table 13-2 Standard RS-232 data rates.

Data rate
110 bps
150 bps
300 bps
600 bps
1200 bps
2400 bps
4800 bps
9600 bps
19200 bps
38400 bps
57600 bps
115200 bps

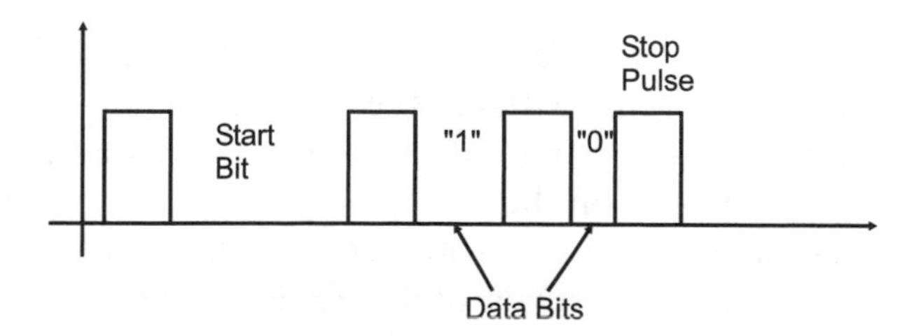

Fig. 13-18. Manchester data encoding.

Manchester encoding is well suited for situations where data can be easily interrupted or there is a conflict in the middle of data reception. Because of this, it is the primary method of data transmission for infrared control (such as used in your TV's remote control).

RS-232 Electrical Standards

In the previous section, I didn't tell the whole story about RS-232 asynchronous serial communications. I left out one very important point – signals do not travel at the same voltage levels as what we've discussed so far in the book. When RS-232 was first developed into a standard, computers and the electronics that drive them were still very primitive and unreliable. Because of that, we've got a couple of legacies to deal with and this can complicate connecting digital electronics circuits to another device using RS-232.

The biggest concern is the voltage levels of the data. A "Mark" ("1") is specified to be −3 volts to −12 volts and a "Space" ("0") is +3 volts to +12 volts. This means that there must be some kind of voltage level conversion when passing RS-232 to or from digital electronic devices. There are a number of ways of doing this.

Before working with the voltage levels, I just want to say a few words about the "handshaking" signals built into RS-232 (these are all the RS-232 connections other than RX, TX and GND in Fig. 13-17). These six additional lines (which are at the same logic levels as the transmit/receive lines) are used to interface between devices and control the flow of information between computers.

The "Request To Send" ("RTS") and "Clear To Send" ("CTS") lines are used to control data flow between the computer ("DCE") and the modem ("DTE" device). The "Data Transmitter Ready" ("DTR") and "Data Set Ready" ("DSR") lines are used to establish communications. There are two more handshaking lines that are available in the RS-232 standard that you should be aware of, even though chances are you will never connect anything to them. The first is the "Data Carrier Detect" ("DCD"), which is asserted when the modem has connected with another device (i.e. the other device has "picked up the phone"). The "Ring Indicator" ("RI") is used to indicate to a PC whether or not the phone on the other end of the line is ringing or if it is busy. Few of these lines are used in modern RS-232 applications and, as shown in Fig. 13-19, the DTR/DSR and CTS/RTS pairs are often simply shorted together to avoid any "hardware handshaking" issues with the PC.

There is a common ground connection between the DCE and DTE devices. This connection is critical for the RS-232 level converters to determine the actual incoming voltages. The ground pin should never be connected to a chassis or shield ground (to avoid large current flows or be shifted and prevent accurate reading of incoming voltage signals). Incorrect grounding of an application can result in the computer or the device it is

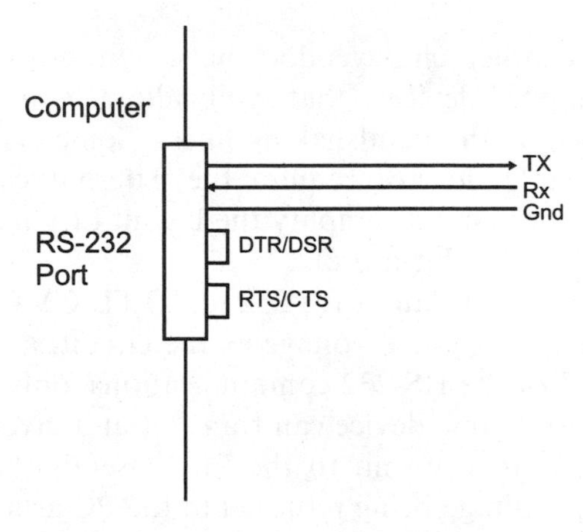

Fig. 13-19. Three wire RS-232 connection.

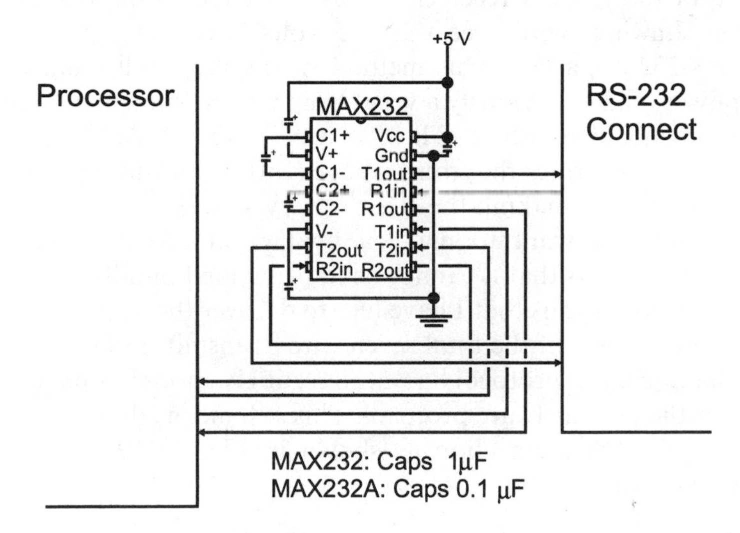

Fig. 13-20. MAX232 RS-232 level shirting chip and schematic diagram.

interfacing having to reset or have they power supplies blow a fuse or burn out. The latter consequences are unlikely, but I have seen it happen in a few cases. To avoid these problems make sure that chassis and signal grounds are separate or connected by a high value (hundreds of k Ω) resistor.

The most popular method for converting TTL/CMOS Logic signals to RS-232 levels is to use the MAXIM MAX232 (see Fig. 13-20,) which has a built in charge-pump voltage generator. This chip is ideal for implementing three-wire RS-232 interfaces (or to add a simple DTR/DSR or RTS/CTS handshaking interface). Ground for the incoming signal is normally connected to the digital electronics ground. Along with the MAX232,

MAXIM and some other chip vendors have a number of other RS-232 charge-pump equipped devices that will allow you to handle more RS-232 lines (to include the handshaking lines). Some charge-pump devices that are also available do not require the external capacitors that the MAX232 chip does, which will simplify the layout of your circuit (although these chips do cost quite a bit more).

The next method of translating RS-232 and TTL/CMOS voltage levels is to use the transmitter's negative voltage as the circuit, as Fig. 13-21 shows. This circuit relies off of the RS-232 communications, only running in "Half-Duplex" mode (i.e. only one device can transmit at a given time). When the external device wants to transmit to the PC, it sends the data either as a "Mark" (leaving the voltage being returned to the PC as a negative value) or as a "Space" by turning on the transistor and enabling the positive voltage output to the PC's receivers. If you go back to the RS-232 voltage specification drawing, you'll see that +5 volts is within the valid voltage range for RS-232 "Spaces". This method works very well (consuming just about no power) and is obviously a very cheap way to implement a three-wire RS-232 bi-directional interface. There is a chip, the Dallas Semiconductor DS275, which incorporates the circuit above (with a built-in inverter) into the single package shown, making the RS-232 very simple.

Before finishing, I want to make a final point on three wire RS-232 connections. The first is that it cannot be implemented blindly; in about 20% of the RS-232 applications that I have had to do over the years, I have had to implement some subset of the total seven wire (transmit, receive, ground and four handshaking lines) protocol lines. Interestingly enough, I have never had to implement the full hardware protocol. This still means that four out of five times if you wire the connection as shown in Fig. 13-19, the application would have worked.

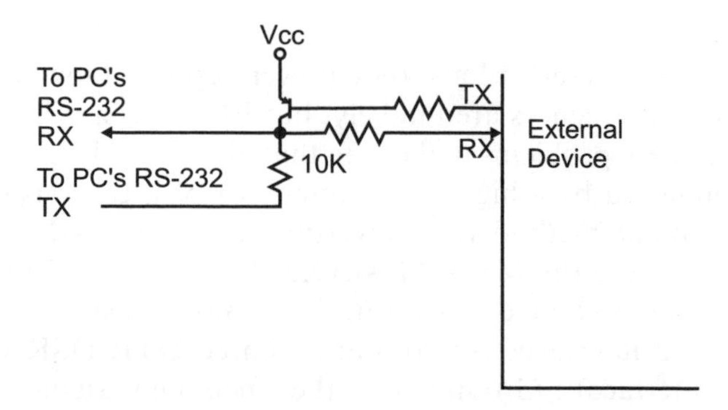

Fig. 13-21. "Voltage stealing" RS-232 interface.

Quiz

1. Which type of memory responds in the least amount of time?
   (a) Processor cache
   (b) All three types respond equally fast
   (c) Main Memory (S)DRAM
   (d) Processor ROM

2. What is not a characteristic of a computer bus?
   (a) Controlled by a central processor
   (b) The ability of peripherals on the bus to request information
   (c) Data can be read/written to peripheral busses
   (d) Keeping a log of the processor bus accesses

3. The minimum timing between signal edges in the parallel port is:
   (a) 0.5 μs
   (b) 0.5 ms
   (c) 0.5 ns
   (d) There are no delays; the port works completely asynchronously

4. The raster display on a video display looks like:
   (a) A series of parallel lines going across the display
   (b) Parallel lines that loop back like plough lines in a farmer's field
   (c) A series of small dots drawn on the video display
   (d) A series of brush strokes on the video display

5. "Horizontal synch" in a video display:
   (a) Moves the electron beam in the CRT to the left
   (b) Moves the electron beam in the CRT to the top
   (c) Starts the raster going side to side
   (d) Is needed for CRTs laid down on their side

6. Devices on a synchronous serial bus are normally referred to as:
   (a) "Dominant" and "Recessive"
   (b) "Master" or "Slave"
   (c) "1" or "0"
   (d) "Transmitter" or "Receiver"

7. The maximum speed of the "SPI" synchronous serial protocol is:
   (a) 100 kpbs (thousands of bits per second)
   (b) 1 Mbps (millions of bits per second)

(c) 2 Mbps
(d) 3 Mbps

8. To connect to a modern PC's RS-232 connector, you will need:
   (a) Fiber optic cable
   (b) 75 Ohm Coax and BNC connector
   (c) RJ-45 cable and crimper
   (d) A 9 pin male D-Shell connector

9. The RS-232 voltage levels of −12 volts for a "1" and +12 volts for "0" means that:
   (a) You can connect digital electronics devices directly to RS-232 connections
   (b) You can connect TTL/CMOS Logic drivers directly to RS-232
   (c) You can connect TTL/CMOS Logic inputs directly to RS-232
   (d) You cannot connect digital electronics devices directly to RS-232 connections

10. The MAX232:
    (a) Uses a charge pump to power the circuit
    (b) Requires +12 volts and −12 volts to convert TTL/CMOS Logic signals
    (c) Produces correct RS-232 voltages and converts TTL/CMOS Logic signals
    (d) Will allow a full RS-232 connection with only one chip

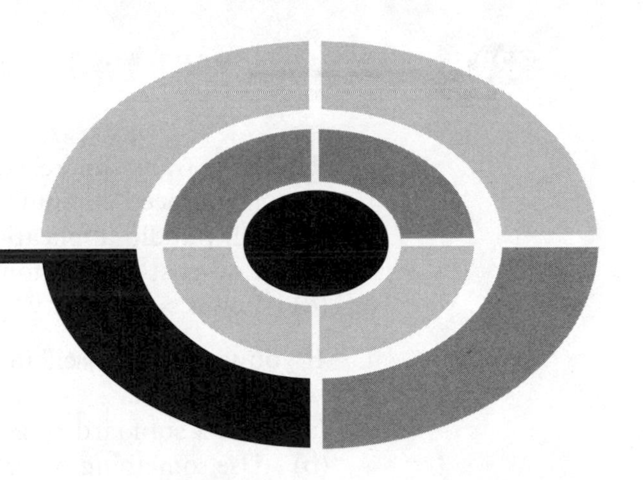

Test for Part Two

Do not refer to the text when taking this test. You may draw diagrams or use a calculator if necessary. A good score is at least 38 correct answers. Answers are in the back of the book. It's best to have a friend check your score the first time so you won't memorize the answers if you want to take the test again.

1. A signal with a 100 ns time "On" and a 230 ns time "Off" has a frequency of:
   (a) 30.3%
   (b) 3 GHz
   (c) 3 MHz
   (d) 3 kHz
2. A toggle flip flop will divide a clock frequency by:
   (a) The value selected in its inputs
   (b) 2
   (c) 1.5, but the output duty cycle is 50%
   (d) The toggle flip flop is used for debouncing switch inputs

3. The NPN transistor relaxation oscillator is well suited for:
   (a) High-speed computer applications
   (b) Only audio applications
   (c) Low-cost applications where accuracy isn't important
   (d) Power supply duty cycle generators

4. If you wanted an oscillator that produced a period of 4 gate delays, you would:
   (a) Use a standard ring oscillator
   (b) Use something other than a ring oscillator
   (c) Use two three inverter ring oscillators and XOR their outputs
   (d) Use a two inverter ring oscillator and divide the output frequency by two using a toggle D flip flop

5. The duty cycle of a ring oscillator is:
   (a) Always 50%
   (b) Equal to the number of inverters used in the ring oscillator multiplied by the gate delay
   (c) Dependent on the technology used in the inverters
   (d) Always 100%

6. Why are TTL inverters not recommended for relaxation oscillators?
   (a) They are too costly
   (b) They do not operate at a fast enough speed
   (c) Their current controlled inputs will affect the operation of the oscillator
   (d) They may short out internally when used as part of an oscillator.

7. A relaxation oscillator has an R1 value of 1k, C of 0.1 μF and R2 equal to 10 k. What frequency will it oscillate at?
   (a) It won't oscillate
   (b) 4.54 kHz
   (c) 4.54 MHz
   (d) 4.54 Hz

8. Crystals used in oscillators are:
   (a) Relatively high cost but very accurate
   (b) Very robust and immune to shock damage
   (c) Low cost with an accuracy that is similar to that of a "canned oscillator"
   (d) Mined in South America

9. The crystal inverter circuit requires the following parts:
   (a) Crystal and inverter
   (b) Crystal, inverter, two capacitors and two resistors
   (c) Crystal, inductor, three resistors, two capacitors and an NPN transistor
   (d) Crystal, diode, two capacitors, hand wound coil, two diodes

10. Changing the voltage at the "Vcontrol" pin of a 555 astable oscillator:
    (a) Will change the period of the oscillator output
    (b) Won't change anything
    (c) Will lower the amount of power consumed
    (d) Will add unnecessary costs to the circuit

11. A 555 astable oscillator with $R1 = 47\,k$, $R2 = 100\,k$ and $C = 4.7\,\mu F$ will oscillate at:
    (a) 805 Hz
    (b) 0.805 Hz
    (c) 1.24 Hz
    (d) 1.24 kHz

12. Decreasing value of a resistor or capacitor in a 555 monostable will:
    (a) Lower its operating frequency
    (b) Raise its operating frequency
    (c) Increase the pulse width
    (d) Decrease the pulse width

13. A chip with two 555 timers built in has the part number:
    (a) 75555
    (b) 556
    (c) No such chip exists
    (d) 2x555

14. The delay in a "canned delay line" is produced by:
    (a) Elves
    (b) A piece of quartz, cut to provide a specified delay
    (c) A long piece of copper wire, formed as a coil
    (d) A large array of simple gates

15. Why is reset passed to the "Output Formatter" in the sequential circuit block diagram?
    (a) To control the operation of tri-state drivers built into the "Output Formatter"

(b) To synchronize outputs with the operation of the chip
(c) To minimize power usage when the chip is reset
(d) To speed up operation of the sequential circuit

16. To create a cascaded counter from multiple chips, what signal(s) are passed between the chips?
(a) The most significant two bits
(b) The "Carry" from the least significant counter and the "Borrow" from the more significant counter
(c) Just the carry from the least significant counter to the clock input of the more significant counter
(d) The clock and chip enable signals to be shared between the two counters

17. What are the disadvantages of serial data transmission over parallel data transmission?
(a) Increased product chip count
(b) Slower data transmission
(c) Increased product power requirements
(d) Increased data error rate

18. The linear feedback shift register polynomial defines:
(a) The XOR "taps" used to modify the incoming data
(b) The execution order of magnitude
(c) The value needed to decode any value run through the LFSR
(d) The relationship between the number of bits and speed of the LFSR's operation

19. The number of states and inputs in a hardware state machine are a function of:
(a) The fanout capabilities of the chips used
(b) The number of discrete states required
(c) The requirements of the application
(d) The size of the ROM used

20. Since commands to a Hitachi 44780 controlled LCD require 160 μs or more to execute, what is an effective way of waiting the suggested delay before sending the next command?
(a) Using a 555 timer wired as a monostable
(b) Passing requests to the LCD through an RC delay
(c) Polling the busy flag
(d) Using a clock with a 5 ms period

21. What signal is not included in a typical sequential logic circuit interface bus?
    (a) _Read
    (b) Data Read Ready
    (c) _Write
    (d) Address Bit 0

22. A "three to eight" decoder:
    (a) Cannot be implemented in TTL or CMOS Logic
    (b) Converts three data selects into eight device read requests
    (c) Converts three clock cycles into eight state machine addresses
    (d) Converts three binary bits into eight interface device selects

23. The segments to be turned on in a seven-segment LED display is determined by:
    (a) A circuit which enables the common cathodes or common anodes in the LED displays
    (b) A microprocessor built into each seven-segment LED display
    (c) A combinatorial circuit performing Boolean arithmetic operations on the display bits
    (d) A sequential circuit that compares the display bits to expected values

24. A PWM with a period of 10 μs and a "high" signal of 4 μs has a duty cycle of:
    (a) Not enough information is available to determine the duty cycle
    (b) 60%
    (c) 0.4
    (d) 40%

25. To produce a PWM output signal of 20 kHz and have 6 bits of accuracy, what input clock frequency should be used?
    (a) 20 kHz
    (b) 64 Hz
    (c) 1.28 MHz
    (d) 5.1 MHz

26. The two inverter debounce circuit is best suited for what kind of input devices?
    (a) Push buttons
    (b) 40%

(c) 0.4
(d) Double throw switches

27. Hysteresis causes logic inputs to be registered:
    (a) Faster than normal inputs
    (b) At lower thresholds than devices with standard inputs
    (c) At more extreme thresholds than devices with standard inputs
    (d) When there is an inductor on the line

28. A switch matrix keypad is:
    (a) Used by Trinity and Neo to access the Matrix
    (b) A two-dimensional array of switches used to provide multiple button inputs
    (c) A series of switches that change the output signal of a sequential circuit
    (d) The natural evolution of a single button to multiple buttons

29. The Princeton architecture was criticized because:
    (a) The single memory interface was felt to be a performance bottleneck
    (b) It relied too heavily on vacuum tube technology
    (c) The design was taken from other, earlier computer designers
    (d) The amount of space required for it seemed to be prohibitive

30. The first digit to the right of the decimal point in a binary floating point number has a value of:
    (a) 0.1 decimal
    (b) 0.25 decimal
    (c) 0.5 decimal
    (d) 1.0 decimal

31. Floating point numbers are stored in data size except:
    (a) 2 bytes
    (b) 4 bytes
    (c) 8 bytes
    (d) 10 bytes

32. SRAM memory is best suited for applications which require:
    (a) Data stacks
    (b) Gigabytes of memory
    (c) Mass storage of data
    (d) Fast data access for a relatively large memory area

33. Stacks cannot be modeled by:
    (a) Trays in a lunchroom
    (b) Pieces of paper on a person's desk
    (c) Stacks of paper produced by a computer processor
    (d) A jelly bean jar

34. The operation of the Zener regulator is analogous to:
    (a) A car's carburetor
    (b) A bowl with a hole cut in the bottom
    (c) A PWM output driven by a comparator
    (d) A comparator output driven by sensor

35. The operation of the linear regulator is analogous to:
    (a) A car's carburetor
    (b) A bowl with a hole cut in the bottom
    (c) A PWM output driven by a comparator
    (d) A comparator output driven by sensor

36. A 10:1 transformer has an input of 110 volts AC. What is its output?
    (a) 11 volts DC
    (b) 11 volts AC
    (c) 110 volts DC
    (d) 11 amps AC

37. "PAL" I/O pins are:
    (a) Input and Output
    (b) Programmable Input or Output
    (c) Input only
    (d) Output only

38. What is not a "local memory" device:
    (a) Processor cache
    (b) PCI status registers
    (c) Main memory (S)DRAM
    (d) Processor ROM

39. The Logic levels for the PC's parallel port are?
    (a) 3.3 volt CMOS Logic compatible
    (b) 5 volt TTL/CMOS Logic compatible
    (c) 5 volt "HCT" Logic compatible
    (d) Current controlled, voltage levels are not considered

40. The inventor of the parallel port pinout standard was:
    (a) IBM
    (b) Apple Computers
    (c) The Centronics Corporation
    (d) Xerox FARC

41. The "Data" bits of the PC's parallel port have what kind of outputs?
    (a) Discrete transistors
    (b) CMOS totem pole outputs
    (c) Pulled up Open Collectors that can be changed by external devices
    (d) TTL tri-state buffers

42. "Vertical synch" in a video display:
    (a) Moves the electron beam in the CRT to the left
    (b) Moves the electron beam in the CRT to the top
    (c) Starts the raster going up and down
    (d) Is needed for CRTs pointing upwards

43. The "NTSC" video standard was first developed for:
    (a) Military radar screens
    (b) Portable news gathering services
    (c) Electronic instruments
    (d) Televisions

44. The "Microwire" synchronous serial protocol sends:
    (a) A word consisting of 16 bits of data
    (b) A word consisting of 8 bits of data
    (c) Status Information before Data
    (d) Command Information before Data

45. I2C's "Acknowledge" bit:
    (a) Is used as a "Parity Bit" for Received Data
    (b) Indicates the Received Data was Valid
    (c) Is a placeholder in case 9 bits of data is required
    (d) Is not currently implemented, but will be in the future

46. Baudot asynchronous serial communications was developed for:
    (a) Networking
    (b) Televisions
    (c) Teletypes
    (d) Telegraphs

47. What are the differences between Baudot asynchronous serial communications and NRZ serial communications?
    (a) None
    (b) The number of bits transmitted
    (c) The speed at which data is transmitted
    (d) The number of "overhead" bits

48. The primary difference between Manchester Encoding and NRZ is:
    (a) The media used to transmit the data
    (b) The bit timing for "0" and "1" change in Manchester Encoding
    (c) The data rates used to transmit the data
    (d) The companies that hold the original patents on the technologies

49. To implement a three wire RS-232 connection, you will have to tie together:
    (a) RTS/CTS and DSR/DTR
    (b) RI/CTS and DSR/RTS
    (c) DCD/DSR and RTS/CTS
    (d) DCD/CTS and RI/DSR

50. The RS-232 to TTL/CMOS level translator shown in Fig. Test 2.1:
    (a) Is an inexpensive, full capability RS-232 interface
    (b) Sends data in "half duplex" mode only
    (c) Can be used for handshaking lines as well as RX and TX
    (d) Is limited by the speed it can operate at

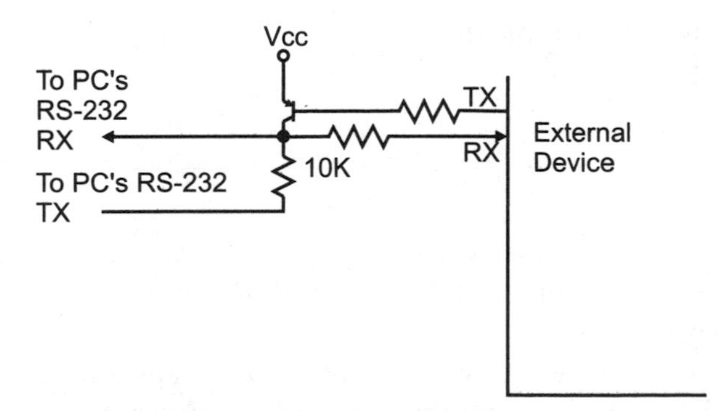

Fig. Test 2-1.

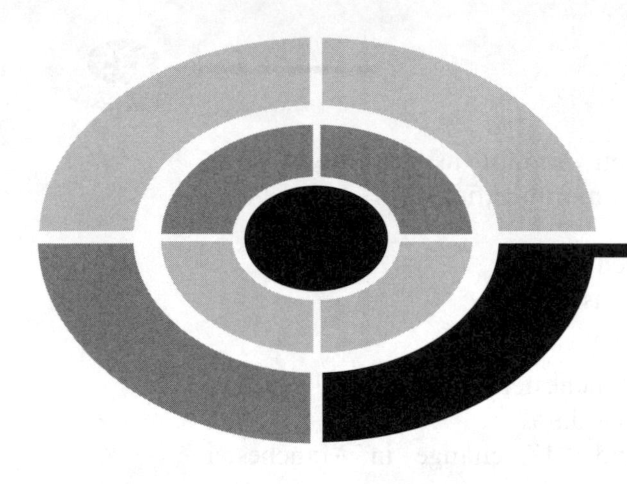

Final Exam

Do not refer to the text when taking this test. You may draw diagrams or use a calculator if necessary. A good score is at least 75 correct answers. Answers are in the back of the book. It's best to have a friend check your score the first time so you won't memorize the answers if you want to take the test again.

1. The NAND Gate is the base technology for:
   (a) TTL logic circuits
   (b) Your TV set
   (c) Computer chips
   (d) Symbolic logic

2. What is not an indicator for an Inverted bit?
   (a) “!”
   (b) “_”
   (c) “^”
   (d) “-”

3. Combinatorial logic is used:
   (a) In a DJ's mixing table to combine audio inputs together with special effects
   (b) To produce a desired output from a specific set of inputs
   (c) To express the mathematics of cryptography
   (d) To collate data from different sources

4. The following truth table represents the following basic gate:

Input "A"	Input "B"	Output
0	0	0
0	1	1
1	1	0
1	0	0

(a) AND
(b) OR
(c) NOT
(d) None

5. The following truth table can be represented as the sum of products:

Input "A"	Input "B"	Input "C"	Output
0	0	0	0
0	0	1	0
0	1	1	1
0	1	0	0
1	1	0	0
1	1	1	0
1	0	1	0
1	0	0	1

(a) $(A \cdot B \cdot C) + (!A + !B + !C)$
(b) $(!A \cdot B \cdot C) + (!A + !B + !C)$
(c) $(!A \cdot B \cdot C) + (A + !B + !C)$
(d) $(A \cdot !B \cdot C) + (A + !B + !C)$

6. The following truth table can be represented as the product of sums:

Input "A"	Input "B"	Input "C"	Output
0	0	0	0
0	0	1	1
0	1	1	1
0	1	0	1
1	1	0	1
1	1	1	0
1	0	1	1
1	0	0	1

(a) (!A + !B + !C) · (A + B + C)
(b) (!A + !B + C) · (A + B + C)
(c) (!A + !B + !C) · (A + !B + C)
(d) (!A + B + !C) · (A + B + C)

7. Oscilloscopes are used for:
(a) Testing the impedance of a circuit trace
(b) Display the state of multiple signals
(c) Displaying the current voltage of individual signals
(d) Displaying the analog voltage of individual signals

8. According to the AND associative law, which equation is true?
(a) (A + B) + C = A + (B + C) = A + B + C
(b) A · (B + C) = (A · B) + (A · C)
(c) A + (B · C) = (A + B) · (A + C)
(d) !(A + B) = !A · !B

9. A CMOS logic technology has a gate delay of 25 ns and an AND statement built from it requires two gate delays for the signal to pass through it. How long will it take a signal to pass through the AND gate?
(a) 0 ns
(b) 25 ns

(c) 50 ns
(d) Indeterminate

10. The first step in reducing the following truth table is:

Input "A"	Input "B"	Input "C"	Output
0	0	0	0
0	0	1	0
0	1	1	1
0	1	0	1
1	1	0	0
1	1	1	1
1	0	1	1
1	0	0	0

(a)

Input "A"	Input "B"	Input "C"	Output
0	0	0	0
0	0	1	0
0	1	x	1
1	1	0	0
1	x	1	1
1	0	0	0

(b)

Input "A"	Input "B"	Input "C"	Output
0	0	0	0
0	0	1	0
0	1	x	1
1	1	0	0
1	1	1	1
1	0	1	1
1	0	0	0

(c)

Input "A"	Input "B"	Input "C"	Output
0	0	0	0
0	0	1	0
0	1	1	1
x	1	0	1
1	1	1	1
1	0	1	1
1	0	0	0

(d)

Input "A"	Input "B"	Input "C"	Output
0	0	0	0
0	0	1	0
0	1	1	1
0	1	0	1
1	1	0	0
x	1	1	1
1	0	0	0

11. Using De Morgan's NAND Theorem. "!(!A · !B)" is equal to:
    (a) A + B
    (b) !A + !B
    (c) !(A · B)
    (d) A · B

12. The sum of products logic equation

    `Output` = (`!A` · `B` · `!C`) + (`!A` · `!B` · `!C`)

    can be reduced to:

    (a) A · C
    (b) !A · !C
    (c) !C · CB
    (d) C

13. A Karnaugh map has three bits equal to 1, can one circle (and one AND statement) be used to represent the logic function?
    (a) Yes
    (b) Yes, if inverters can be used
    (c) Yes, if an XOR gate could be used
    (d) No

14. The four bit Karnaugh map

	AB			
	00	01	11	10
CD-00	0	1	1	1
01	0	1	0	0
11	0	1	0	0
10	0	1	1	1

has the optimized sum of product equation:

(a) Output = (!B · !D) + (!A · !B)
(b) Output = (A · B) + (A · D)
(c) Output = (!B · A) + (B · D)
(d) Output = (!A · B) + (A · !D)

15. The NAND equivalent to an OR gate is:
(a) Built from two NAND gates and requires three gate delays for a signal to pass through
(b) Built from two NAND gates and requires two gate delays for a signal to pass through
(c) Built from three NAND gates and requires two gate delays for a signal to pass through
(d) Built from one NOR gate as well as a NOT gate and requires two gate delays for a signal to pass through

16. The voltage drop across a resistor in a parallel circuit:
(a) Is the same as the applied voltage
(b) Is proportional to the power dissipated in the circuit
(c) Is always zero
(d) Is proportional to the resistor's value relative to the total resistance in the circuit multiplied by the applied voltage

17. A 47μF capacitor:
(a) Has a value of 47 trillionths of a farad.
(b) Weighs 47 micrograms
(c) Probably uses polyester as a dielectric
(d) Has a value of 0.000047 farads

18. The equivalent resistance of a 10 ohm resistor in parallel with 10 ohm and 20 ohm resistors in series is:
    (a) 40 ohms
    (b) 30 ohms
    (c) 7.5 ohms
    (d) 6.7 ohms

19. In a PNP bipolar transistor, collector current:
    (a) Flows into the emitter through to the collector
    (b) Is dependent on the emitter current
    (c) Is negative with respect to the power source
    (d) Like in the NPN bipolar transistor, does not include the base current

20. CMOS Logic is:
    (a) Sound controlled
    (b) Resistor controlled
    (c) Current controlled
    (d) Voltage controlled

21. "Breadboards" allow:
    (a) Delivery of bagels and other breakfast foods without any garbage
    (b) Simple creation of electronic products that can be delivered to consumers
    (c) The method that provides the fastest signal paths for digital electronics circuits
    (d) Simple creation of prototype circuits

22. The holes in a breadboard:
    (a) Allow effective cooling of the breadboard circuitry
    (b) Allow components to be mounted and interconnected
    (c) Are there simply for aesthetics
    (d) Minimize the resistance encountered by signals

23. A circuit that would provide valid TTL and CMOS logic inputs would:
    (a) Both source current as well as provide high and low voltages
    (b) Both provide high voltage as well as sink and source current
    (c) Both provide low voltage as well as sink and source current
    (d) Both sink current as well as provide high and low voltages

24. The four bit binary number B'1010 1101' expressed as a hexadecimal number is:
    (a) 171
    (b) 0x0AB
    (c) 0x02231
    (d) 0x0171

25. The most significant digit in the decimal number "1234" is:
    (a) 1
    (b) 2
    (c) 3
    (d) 4

26. The decimal number "42" in hexadecimal is:
    (a) 0x042
    (b) B'0010 0100'
    (c) 2A
    (d) 0x02A

27. The four bit hexadecimal number 0x0F4EA expressed in decimal is:
    (a) −2,838
    (b) 62,698
    (c) B'1111 0100 1110 1010'
    (d) 29.930

28. B'0101' in binary, using the formula Gray code = Binary ^ (Binary≫1) can be converted to the Gray code:
    (a) B'1010'
    (b) B'0110'
    (c) B'0111'
    (d) B'1110'

29. The Gray code B'1010' corresponds to the binary value
    (a) B'1111'
    (b) B'1110'
    (c) B'1101'
    (d) B'1010'

30. A "Carry Look-Ahead" adder:
    (a) Is the only way addition should be carried out in digital electronics
    (b) Is often needlessly complex for most applications
    (c) Produces its result in an amount of time independent of the number of bits

(d) Produces its result in an amount of time dependent on the number of bits

31. In a universe where infinity (the highest possible number) is one thousand (1,000); "−34" could be represented as:
    (a) Only −34
    (b) 966
    (c) 976
    (d) 34

32. Converting the eight bit, two's complement value "−124" to a positive number by inverting each bit and incrementing the result produces the bit pattern:
    (a) B'0111 1100'
    (b) Which is seven bits long and is invalid
    (c) B'1000 0100'
    (d) B'01111 110'

33. Multiplying two eight bit numbers by repeated addition:
    (a) Will require up to 8 addition operations
    (b) Will require up to 255 addition operations
    (c) Cannot be implemented in digital electronics
    (d) Is the fastest way of performing binary multiplication

34. Multiplying a binary number by 64 can be accomplished by:
    (a) Clearing the least significant six bits
    (b) Shifting left six bits
    (c) Shifting right six bits
    (d) Setting the least significant six bits

35. What is the best method of multiplying binary number B'0110 1101' by B'0011 1110'?
    (a) Boothe's algorithm: shift B'0110 1101' to the left by 7 and subtract B'0110 1101 from the result
    (b) Shifting left seven bits according to Boothe's algorithm
    (c) Shifting right six bits according to Boothe's algorithm
    (d) Boothe's algorithm: shift B'0110 1101' to the right by 7 and subtract B'0110 1101 from the result

36. If you looked up the "gate delay" specification for the technology used to build an AND gate chip, what would the chip's gate delay be?
    (a) Exactly what is listed for the technology used by the AND gate chip
    (b) Dependent on whether or not the chip technology is NOR or NAND gate based
    (c) Twice the listed gate delay
    (d) Three times the listed gate delay

37. What tools should *not* be used to find a possible race condition in a circuit?
    (a) An oscilloscope
    (b) A logic analyzer
    (c) A logic simulator
    (d) A time delay reflectometer

38. A chip fanout of 6 means:
    (a) The chip's output can drive six inputs
    (b) The chip provides six outputs for a single input
    (c) There can be six chips connected to the various outputs
    (d) The chip provides six outputs, each one a variation on the inputs

39. When working with TTL, the Mickey Mouse "AND" circuit should have a resistor value of:
    (a) The Mickey Mouse AND gate will not work with TTL
    (b) 4 watts
    (c) 10 k
    (d) 470 Ω

40. Can a pulled up button (which pulls to ground) be attached to a dotted AND bus?
    (a) Yes, if the signal passes through an open collector transistor
    (b) Yes, if the signal passes through a gate with an open collector driver
    (c) Unconditionally, yes
    (d) No

41. Can a totem pole driver be attached to a dotted AND bus?
    (a) Yes, if the power to the chip with the totem pole driver is the same as the chips on the dotted AND bus
    (b) Yes, if care is taken to ensure grounds are kept at the same voltage level

(c) Unconditionally, yes
(d) No

42. When the TTL input is pulled low, how much current flows out of the pin?
(a) 1 μA
(b) 100 μA
(c) 1 mA
(d) No current at all

43. The most popular output driver type found in logic chips is:
(a) The totem pole driver
(b) Tri-state driver
(c) Open collector
(d) Open drain

44. A commercial microprocessor reset chip will:
(a) Pull the microprocessor's _Reset line low at power up and when its input voltage falls below a set value
(b) Drive the microcontroller's _Reset line high at power up and when its input voltage falls below a set value
(c) Pull the microprocessor's _Reset line low at power up.
(d) Turn on a LED when its input voltage falls below a set value to indicate a power supply problem

45. A small triangle on the input of a sequential device indicates:
(a) The pin is buffered before being processed within the chip
(b) The signal is inverted before being processed within the chip
(c) The signal is a clock to the chip's internal flip flops
(d) The circuitry connected to the chip was either not completed or qualified and it should be used with caution

46. Which formula approximates the time required for an RC network to change state due to a sudden voltage?
(a) $t = 2.2 \times R \times C$
(b) $t = -RC \times \ln(Vcc)$
(c) $t = -RC \times \ln(Vcc - (Vcc \times e^{-1/RC}))$
(d) $t = V \times i$

47. A switch bounces:
(a) Never
(b) Most of the time, but this can be filtered out with a capacitor
(c) Only if it is not feeding the input of a flip flop
(d) Always

48. The NOR "RS" flip flop saves data when the inputs are:
    (a) R = 0, S = 0
    (b) R = 0, S = 1
    (c) R = 1, S = 1
    (d) R = 1, S = 0

49. The NAND "RS" flip flop sets the output (drives out "1") when the inputs are:
    (a) R = 0, S = 0
    (b) R = 0, S = 1
    (c) R = 1, S = 1
    (d) R = 1, S = 0

50. Reset circuitry for sequential circuits should be used:
    (a) Always
    (b) Only for digital clocks to avoid the clock starting up at a random time
    (c) In situations where the start up values are critical
    (d) Never

51. A signal with a 100 ns time "On" and a 230 ns time "Off" has a frequency of:
    (a) 30.3%
    (b) 3 GHz
    (c) 3 MHz
    (d) 3 kHz

52. A toggle flip flop will divide a clock frequency by:
    (a) The value selected in its inputs
    (b) 2
    (c) 1.5, but the output duty cycle is 50%
    (d) The toggle flip flop is used for debouncing switch inputs

53. The NPN transistor relaxation oscillator is well suited for:
    (a) High-speed computer applications
    (b) Only audio applications
    (c) Low-cost applications where accuracy isn't important
    (d) Power supply duty cycle generators

54. If you wanted an oscillator that produced a period of 4 gate delays, you would:
    (a) Use a standard ring oscillator
    (b) Use something other than a ring oscillator
    (c) Use a two three inverter ring oscillators and XOR their outputs

(d) Use a two inverter ring oscillator and divide the output frequency by two using a toggle D flip flop

55. The duty cycle of a ring oscillator is:
    (a) Always 50%
    (b) Equal to the number of inverters used in the ring oscillator multiplied by the gate delay
    (c) Dependent on the technology used in the inverters
    (d) Always 100%

56. Why are TTL inverters not recommended for relaxation oscillators?
    (a) They are too costly
    (b) They do not operate at a fast enough speed
    (c) Their current controlled inputs will affect the operation of the oscillator
    (d) They may short out internally when used as part of an oscillator

57. A relaxation oscillator has an R1 value of 1 k, C of 0.1 μF and R2 equal to 10 k. What frequency will it oscillate at?
    (a) It won't oscillate
    (b) 4.54 kHz
    (c) 4.54 MHz
    (d) 4.54 Hz

58. Crystals used in oscillators are:
    (a) Relatively high cost but very accurate
    (b) Very robust and immune to shock damage
    (c) Low cost with an accuracy that is similar to that of a "canned oscillator"
    (d) Mined in South America

59. The crystal inverter circuit requires the following parts:
    (a) Crystal and inverter
    (b) Crystal, inverter, two capacitors and two resistors
    (c) Crystal, Inductor, three resistors, two capacitors and an NPN transistor
    (d) Crystal, diode, two capacitors, hand wound coil, two diodes

60. Changing the voltage at the "Vcontrol" pin of a 555 astable oscillator:
    (a) Will change the period of the oscillator output
    (b) Won't change anything

(c) Will lower the amount of power consumed
(d) Will add unnecessary costs to the circuit

61. A 555 astable oscillator with R1 = 47 k, R2 = 100 k and C = 4.7 μF will oscillate at:
    (a) 805 Hz
    (b) 0.805 Hz
    (c) 1.24 Hz
    (d) 1.24 kHz

62. Decreasing value of a resistor or capacitor in a 555 monostable will:
    (a) Lower its operating frequency
    (b) Raise its operating frequency
    (c) Increase the pulse width
    (d) Decrease the pulse width

63. A chip with two 555 timers built in has the part number:
    (a) 75555
    (b) 556
    (c) No such chip exists
    (d) 2 × 555

64. The delay in a "canned delay line" is produced by:
    (a) Elves
    (b) A piece of quartz, cut to provide a specified delay
    (c) A long piece of copper wire, formed as a coil
    (d) A large array of simple gates

65. Why is reset passed to the "Output Formatter" in the sequential circuit block diagram?
    (a) To control the operation of tri-state drivers built into the "Output Formatter"
    (b) To synchronize outputs with the operation of the chip
    (c) To minimize power usage when the chip is reset
    (d) To speed up operation of the sequential circuit

66. To create a cascaded counter from multiple chips, what signal(s) are passed between the chips?
    (a) The most significant two bits
    (b) The "Carry" from the least significant counter and the "Borrow" from the more significant counter
    (c) Just the carry from the least significant counter to the clock input of the more significant counter

(d) The clock and chip enable signals to be shared between the two counters

67. What are the disadvantages of serial data transmission over parallel data transmission?
    (a) Increased product chip count
    (b) Slower data transmission
    (c) Increased product power requirements
    (d) Increased data error rate
68. The linear feedback shift register polynomial defines:
    (a) The XOR "taps" used to modify the incoming data
    (b) The execution order of magnitude
    (c) The value needed to decode any value run through the LFSR
    (d) The relationship between the number of bits and speed of the LFSR's operation
69. The number of states and inputs in a hardware state machine are a function of:
    (a) The fanout capabilities of the chips used
    (b) The number of discrete states required
    (c) The requirements of the application
    (d) The size of the ROM used
70. A constant delay of 100 μs can be used for instruction and data writes to the 44780 controlled LCD except when:
    (a) Writing the "Cursor Move" instruction
    (b) When scrolling the LCD upwards
    (c) When writing the "Screen Clear" and "Cursor Home" commands
    (d) Backspacing from the start of the current line to the start of the next
71. What signal is not included in a typical sequential logic circuit interface bus?
    (a) _Read
    (b) Data Read Ready
    (c) _Write
    (d) Address Bit 0
72. A "three to eight" decoder:
    (a) Cannot be implemented in TTL or CMOS Logic
    (b) Converts three data selects into eight device read requests
    (c) Converts three clock cycles into eight state machine addresses
    (d) Converts three binary bits into eight interface device selects

73. The segments to be turned on in a seven-segment LED display is determined by:
    (a) A circuit which enables the common cathodes or common anodes in the LED displays
    (b) A microprocessor built into each seven-segment LED display
    (c) A combinatorial circuit performing Boolean arithmetic operations on the display bits
    (d) A sequential circuit that compares the display bits to expected values

74. A PWM with a period of 10 μs and a "high" signal of 4 μs has a duty cycle of:
    (a) Not enough information is available to determine the duty cycle
    (b) 60%
    (c) 0.4
    (d) 40%

75. To produce a PWM output signal of 20 kHz and have 6 bits of accuracy, what input clock frequency should be used?
    (a) 20 kHz
    (b) 64 Hz
    (c) 1.28 MHz
    (d) 5.1 MHz

76. The two inverter debounce circuit is best suited for what kind of input devices?
    (a) Push buttons
    (b) 40%
    (c) 0.4
    (d) Double throw switches

77. Hysteresis causes logic inputs to be registered:
    (a) Faster than normal inputs
    (b) At lower thresholds than devices with standard inputs
    (c) At more extreme thresholds than devices with standard inputs
    (d) When there is an inductor on the line

78. A switch matrix keypad is:
    (a) Used by Trinity and Neo
    (b) A two-dimensional array of switches used to provide multiple button inputs

(c) A series of switches that change the output signal of a sequential circuit
(d) The natural evolution of a single button to multiple buttons

79. The Princeton architecture was criticized because:
(a) The single memory interface was felt to be a performance bottleneck
(b) It relied too heavily on vacuum tube technology
(c) The design was taken from other, earlier computer designers
(d) The amount of space required for it seemed to be prohibitive

80. The first digit to the right of the decimal point in a binary floating point number has a value of:
(a) 0.1 decimal
(b) 0.25 decimal
(c) 0.5 decimal
(d) 1.0 decimal

81. Floating point numbers are stored in data size except:
(a) 2 bytes
(b) 4 bytes
(c) 8 bytes
(d) 10 bytes

82. SRAM memory is best suited for applications which require:
(a) Data stacks
(b) Gigabytes of memory
(c) Mass storage of data
(d) Fast data access for a relatively large memory area

83. Stacks cannot be modeled by:
(a) Trays in a lunchroom
(b) Pieces of paper on a person's desk
(c) Stacks of paper produced by a computer processor
(d) A jelly bean jar

84. The operation of the Zener regulator is analogous to:
(a) A car's carburetor
(b) A bowl with a hole cut in the bottom
(c) A PWM output driven by a comparator
(d) A comparator output driven by sensor

85. The operation of the linear regulator is analogous to:
(a) A car's carburetor

(b) A bowl with a hole cut in the bottom
(c) A PWM output driven by a comparator
(d) A comparator output driven by sensor

86. A 10:1 transformer has an input of 110 volts AC. What is its output?
(a) 11 volts DC
(b) 11 volts AC
(c) 110 volts DC
(d) 11 amps AC

87. "PAL" I/O pins are:
(a) Input and Output
(b) Programmable Input or Output
(c) Input only
(d) Output only

88. What is not a "local memory" device:
(a) Processor cache
(b) PCI status registers
(c) Main memory (S)DRAM
(d) Processor ROM

89. The Logic levels for the PC's parallel port are?
(a) 3.3 volt CMOS Logic compatible
(b) 5 volt TTL/CMOS Logic compatible
(c) 5 volt "HCT" Logic compatible
(d) Current controlled, voltage levels are not considered

90. The inventor of the parallel port pinout standard was:
(a) IBM
(b) Apple Computers
(c) The Centronics Corporation
(d) Xerox PARC

91. The "Data" bits of the PC's parallel port have what kind of outputs?
(a) Discrete transistors
(b) CMOS totem pole outputs
(c) Pulled up Open Collectors that can be changed by external devices
(d) TTL tri-state buffers

92. "Vertical synch" in a video display:
    (a) Moves the electron beam in the CRT to the left
    (b) Moves the electron beam in the CRT to the top
    (c) Starts the raster going up and down
    (d) Is needed for CRTs pointing upwards

93. The "NTSC" video standard was first developed for:
    (a) Military radar screens
    (b) Portable news gathering services
    (c) Electronic instruments
    (d) Televisions

94. The "Microwire" synchronous serial protocol sends:
    (a) A word consisting of 16 bits of data
    (b) A word consisting of 8 bits of data
    (c) Status Information before Data
    (d) Command Information before Data

95. I2C's "Acknowledge" bit:
    (a) Is used as a "Parity Bit" for Received Data
    (b) Indicates the Received Data was Valid
    (c) Is a placeholder in case 9 bits of data is required
    (d) Is not currently implemented, but will be in the future

96. Baudot asynchronous serial communications was developed for:
    (a) Networking
    (b) Televisions
    (c) Teletypes
    (d) Telegraphs

97. What are the differences between Baudot asynchronous serial communications and NRZ serial communications?
    (a) None
    (b) The number of bits transmitted
    (c) The speed at which data is transmitted
    (d) The number of "overhead" bits

98. The primary difference between Manchester Encoding and NRZ is:
    (a) The media used to transmit the data
    (b) The bit timing for "0" and "1" change in Manchester Encoding
    (c) The data rates used to transmit the data

99. To implement a three wire RS-232 connection, you will have to tie together:
   (a) RTS/CTS and DSR/DTR
   (b) RI/CTS and DSR/RTS
   (c) DCD/DSR and RTS/CTS
   (d) DCD/CTS and RI/DSR

100. The RS-232 to TTL/CMOS level translator shown in Figure Final Exam 1:
   (a) Is an inexpensive, full capability RS-232 interface.
   (b) Send data in "half duplex" mode only
   (c) Can be used for handshaking lines as well as RX and TX
   (d) Is limited by the speed it can operate at

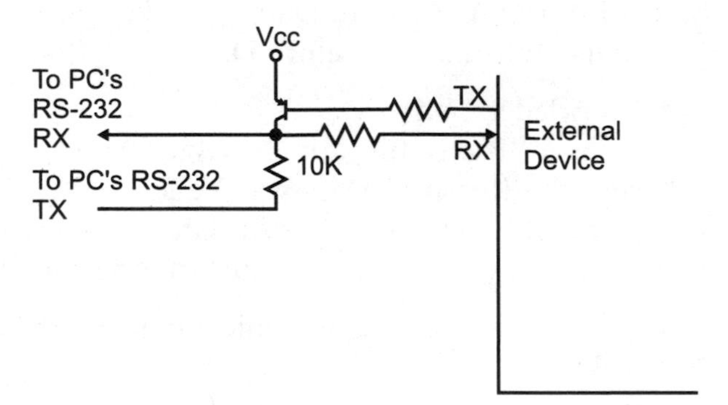

Fig. Final Exam-1.

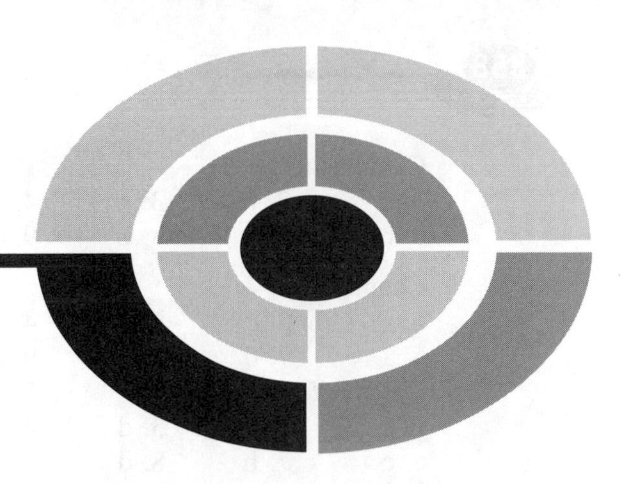

Appendix: Answers to Quiz, Tests, and Final Exam

Chapter 1:

1. b	2. a	3. c	4. a	5. d
6. c	7. c	8. b	9. c	10. a

Chapter 2:

1. c	2. a	3. a	4. d	5. d
6. b	7. c	8. d	9. a	10. b

Chapter 3:

1. d	2. b	3. c	4. c	5. d
6. c	7. c	8. d	9. c	10. a

Chapter 4:

1. c	2. b	3. b	4. d	5. d
6. b	7. c	8. b	9. c	10. a

Chapter 5:

1. a	2 .c	3. a	4. c	5. d
6. a	7. d	8. b	9. b	10. d

Chapter 6:

1. a	2. c	3. d	4. d	5. c
6. c	7. b	8. d	9. b	10. d

Chapter 7:

1. c	2. b	3. b	4. c	5. c
6. d	7. a	8. b	9. c	10. c

Test for Part One

1. d	2. b	3. c	4. a	5. b	6. d	7. b	8. c
9. c	10. b	11. a	12. a	13. a	14. b	15. d	16. b
17. b	18. b	19. a	20. c	21. d	22. a	23. a	24. c
25. c	26. a	27. b	28. d	29. c	30. c	31. a	32. c
33. c	34. b	35. c	36. b	37. c	38. b	39. c	40. a
41. b	42. d	43. a	44. a	45. d	46. d	47. b	48. c
49. d	50. b						

Chapter 8:

1. a	2. d	3. a	4. d	5. a
6. d	7. c	8. d	9. a	10. b

Chapter 9:

1. b	2. d	3. a	4. d	5. a
6. d	7. c	8. b	9. a	10. b

Chapter 10:

1. b	2. a	3. c	4. a	5. b
6. b	7. a	8. a	9. d	10. a

Chapter 11:

1. c	2. d	3. b	4. c
5. b	6. a	7. d	8. a

Chapter 12:

1. b	2. d	3. b	4. c	5. c
6. b	7. a	8. d	9. c	10. d

Chapter 13:

1. a	2. d	3. a	4. a	5. a
6. b	7. d	8. d	9. d	10. c

Test for Part Two

1. c	2. b	3. c	4. b	5. a	6. c	7. b	8. a
9. b	10. a	11. c	12. d	13. b	14. c	15. a	16. c
17. b	18. a	19. d	20. c	21. b	22. d	23. c	24. c
25. c	26. d	27. c	28. b	29. a	30. c	31. a	32. d
33. c	34. b	35. a	36. b	37. a	38. b	39. b	40. c
41. c	42. b	43. d	44. a	45. b	46. c	47. a	48. b
49. a	50. b						

Final Exam:

1. a	2. c	3. b	4. d	5. c	6. a	7. d	8. b
9. c	10. a	11. a	12. b	13. d	14. d	15. c	16. a
17. d	18. c	19. a	20. d	21. d	22. b	23. d	24. b
25. a	26. d	27. b	28. c	29. a	30. c	31. b	32. a
33. b	34. b	35. a	36. b	37. c	38. a	39. d	40. c
41. d	42. c	43. a	44. a	45. c	46. a	47. d	48. a
49. d	50. a	51. c	52. b	53. c	54. b	55. a	56. c
57. b	58. a	59. b	60. a	61. c	62. d	63. b	64. c
65. a	66. c	67. b	68. a	69. d	70. c	71. b	72. d
73. c	74. c	75. c	76. d	77. c	78. b	79. a	80. c
81. a	82. d	83. c	84. b	85. a	86. b	87. a	88. b
89. b	90. c	91. c	92. b	93. d	94. a	95. b	96. c
97. a	98. b	99. a	100. b				

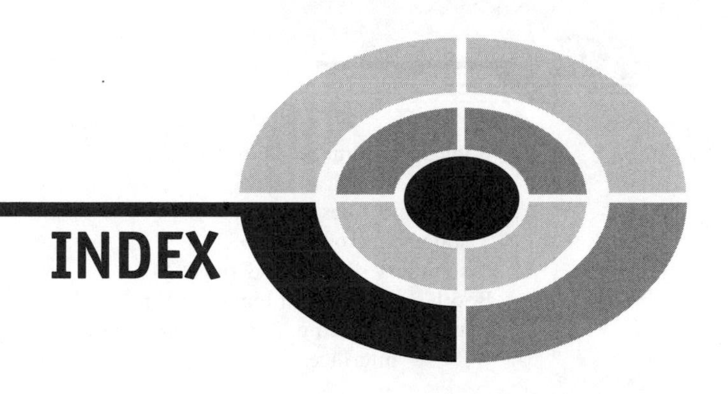

INDEX

Page numbers for illustrations are shown in **bold** type

ABOUT THE AUTHOR

Myke Predko is New Technologies Test Engineer at Celestica in Toronto, Canada. He is the author of McGraw-Hill's *Programming and Customizing PICMicro Microcontrollers*, Second Edition, and is a co-designer of both *TAB Electronics Build Your Own Robot Kits*.